Pressure Regimes in Sedimentary Basins and Their Prediction

An outgrowth of the international forum sponsored by the Houston chapter of the American Association of Drilling Engineers Houston, Texas, September 2–4, 1998

Edited by

Alan R. Huffman
Glenn L. Bowers

AAPG MEMOIR 76

Published jointly by
The American Association of Petroleum Geologists
Tulsa, Oklahoma
and
The Houston Chapter of The American Association of Drilling Engineers
Houston, Texas

Printed in the U.S.A.

ISBN 0-89181-357-8

AAPG Association Editor: Neil F. Hurley, 1997–2001; John C. Lorenz, 2001–2004
AAPG Geoscience Director: Robert C. Millspaugh
AAPG Bulletin and Book Editor: Carol E. Christopher
Copyediting and Production: Impressions Book and Journal Services, Inc.
Printed by Walsworth Publishing Company

This book and other AAPG publications are available from

The AAPG Bookstore
P.O. Box 979
Tulsa, OK 74101-0979
U.S.A.
Tel 1-918-584-2555 or 1-800-364-AAPG (U.S.A.)
Fax 1-918-584-0469 or 1-800-898-2274 (U.S.A.)
http://bookstore.aapg.org

Geological Society Publishing House
Unit 7, Brassmill Enterprise Centre
Brassmill Lane, Bath BA1 3JN
U.K.
Tel +44-1225-445046
Fax +44-1225-442836
www.geolsoc.org.uk

Canadian Society of Petroleum Geologists
#160, 540-Fifth Avenue S.W.
Calgary, Alberta
Canada T2P 0M2
Tel 1-403-264-5610
Fax 1-403-264-5898
www.cspg.org

Affiliated East-West Press Private Ltd.
G-1/16 Ansari Road Darya Ganj
New Delhi 110 002
India
Tel +91-11-3279113
Fax +91-11-3260538
e-mail: affiliat@nda.vsnl.n

Acknowledgments

AAPG and AADE thank the following for their generous contributions to *Pressure Regimes in Sedimentary Basins and Their Prediction*

BP America Inc.

Knowledge Systems, Inc.

Minerals Management Service

Unocal Corporation

Contributions are applied toward the production costs of publication, thus directly reducing the book's purchase price and making the volume available to a larger readership.

Cover

The cover images (©2002, Conoco, Inc.) were generated with the GEOPRESS™ program, Conoco Inc.'s workstation application for predicting pore-fluid pressure and other attributes from seismic velocities. Seismic data are used with permission of WesternGeco.

The background image shows a seismic section with the predicted pore-fluid pressure gradient as the color underlay. In general, the gradient increases with depth. The white color at the top of the section indicates a normal hydrostatic gradient. The purple at depth indicates a gradient of about 15 ppg effective mud weight. The vertical, red dotted line is the location of a control well. The red solid curved line indicates the top of salt. The white color below the top of the salt horizon does not indicate normal pressure. No pressure prediction was made in the salt. The green line is a fault.

The blue markers in the graph on the lower right are crossplotted velocity–effective stress data pairs. The effective stress values are based on pore-fluid pressures from a model of a normally pressured section. The velocities are based on a regularly sampled, smoothed seismic velocity function. The red curve is a power law function fit to the crossplotted values.

The blue stair-step line in the graph in the top left corner is a velocity function. The original stacking velocity function was smoothed and resampled to a regular time interval. The red and green curves are predicted normal velocity trends. The normal trends are based pore-fluid pressures from a hydrostatic model. The two curves are different because each one used a different equation to predict velocity from effective stress. H-1 and Top Salt are horizons. The depth datum is the sea floor or mudline. The velocities in this track are the same as those in the velocity-effective stress crossplot. The red normal velocity trend in this track is based on the red velocity-effective stress curve in the crossplot.

Contents

About the Editors

Alan R. Huffman received a bachelor's degree in geology from Franklin and Marshall College in 1983 and a Ph.D. in geophysics from Texas A&M University in 1990. His main research focus was rock physics, rock mechanics, and crustal seismology. His dissertation research on shock deformation of silicates led to numerous publications and to his being recognized as a leading expert in the area of high strain rate deformation of natural materials. In 1989, he was appointed science manager of DOSECC, the United States Continental Drilling Program. In 1990, Huffman went to work in the Offshore/Alaska Division of Exxon, and he transferred into the newly formed Exxon Exploration Company in 1992. During his tenure at Exxon, Huffman was involved in the development of proprietary technology, including three-dimensional amplitude vs. offset (3-D AVO), pore-pressure prediction, seismic attribute technology, and specialist software for well log and physical properties analysis and quantitative attribute analysis. Huffman was also actively involved in the exploration and field development process from the acquisition of 3-D seismic, to final interpretation, and to the planning and drilling of wells in the United States, the Far East, and west Africa.

In March 1997, Huffman joined Conoco as manager of the Seismic Imaging Technology Center (SITC). In this role, Huffman manages the development and application of high-end geophysical technology for Conoco's upstream businesses. Huffman is active in industry and professional affairs, having chaired numerous technical conferences and having served on organizing committees for the Society of Exploration Geophysicists, the American Association of Drilling Engineers, AAPG, and the Society of Petroleum Engineers. He is a member of the Society of Exploration Geophysicists, AAPG, AEG, the Geological Society of America, and the American Geophysical Union. He is a director and executive committee member of the Midcontinent Oil and Gas Association and also a director of the United States Oil and Gas Association.

Glenn L. Bowers is president of Applied Mechanics Technologies (AMT) in Houston, Texas, which provides consulting services in abnormal pore pressure, wellbore stability, wellbore fracturing, and general rock mechanics. Prior to starting AMT in 1994, he spent 12 years at Exxon Production Research Company, where he performed research in these same areas.

Bowers holds B.S. and M.S. degrees in mechanical engineering from the University of Akron and a Ph.D. in theoretical and applied mechanics from the University of Illinois. He is a member of the Society of Petroleum Engineers (SPE) and the American Association of Drilling Engineers (AADE) and was founder and chairman of the AADE's committee on abnormal pressure drilling technology. Bowers was a member of the organizing committee for the 1993 SPE Forum Series in North America workshop "Pore Pressure Estimation and Detection," served as co-organizer for the AADE's 1998 international conference "Pore Pressures in Sedimentary Basins and Their Prediction," and was co-organizer of the special session "Geophysical Methods in Pore Pressure Estimation" that was held at the 2001 Offshore Technology Conference.

Preface

Pore pressure and pressure regimes have always been recognized as factors of critical importance in the evolution of hydrocarbon provinces. Until recently, however, the ability to measure and predict pressures has been limited by two critical factors: our limited knowledge of those regimes and their physics, and the lack of sufficiently robust technology for predicting pressures in the subsurface. Historically, pressure prediction has used basic data such as mud weights, repeat formation tests (RFT), and leak-off Tests (LOT) from offset well control, along with seismic velocities and well logs, to predict pressure. One of the biggest challenges for the pressure-prediction community has been to educate each other across disciplinary boundaries. This challenge was a formidable one, considering that the geophysicist and geologist commonly speak very different technical languages than the reservoir engineer and drilling engineer.

To address this cross-disciplinary educational issue, the Drilling Information Standing Committee on Abnormal Pressure Drilling Technology of the Houston chapter of the American Association of Drilling Engineers (AADE) sponsored an international forum on the topic of Pressure Regimes in Sedimentary Basins and Their Prediction. This conference was held September 2–4, 1998 at the Del Lago Resort and Conference Center on Lake Conroe just north of the Houston, Texas, metropolitan area. The goal of this conference was to bring together experts in pore-pressure prediction from the geological, geophysical, reservoir, and drilling communities and to share the state-of-the-art from each discipline with the attendees. The meeting was attended by 159 delegates from 14 countries and represented 25 operating companies, 10 service companies, 5 drilling contractors, 15 universities and industry research laboratories, and 4 government agencies (United States and Canadian). The balance of geoscientists and engineers confirmed the interdisciplinary interest in the topic. The meeting was built around 35 invited presentations in five half-day topical sessions on (1) shale mechanics, (2) overpressure mechanisms, (3) pore pressure and fracture gradient prediction methods, (4) pressure regimes at the prospect and basin scales, and (5) pressure management while drilling and frontier issues. Each session also had five or six impromptu presentations to stimulate discussion on each topic. The meeting was sponsored by Conoco, the U.S. Department of Energy, The Gas Research Institute, Mobil, and Chevron. A meeting summary can be found in the April 1999 edition of the Society of Exploration Geophysicists *Leading Edge* magazine.

Following the forum, the presenters were invited to submit papers for a special AAPG publication. Nineteen papers were submitted from the original 35 invited papers, and these comprise the chapters of the current volume. The chapters have been organized into three groups.

The first seven chapters are oriented toward rock physics and pore-pressure theory. Chapter 1 (Swarbrick et al.) reviews the basic types of overpressure mechanisms, and examines the potential excess pressure each can produce. Chapter 2 (Miller et al.) addresses differences between in-situ stress states associated with externally vs. internally generated overpressure. In Chapter 3, Holbrook discusses relationships between sediment compaction behavior and the properties of the sediment grains. Chapter 4 (Dvorkin and Nur) reviews the concept of critical porosity, and presents equations for velocity as a function of porosity and critical porosity. Bowers and Katsube (Chapter 5) discuss the role shale pore structure plays in wire-line logs having different sensitivities to different overpressure mechanisms. Chapter 6 (Lahann) looks at compaction trend changes caused by smectite diagenesis as a potential source of overpressure. In Chapter 7, Katahara and Corrigan consider the effect pore fluid compressibility has on pore pressure; in particular, how this dictates whether erosion will lead to overpressure.

Chapters 8 through 13 focus on pore pressure and fracture gradient prediction in different geologic environments. Chapter 8 (Addis and Yassir) looks at how different overpressure mechanisms and different tectonic settings impact overpressure detection and estimation. Hennig et al. (Chapter 9) continue this theme with results from a Papua New Guinea study on pore-pressure prediction in an active thrust region. Chapter 10 (Smith) summarizes data accumulated by the Minerals and Management Service (MMS) on the distribution of overpressure in the deepwater Gulf of Mexico. In Chapter 11, Holasek presents equations for estimating overburden stresses and fracture gradients in the Gulf of Mexico. Chapter 12, by Singh and Emery, discusses a technique for extrapolating fracture gradient data from shales to sands, and from virgin to depleted reservoirs. Stump and Flemings (Chapter 13) then present permeability, current, and maximum past effective stress data derived from laboratory tests on shale samples from the Eugene Island 330 field, offshore Louisiana.

The last six chapters cover a variety of topics ranging from geophysics to drilling technology. In Chapter 14, Holbrook reviews a methodology for estimating pore pressures and fracture gradients from logging measurements in multiple lithologies. Chapter 15 (Rasmus) looks at how pore water property changes during burial can influence the response of logging measurements to overpressure. Dutta et al. (Chapter 16) discuss an integrated approach for real-time pore-pressure analysis using the drill bit as a seismic source. Chapter 17, by Smith and Gault, describes new technology that could potentially widen the narrow mud-weight operating ranges that characterize deep-water drilling. In Chapter 18, Bell reviews the fundamentals of seismic velocity analysis, and discusses the potential pitfalls and processing strategies that need to be considered when interpreting velocities for predrill pore-pressure predictions. Huffman (Chapter 19) then closes with an assessment of what new breakthroughs in our understanding of basin loading history, seismic processing, and velocity analysis are critical to advancing the state-of-the-art of pore-pressure prediction.

We would like to thank the AAPG for supporting this effort from a sister society, and we would also like to thank the authors that persevered through the writing and editorial process.

Alan R. Huffman and Glenn L. Bowers, Editors

September 30, 2000

1

Comparison of Overpressure Magnitude Resulting from the Main Generating Mechanisms

Richard E. Swarbrick
University of Durham, Durham, England

Mark J. Osborne
University of Durham, Durham, England;
BP Exploration, Sunbury on Thames, England

Gareth S. Yardley
Heriot-Watt University, Edinburgh, Scotland

ABSTRACT

Overpressure is created by two main processes: (1) stress applied to a compressible rock and (2) fluid expansion. Both processes are most effective in fine-grained lithologies, such as mudrocks and chalks. Both processes involve ineffective fluid expulsion to create pressures in excess of hydraulic equilibrium, emphasizing the importance of permeability (a poorly known rock property in fine-grained sedimentary rocks) in controlling pore pressure in the subsurface. Overpressure generation and fluid expulsion can be modeled assuming Darcy flow though a pore matrix. The basin conditions favoring high-magnitude overpressure from stress are a high sedimentation (loading) rate and/or strong lateral compressive forces. A high sedimentation rate, as a means to create rapid increase in temperature, also favors high-magnitude overpressure from fluid expansion mechanisms. An alternative method to achieve a rapid increase in temperature is a thermal pulse associated with tectonic or magmatic processes.

Magnitude of overpressure from applied stress is controlled by the rate of increase in stress, sediment and fluid compressibility, and the rate of fluid expulsion. Continuous deposition of fine-grained sedimentary rocks leads to onset of overpressure at the fluid retention depth (FRD), the point at which fluid expulsion is no longer fully effective. Overpressure magnitude in this setting commonly increases at approximately 12.0–12.6 MPa/km (0.95–1.0 psi/ft), that is, along a gradient subparallel with the lithostatic stress gradient, implying only minor fluid expulsion. The variable characteristic of each lithology creates differences in the magnitude of overpressure.

Magnitude of overpressure from fluid expansion mechanisms is controlled by the rate of volume change, which can be shown to be slow for the burial rates and temperature gradients found in most basins. In addition the volume increase in all reactions except gas generation is small. Although gas generation has the capacity to create high-magnitude overpressure locally (up to tens of MPa), the magnitude is diluted where a large connected reservoir volume is involved. The process is therefore most likely effective only within gas-generative source rocks and in thin intraformational reservoirs where oil cracks to gas.

Swarbrick, Richard E., Mark J. Osborne, and Gareth S. Yardley, 2002, Comparison of Overpressure Magnitude Resulting from the Main Generating Mechanisms, in A. R. Huffman and G. L. Bowers, eds., Pressure regimes in sedimentary basins and their prediction: AAPG Memoir 76, p. 1–12.

Chemical processes involving fabric change, dissolution/reprecipitation, and solid to liquid transfer may also play a part in creating conditions favoring overpressure through fluid retention, but these cannot be quantified with existing data. Finally, overpressure related to hydraulic head and hydrocarbon buoyancy effects should not be ignored, but the magnitude of overpressure can be easily assessed.

INTRODUCTION

The phenomenon of overpressure in sedimentary basins has been attributed to a wide range of mechanisms that can be related to the following processes: increase in stress applied to a compressible rock, fluid expansion within a restricted pore space, fluid movement, buoyancy, diagenesis, and osmosis (Osborne and Swarbrick, 1997). The ability for each of these mechanisms to generate pressures above hydrostatic pressure depends on the rock and fluid properties of the sedimentary rocks and their rate of change under the normal range of basin conditions.

The magnitude of overpressure varies from basin to basin. Present-day pressure distribution can be assessed from direct measurement in permeable units (e.g., RFT, DST pressures). The pressures in low-permeability lithologies cannot be measured directly but can be inferred from indirect measurements. Most of the techniques are linked to porosity and assume that the porosity is controlled by the maximum effective stress the sediment has experienced. An upper limit to overpressure is the minimum stress plus tensile strength of the sediment (sometimes referred to as the fracture pressure) (Figure 1). In some basins (for example, North Sea, mid-Norway, Scotian Shelf) overpressure is in excess of 50 MPa (7250 psi) and close to the fracture pressure at drilling depths of 5.0 km. The purpose of this chapter is to review each of the principal mechanisms for creating overpressure and to evaluate their effectiveness to create the magnitude of overpressure found in typical basins, assuming realistic conditions of basin evolution.

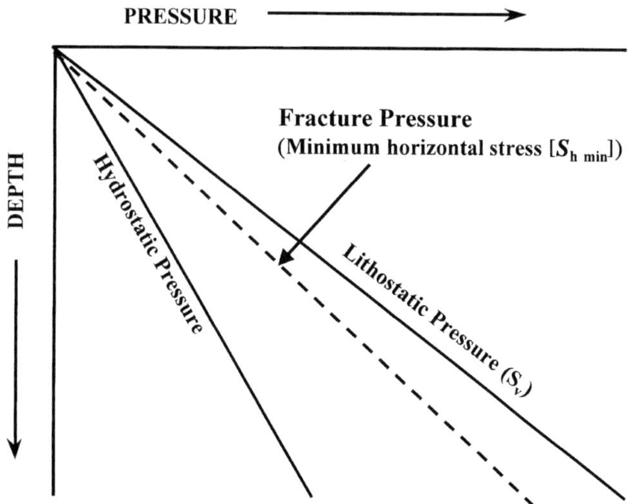

Figure 1. Schematic pressure vs. depth profile showing fracture pressure derived from leak-off tests.

PRINCIPAL PROCESSES

Stress-Related: Vertical

Increase in vertical stress during loading can lead to incomplete dewatering of the sediment when part of the weight of the load is added to the pore-fluid pressure. The mechanism is commonly termed "disequilibrium compaction," and the physical manifestation in the bulk rock is excess pore pressure and a higher porosity relative to the normally pressured and fully compacted rock at the same depth. Disequilibrium compaction overpressure commences at the depth where the permeability becomes too low to allow complete dewatering (fluid retention depth [FRD]) (Figure 2). The onset of overpressure is controlled by the loading rate and the porosity and permeability evolution of the sediment during burial. The FRD is shallower in low-permeability and/or more compressible sedimentary rocks such as mudrocks and deeper in more permeable and less compressible rocks such as silts and sands, assuming the same sedimentation rate. Conversely, for the same sediment the FRD is shallow at high sedimentation rates and deep at slow sedimentation rates. Luo and Vasseur (1992) show that the main controls on overpressure from disequilibrium compaction are loading rate, compaction coefficient (a method of expressing rock compressibility), temperature, and permeability, which controls the rate of fluid expulsion.

When modeling vertical loading stress, the loading rates are derived from reconstruction of the burial history and input of rock properties (for example, permeability and rock and fluid compressibility for the main sediment types), and their evolution during burial is required. Modeling provides a useful guide to the magnitude of overpressure where the parameters are carefully chosen and where mechanical compaction processes govern porosity reduction and fluid

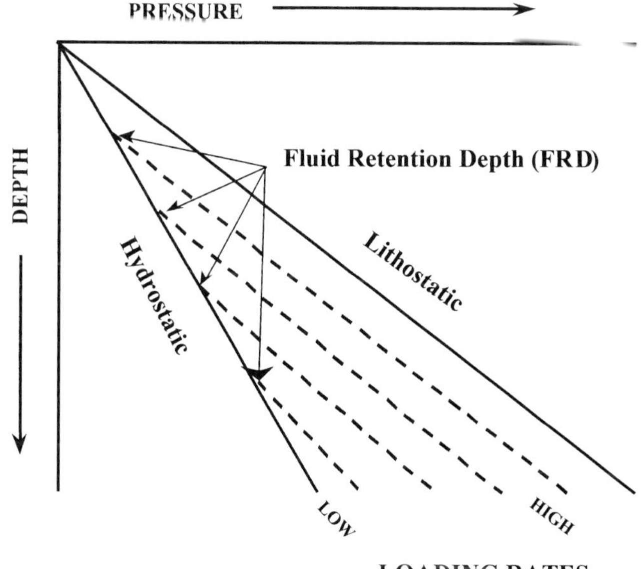

Figure 2. Pressure-depth profiles for overpressured fine-grained rocks created by disequilibrium compaction. Sedimentation rate and sediment permeability are the main processes that govern the magnitude of overpressure.

expulsion. Most basin modeling software uses the Terzaghi effective stress relationship,

$$\sigma_V = S_V - P_f \quad (1)$$

where σ_V is vertical effective stress, S_V is vertical load (overburden), and P_f is pore pressure. Goulty (1998) has pointed out that pore pressure is controlled by mean effective stress (S_m), which may be determined from estimation of both vertical and horizontal stresses. Use of the vertical effective stress in extensional basins (where S_V is the largest of the three principal stresses) can lead to an underestimation of pore pressure in mudrocks (see Harrold et al. [1999] for discussion).

When modeling overpressure from disequilibrium compaction, a full three-dimensional description of the subsurface is required ideally because connectivity of high-permeability strata leads to rapid redistribution of the fluids and pressures, generally from a downdip location where differential subsidence has created higher magnitude overpressure. This lateral transfer, whereby higher than expected overpressure exists at the crest of tilted reservoirs, is another cause of underestimation of overpressure using one-dimensional models. The effect is most extreme when tilting accompanies burial (for example, during active growth faulting, such as in the offshore Gulf of Mexico) (Yardley and Swarbrick, 2000).

Stress-Related: Lateral

Lateral compression is an alternative way in which stress can be imposed on a body of sedimentary rocks to cause overpressure. Areas of thrusting and folding typically contain overpressured rocks, and the magnitude of overpressure in these regions relates to both the amount of stress and strain in the rocks and their physical properties. Examples of high-magnitude overpressure in compression include the Barbados Accretionary Prism (Fisher and Zwart, 1996), the Andes, and Papua New Guinea (Henning et al., 1998).

Fluid Expansion

Overpressure is created by fluid expansion in low-permeability rocks, where pore fluid volume increases with minimal change in porosity and at a rate that does not permit effective dissipation of fluid. Causes of fluid expansion are cited in the literature as clay dehydration, smectite-illite transformation, maturation of source rocks to oil and gas, gas cracking, and mineral precipitation/cementation reactions (see Osborne and Swarbrick, 1997). In addition, fluid and rock expansion occurs because of a temperature increase during burial, leading to aquathermal pressuring (Barker, 1972).

In all the preceding mechanisms, the magnitude of overpressure is controlled by the rate of volume change, as well as the properties of the sedimentary rocks in which the change occurs. Using fluid-flow relationships governed by the Darcy flow law, the magnitude of overpressure for each fluid expansion mechanism can be determined if the rate of volume change is known or can be estimated. Realistic dimensions for the part of the rock in which the change occurs (compartment) and the seal must be assumed, and the permeability of the seal needs to be assigned. Vertical flow through the rock matrix is also assumed, whereas flow through microfractures is also considered likely in such low-permeability sedimentary rocks (Palciauskas and Domenico, 1980; Capuano, 1993). Ignoring microfracture permeability overestimates the magnitude of overpressure created by in-situ volume expansion.

Overpressure from fluid expansion mechanisms has received widespread commentary in the literature but little quantification. Many of the processes of fluid expansion are poorly constrained in relation to the magnitude of fluid change and its rate. Osborne and Swarbrick (1997) examined the processes and the rate of some of the reactions. What must be defined are the volume change and the rate at which the volume change occurs under a typical range of basin conditions. We assume in this chapter that Darcy's flow equation is an applicable description of subsurface fluid flow. To calculate the comparative magnitude of overpressure under the same range of basin conditions, we determined fluid flux from the total volume change (vol. %) for the time interval over which the change occurs (m.y.) for each fluid expansion mechanism. The fluid flux (Q) is calculated for a compartment 10 m thick overlain by a seal 100 m thick (L), whose permeability (k) is assumed to be 10^{-21} m^2 (10^{-6} md). Typical permeability for argillaceous rocks ranges from 10^{-17} to 10^{-22} m^2 (10^{-2}–10^{-7} md) (Neuzil, 1994), and an average permeability of 10^{-21} m^2 (10^{-6} md) for the entire seal thickness is considered to represent the lower bound of likely sealing conditions in the absence of extensive diagenetic cementation.

Darcy's flow equation relates volumetric rate of flow to the effective permeability of the rock, the flow path dimensions, and the pressure drop along the length of the flow path (seal thickness):

$$Q = \frac{kA\Delta P}{\mu L} \qquad (2)$$

where Q is the flow rate (cm^3/s), k is the permeability (darcy), A is the cross sectional area of the flow path (cm^2), ΔP is the pressure drop along the flow path (atmospheres), μ is the fluid viscosity (cP), and L is the flow path length (cm). One darcy is the permeability required for 1.0 cm^3 of fluid with viscosity of 1.0 cP (i.e., fresh water) to flow in one second through a rock mass with 1.0 cm^2 cross sectional area, with a pressure drop of 1.0 atmosphere along a length of 1.0 cm. Formation water of 1.0 g/cm^3 is assumed. The presence of more than one fluid would reduce the permeability and create higher overpressures than calculated. Multiphase fluids are not ubiquitous, however, and are largely restricted to petroleum source rocks, migration pathways, biogenic gas sources, and petroleum traps.

Solving the Darcy flow equation gives the magnitude of the pressure drop (ΔP) across the seal. Consequently, the solution provides the magnitude of overpressure in the compartment. The range of basin conditions shown in Table 1 has been chosen as representative for most basin settings. Each of the fluid expansion mechanisms is exclusively or dominantly controlled by temperature; time is a less important factor than temperature in kinetically controlled reactions, such as smectite to illite transformation. Rapid burial with high geothermal gradients favors the most rapid volume change and produces the largest magnitude overpressure. The range of geothermal gradients from 20°C/km to 40°C/km encompasses most basin settings. The rate of heating is calculated using sedimentation rates that range from 100 m to 2.0 km per m.y. The steady-state thermal conditions we assume tend to overestimate the overpressure. Note that high sedimentation rates are generally associated with low geothermal gradients (e.g., Gulf of Mexico offshore Louisiana, 25°C/km and 1.0–2.0 km/m.y.; Caspian Sea, 16°C/km and 1.25 km/m.y. (Bredehoeft et al., 1988).

Table 1. Modeling Conditions Chosen to Capture the Burial and Thermal Conditions of Most Sedimentary Basins

	High	Low
Geothermal gradients (°C/km)	40	20
Sedimentation Rate (km/m.y.)	2.0	0.01
Permeability (m^2/md)	10^{-21} [10^{-6}]	

MAGNITUDE OF OVERPRESSURE

Stress-Related: Vertical

The amount of vertical stress (loading) controls the maximum pore-pressure increase possible during any period of sediment burial. According to Terzaghi's relationship, the maximum increase in pore pressure equals the total added load, but only if there is no fluid expulsion, the sediment is compressible, and fluid compressibility is also ignored. In reality the sediment does dewater, and as it does, the sediment properties change. Of course, the rock has some strength, and there is some compression of the fluid phase, small for oil and water and larger for gas. Empirical data from several basins experiencing rapid deposition of fine-grained sedimentary rocks, however, show systematic increase in magnitude of overpressure below the FRD (Figure 2). The pressure gradient is approximately 12.0–12.6 MPa/km (1.0 psi/ft). Examples of this type of setting, having a pressure gradient subparallel with the lithostatic gradient, include Gulf of Mexico deep water, Tertiary deposits of the North Sea, and the Nile Delta, Egypt. The gradual increase of overpressure subparallel with the lithostatic pressure gradient, illustrated systematically in Figure 2, is characteristic of a continuous sedimentary column dominated by fine-grained lithologies. By contrast, where high-permeability lithologies are present in the upper part of the sequence, vertical hydraulic equilibrium is achieved. In many deltaic sequences (e.g., Gulf of Mexico [Dickinson, 1953]; Mahakam Delta, Indonesia [Burrus, 1998]) the upper sand-rich compartment is therefore normally pressured, passing into a highly overpressured mud-rich section beneath, with a sharp transition zone. The transition zone is lithology controlled and is not in this instance due to pressure-generating mechanisms other than disequilibrium compaction (Swarbrick and Osborne, 1996).

Profiles of pore-pressure measurements from reservoirs within mudrock-dominated sequences where the increase of pressure with depth is subparallel with the lithostatic pressure gradient have been used in this study to establish the FRD (Table 2; Figure 3). The FRD is determined by intersection of a shale pressure trend, determined from pore-pressure measurements in isolated sands with the hydrostatic pressure curve (Figure 4). The FRD is measured from seabed. A correlation exists between sedimentation rate (log scale) and FRD where the data from shale-dominated regions are compiled (Figure 3). The FRD is shallow where the sedimentation rate is high, leading to high-magnitude overpressure (low effective stress) at depth. Fluid-flow modeling (unpublished) shows that permeability at the FRD is likely to be in the range from 1 to about 10 nd. The scatter in the data is most likely due to natural variations in both permeability and the compaction coefficient of fine-grained sedimentary rocks.

Stress-Related: Lateral

Magnitude of overpressure in sedimentary rocks undergoing lateral stress is poorly known because there has been less petroleum exploration in this setting and therefore few well-documented studies in the literature. The magnitude of overpressure is expected to be higher in tectonically related regions where there is also active vertical stress loading. The upper limit of overpressure remains the same in both cases, namely, the minimum stress. In areas where lateral stress is dominant, the minimum stress is the vertical stress and is equal to the lithostatic stress. In tectonically relaxed areas the minimum stress is the horizontal stress and is lower than the vertical stress.

In summary, high-magnitude stress-related overpressure results from either long periods of rapid loading involving fine-grained sedimentary rocks or extreme tectonic compression. The largest magnitude overpressure created by disequilibrium compaction alone (without considering fluid redistribution by lateral and vertical fluid transfer) is equal to the total stress applied. The shallower the FRD, the larger the potential magnitude of overpressure. Evidence for disequilibrium compaction is found by examination of the porosity-effective stress relationship relative to lithology (Aplin et al., 1995; Harrold et al., 1999). The magnitude of overpressure (and hence effective stress) should correlate with porosity of the fine-grained rocks in which the disequilibrium compaction overpressure is generated. Analysis of porosity trends (or a proxy for porosity, such as sonic velocity, density, or resistivity) forms the basis for several pore-pressure prediction methods (e.g., Equivalent Depth, Eaton) (see Mouchet and Mitchell, 1989, p. 64). Porosity, however, does not always correlate with vertical effective stress and pore pressure, as reviewed by Goulty (1998).

Table 2. Tabulation of Data for Basins That Exhibit Lithostat Parallel Pressure Profiles

	Basin Name	Fluid Retention Depth (km)	Average Sedimentation Rate (m/m.y.)	Reference
1	Flexture trend, Gulf of Mexico	1.3	600	Mann and Mackenzie, 1990
2	Flexture trend, Gulf of Mexico	0.95	3000	Mann and Mackenzie, 1990
3	Northern North Sea	1.2	35	Mann and Mackenzie, 1990
4	Haltenbanken, mid-Norway	1.3	50	Mann and Mackenzie, 1990
5	Nile Delta	0.75	810	Mann and Mackenzie, 1990
6	Malay Basin	0.95	240	Yusoff and Swarbrick, 1994
7	Tertiary, Central North Sea	1.0	55	Holm, 1998
8	Upper Tertiary, Central North Sea	0.7	500*	Swarbrick et al., 1998
9	Southeast Trinidad	0.7	2000	Heppard et al., 1998
10	Gulf of Paris, Trinidad	1.2	200	Heppard et al., 1998
11	Gulf of Mexico shallow silty	1.5	1000	unpublished
12	Gulf of Mexico deep water sey	0.8	1000	unpublished
13	Gulf of Mexico deep water shaley	0.7	2000	unpublished
14	Gulf of Mexico deep wateshaley	0.86	1000	unpublished
15	Gulf of Mexico deep water shaley	0.5	2850	unpublished
16	Gulf of Mexico deep water shaley	0.86	1400	unpublished

*Sedimentation rate as high as 500 m/m.y. in last 3 m.y.

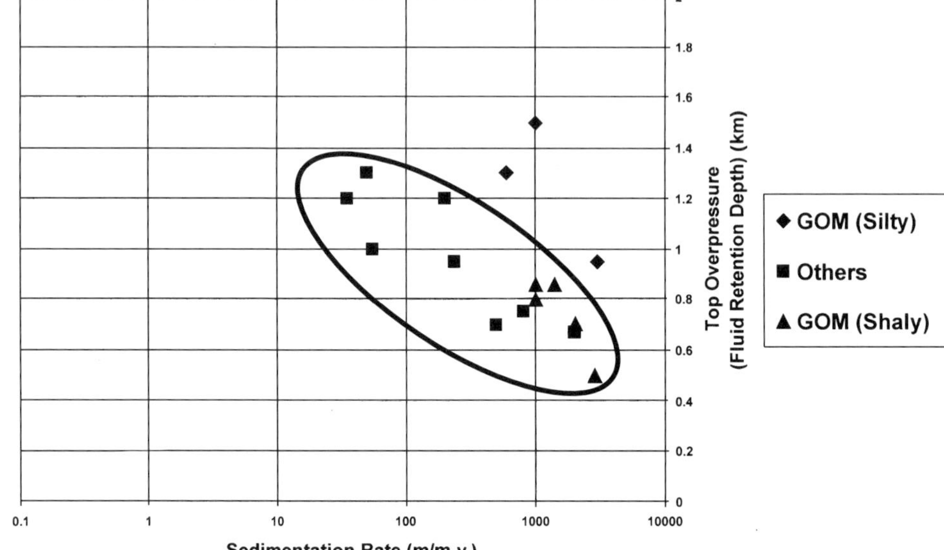

Figure 3. Plot of sedimentation rate vs. FRD. The FRD is estimated from pore-pressure trends, projected to the intersection with the hydrostatic pressure curve, and measured from the seabed. GOM = Gulf of Mexico.

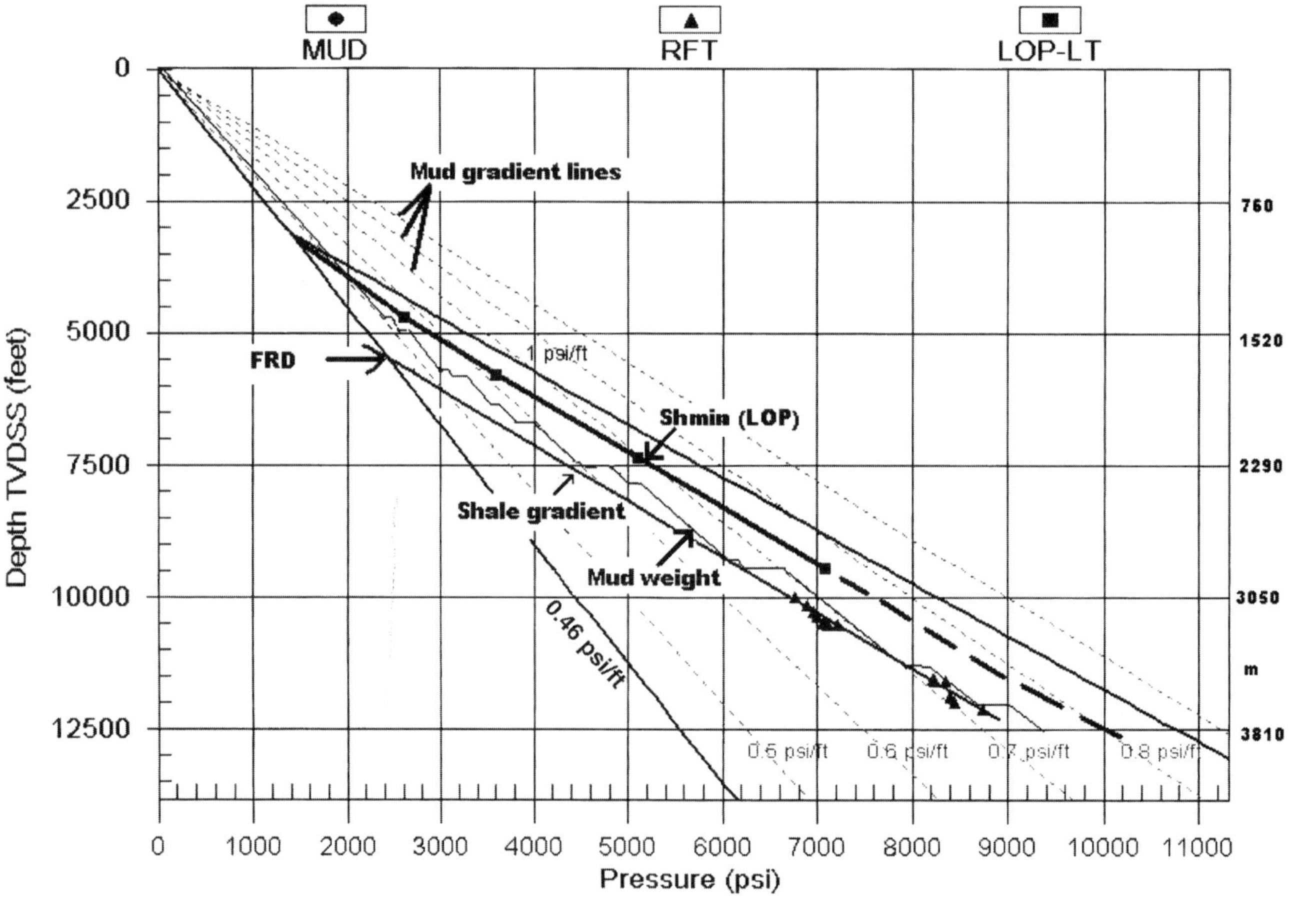

Figure 4. Pressure vs. depth plot for a deep-water well in the Mississippi Canyon area of the Gulf of Mexico. The direct pore-pressure measurements provide a shale pressure trend that intersects the hydrostatic pressure curve at the FRD, 1800 m below sea level, 863 m below seabed. RFT = repeat formation tester; LOP-LT = leak-off pressure; Shmin = minimum horizontal stress.

Fluid Expansion Mechanisms

Aquathermal

Increasing temperature during burial causes both rocks and fluids to expand. The volume expansion of rock is approximately one order of magnitude smaller for rock than for water (Miller, 1995) and therefore can be ignored. The fluid volume change due to aquathermal expansion is 1.65% for an increase in temperature of 40°C (Osborne and Swarbrick, 1997). At a heating rate of 40°C/km and a sedimentation rate of 2.0 km/m.y., the volume expansion occurs in 0.5 m.y., yielding a fluid flux equal to 3.3 vol. %/m.y. Under a typical range of basin conditions (Table 1) this rate of volume change produces approximately 0.7 MPa (100 psi) overpressure, assuming seal permeability of 10^{-21} m^2 (10^{-6} md). Aquathermal expansion does not therefore represent an effective mechanism to create the high-magnitude overpressure (e.g., 60 MPa [8700 psi] at 7.0 km) found in some basins, which is in agreement with the modeling results of Luo and Vasseur (1992).

Smectite Dehydration

The most common hydrated, detrital mineral in fine-grained clastic sedimentary rocks is smectite. Smectite dehydration involves three stages of dewatering, which Osborne and Swarbrick (1997) estimate lead to a total increase of 4.0 vol. % occurring in three stages, each of about 1.3 vol. %. The first two pulses occur at depths of 0.5–1.5 km with negligible overpressure generated, due to the high sediment permeability at shallow depths. For the modeling conditions shown in Table 1, and with a sediment composed entirely of smectite, the range of magnitude of overpressure generated during the third dewatering phase is 0.05–0.7 MPa (7 and 100 psi). This phase would likely occur at depths of 3.0–5.0 km, depending on geothermal gradient.

Smectite to Illite Transformation

The depth at which smectite transforms to illite ranges from 2.3 to 3.3 km on the Gulf of Mexico shelf, from 3.5 to 4.5 km in the Niger Delta (Bruce, 1984), and from 2.4 to 3.5 km in the North Sea (Pearson and Small, 1988). The exact volume changes involved in the smectite to illite transformation reaction are difficult to ascertain. Osborne and Swarbrick (1999) calculated the volume change associated with 10 possible smectite to illite reaction pathways. The volume change ranges from an increase of 4.1% to a decrease of 8.4% depending on reactants and products. Standard kinetic models were used to calculate the cumulative volume change against time for six of these reactions (Figure 5). The geothermal gradient used by Osborne and Swarbrick (1999) was 35°C/km with a sedimentation rate of 50 m/m.y. The maximum rate of volume increase is 0.2 vol. %/m.y. Scaling to a higher heating rate (40°C/km) and a higher sedimentation rate (2.0 km/m.y.) and assuming an initial smectite content of 20% gives a volume increase of 1.82 vol. %/m.y. These conditions lead to an overpressure of 0.8 MPa (112 psi) where the seal permeability is 10^{-21} m^2 (10^{-6} md).

Kerogen Transformation/Gas Generation

Volume changes occur when kerogen transforms to oil and gas and when oil cracks to gas. The volume change depends on the kerogen source and the density and volume of the petroleum products generated during maturation. Meissner (1978) showed schematically that volume change could be as large as 25 vol. % during oil generation from a type II kerogen to more than 100 vol. % during dry gas stage of maturation. Ungerer et al. (1983) advocated a small volume decrease in the early stage of maturation (thermal maturity up to % R_o = 1.3) followed by an increase to more than 50 vol. % at a late stage of maturation (R_o = 2.0) for a type II kerogen. Calculations by the GeoPOP research group (S. R. Larter, 1996, personal communication) confirm an early small volume decrease followed by a progressive volume increase starting at thermal maturity equivalent of % R_o values of 0.7–0.9. The precise volume change depends on the compositional kinetic model chosen. We have used unpublished data, which is in broad agreement with the estimated volume change at high thermal maturity determined by Ungerer et al. (1983). The magnitude of volume increase ranges from 75 to 110 vol. % at % R_o values of 3.7 if a closed system is assumed (i.e., no fluid loss as pressure increases up to fracture pressure). The total volume change is greater (140–200 vol. %) when pressure is maintained at hydrostatic pressure (i.e., when oil or gas leave the source rock during primary migration).

The potential exists, therefore, to create large magnitude overpressure in source rocks during the later stages of oil maturation and on into gas generation and during in-situ oil to gas cracking. Using the preceding range of maximum volume increase from 75 to 140 vol. % that occurs over the time interval for maturation of a type II source rock from % R_o values of 0.7–1.75 and data for Cretaceous–Jurassic vitrinite published in Goff (1983), the estimated temperature interval over which the transformation has occurred is approximately 70°C. For the range of burial rates and geothermal gradients in Table 1, volume increases are estimated in the range of 3%/m.y. to 100%/m.y., for which we calculate overpressures of 0.5–41 MPa (70–6000 psi). These estimates are probably high because migration of petroleum out of the source rocks and/or maturation

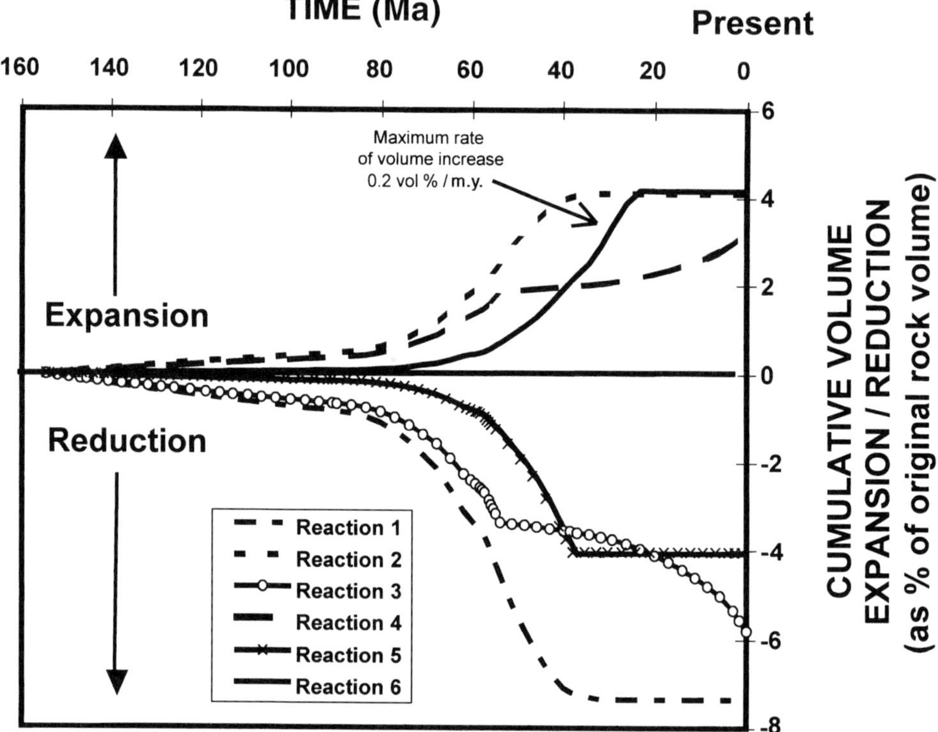

Figure 5. Volume change associated with six reaction pathways for smectite to illite. Maximum rate of change is 4.0 vol. % increase per m.y. (Osborne and Swarbrick, 1999).

retardation at elevated pressures (Carr, 1991, 1999; McTavish, 1998) reduces the overpressure magnitude. A further assumption is that the petroleum occupies all of the available connected pore space, that is, the water phase has been ignored.

The magnitude of overpressure resulting from gas generation is dependent on the location of the transformation. In source rocks, where the petroleum phase may dominate, high-magnitude overpressure could be realized for high petroleum-saturated porosity. Within a petroleum source rock the presence of a second fluid phase also acts to reduce the permeability, making fluid retention more effective. The presence of overpressure in thin reservoirs within the mature Bakken shale, a rich source rock in the Williston basin (Burrus et al., 1996), may be explained by kerogen maturing to oil and gas.

By contrast, in the case of in-situ reservoir oil cracking to gas (expected at temperatures in excess of about 170°C) any overpressure from volume expansion is redistributed through the larger connected pore volume, including the supporting aquifer. If the volume of trapped petroleum is only 10% of the total connected pore volume in the reservoir (an example based on northern North Sea Brent reservoir data) the overpressure is proportionately reduced. Under the most favorable heating rates, we calculate gas generation may account for up to only 4.0 MPa (600 psi) overpressure in the Brent reservoir. Gas generation must therefore be a secondary cause of overpressure because overpressures in excess of 35 MPa (5000 psi) occur at depths of 4.0 km in the northern North Sea (Buhrig, 1989).

Other Mechanisms

Chemical Compaction. A variety of mineral transformations occur during burial. Examples include feldspar to illite, smectite to illite, and kerogen to oil/gas. These reactions involve a set of complex reactions that are largely kinetically controlled. Each of these transformations involves change in rock fabric. Feldspar dissolution creates large secondary pores, potentially removing a part of the load-bearing framework. Smectite to illite transformation creates a much tighter rock fabric with strong alignment of the clays (Colton-Bradley, 1987). Kerogen transformation reduces framework support; Hanebeck et al. (1993) showed experimentally that kerogen in a 10% total organic carbon source rock is load bearing.

Stylolitization (also known as diffusive mass transfer) is another chemical process that alters the framework support. Dissolution (pressure solution) occurs at selected grain contacts. In a closed system concurrent reprecipitation of the same mineral occurs or precipitation of a new mineral. The process of stylolitization is thought to be temperature and stress controlled, although the precise conditions are not well known. Both mineral transformations and stylolitization have the potential to reduce porosity, thereby displacing fluid. Bjorkum and Nadeau (1998) have cited quartz-clay stylolitization and complementary quartz reprecipitation as a principal cause of overpressure in quartz-sandstone reservoirs. We have therefore examined the magnitude overpressure that would result in a reservoir in which quartz dissolution at stylolites and quartz reprecipitation as secondary overgrowths (both porosity reducing processes) were active.

The magnitude of overpressure created by quartz dissolution/reprecipitation has been calculated in the same manner as for fluid expansion mechanisms. Using rates of volume loss up to 3.5 vol. %/m.y. (Osborne and Swarbrick, 1999) the resulting magnitude of overpressure ranges from 0.05 to 1.3 MPa (7.0–200 psi). The mechanism does not therefore explain well the high magnitude of overpressure encountered in many quartz-rich reservoirs. Additionally, if the mechanism was effective as an overpressure-generating mechanism, the amount of quartz should correlate positively with the magnitude of overpressure. In the central North Sea Fulmar Formation the opposite relationship is observed (Osborne and Swarbrick, 1999). There, the amount and rate of quartz cementation (as determined from fluid inclusion microthermometry) is lower in the reservoirs with high overpressure (low effective stress) than in the low-overpressure reservoirs. These observations suggest that effective stress at grain contacts is a principal control on the supply of quartz by dissolution (pressure solution) and its diffusion to sites of reprecipitation.

Chemical compaction processes may, however, be significant in creating overpressure where part of the load-bearing framework is transformed into liquid, as in the case of kerogen to oil/gas, and where the transformation radically alters the compressibility of the framework. Swarbrick and Osborne (1998) gave an example of a rich petroleum source rock with 10 vol. % kerogen and 13% porosity in which kerogen is assumed to have equal load-bearing status with other minerals present. Half of the kerogen is transformed to liquid oil, thereby increasing the porosity to 18%. Assuming a typical porosity–effective stress relationship for shale, the overpressure created is of the order of 7.6 MPa (1100 psi), if there is no fluid loss. Lahann (2002) suggests smectite to illite transformation could create overpressure in a similar manner, as a consequence of load transfer.

Chemical transformations and stylolitization involving solid to liquid transfer and changes in compressibility of the rock may be important as secondary processes and may also help to explain low porosity in highly overpressured reservoirs. Further research is required to establish the impact of chemical reactions on rock strength, particularly in fine-grained sedimentary rocks.

Hydraulic Head. Overpressure can result from a hydraulic head in an adjacent highland area, where there is hydraulic continuity into the subsurface. The magnitude can be easily assessed using the height above datum, fluid densities, and flow rates if the aquifer is active (e.g., Bachu and Underschultz, 1993). A head created by 3.0 km (10,000 ft) of structural relief above datum yields a maximum hydraulic head of 30.0 MPa (4350 psi) pressure.

Hydrocarbon Buoyancy. A pressure differential exists between formation water and petroleum in a petroleum accumulation. The magnitude of overpressure in the hydrocarbon phase is calculated using relevant fluid densities. As an example, a 1.0 km (3300 ft) oil column of 36° API oil creates a buoyancy pressure of 2.0 MPa (290 psi) at the top of the accumulation. The same column of dry gas creates 7.5 MPa (1090 psi) overpressure in the hydrocarbon phase at the top of the accumulation. The overpressure due to hydrocarbon buoyancy is proportional to column height and is in addition to any overpressure in the underlying aquifer.

Osmosis. Salinity variations occur naturally in sedimentary sequences. Salinity contrasts across semipermeable membranes induce osmotic flow that continues as long as the salinity contrast is maintained by recharge. Osborne and Swarbrick (1997) calculated an upper limit to the magnitude of overpressure from osmosis of about 4 MPa (580 psi) for an average shale of North Sea composition. Neuzil (2000) measured osmotically induced overpressures in shallow borehole experiments and further suggested an overpressure magnitude of up to 20 MPa (2900 psi) as an upper limit for a low-porosity, smectite-rich shale.

DISCUSSION AND CONCLUSIONS

The most potent mechanism for creating overpressure in young sedimentary basins is vertical stress during rapid sediment burial. Vertical stress is imposed simultaneously on the entire rock section, and continuation of the process over geological time in rocks with similar properties yields pressure profiles in which overpressure commences at the FRD and continues to increase with increasing depth. The magnitude of overpressure observed can be large and is controlled mainly by burial rate, the permeability evolution of the sediment, and the compressibility of the rock and fluid. Tectonically induced shortening can also create overpressure. Retention of overpressure in older sedimentary rocks in which there was an early phase of rapid burial depends on the permeability evolution of the sediments controlling dissipation.

Of the many fluid expansion mechanisms, only gas generation during the later stages of source rock maturation and/or in-situ cracking of oil to gas create high-magnitude overpressure under the range of conditions considered in this chapter. Volume change is small for other mechanisms, so high-magnitude overpressure does not result, even where permeability is at extremely low values. The magnitude of all overpressure depends critically on the permeability of the sedimentary rocks, both in its creation and in its dissipation. High-magnitude overpressure from gas generation is likely limited to source rocks because the effects are diluted within the connected aqueous pore volume. A diagram comparing the rate of volume change for each of the main fluid expansion mechanisms, plus quartz diagenesis (Figure 6), illustrates the dependence on sediment permeability for the magnitude of overpressure and the way in which gas generation creates an order of magnitude of greater in-situ overpressure.

Two main observations challenge the explanation of high-magnitude overpressure from disequilibrium compaction alone: (1) pressures close to fracture pressure and (2) lower shale porosity than expected for the effective stresses measured. Pressures close to fracture pressure require very shallow FRDs, namely, very low permeabilities early in burial history. One lithology likely to have such low permeability near the surface is evaporites. Very rapid deposition of fine muds, for example in a deep-water, continental slope setting, is also known to create shallow overpressure. In the absence of these untypical lithologies a secondary source of overpressure is required. The principal candidates are gas generation, lateral transfer, and chemical changes, the latter being poorly quantified at present.

Where porosity is lower than expected relative to overpressure from disequilibrium compaction, two main explanations exist. The overpressure may postdate conventional mechanical porosity reduction, that is, the rocks have experienced a higher effective stress in the past, and a later secondary source of overpressure has subsequently reduced the effective stress. Examples of such behavior are described by Bowers (1994) and Hart et al. (1995) from the Gulf of Mexico; by Burrus (1998) from the Mahakam Delta, Indonesia; and by Hermanrud et al. (1998) from mid-Norway. In each case porosity is lower than can be reasonably explained by disequilibrium compaction overpressure, and a secondary source is assumed. An alternative explanation lies in reduction of porosity of the fine-grained sedimentary

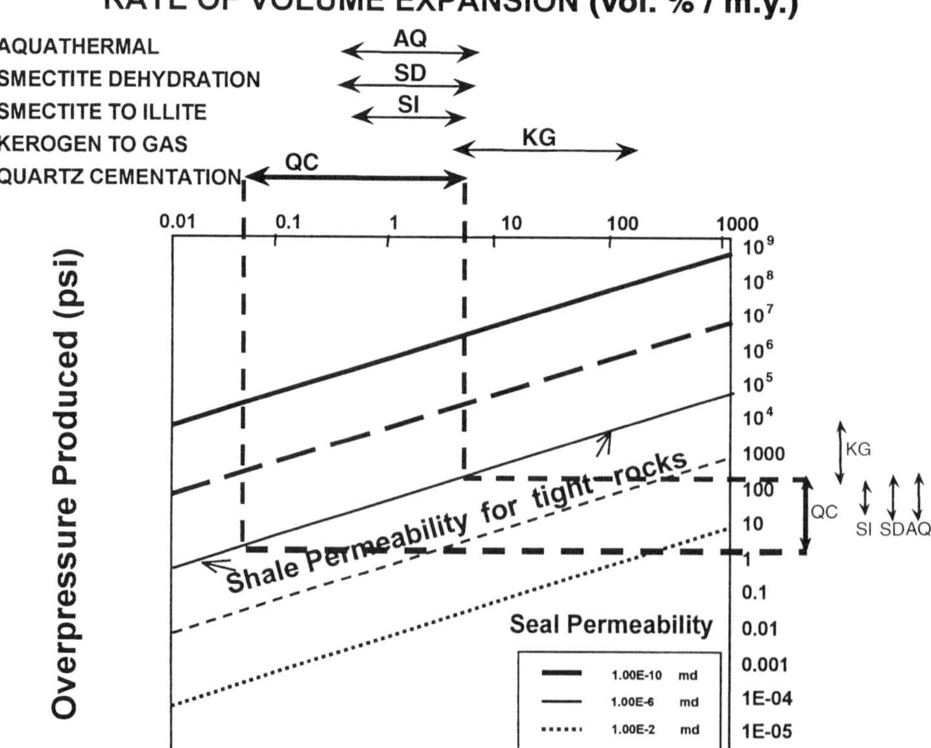

Figure 6. Plot of rate of volume change for the main fluid expansion mechanisms, including quartz cementation, against magnitude of overpressure for a range of seal permeability. Calculations assumed a seal compartment 10 m thick and seal of 100 m. Example for quartz cementation illustrated with dashed lines.

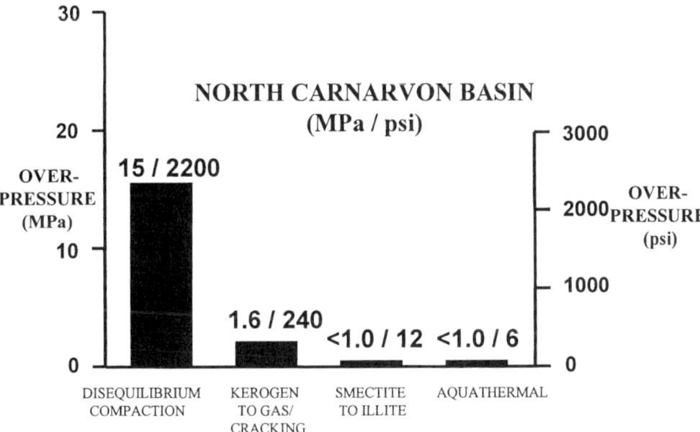

Figure 7. Comparison of overpressure magnitude for stress and volume expansion mechanisms under conditions of burial for the north Carnarvon Basin, northwest Australia. The estimates for fluid expansion (in particular gas generation) consider the in-situ overpressure corrected for redistribution into the accompanying water phase. Assumptions for model data: burial rate = 100 m/m.y.; temperature gradient = 34°C/km; minimum permeability = 10^{-21} m^2 (10^{-6} md) = nanoDarcy (Swarbrick and Hillis, 1999).

rocks independent of stress. Changes in porosity and permeability during both mechanical and chemical compaction processes are poorly known and remain a fruitful area for future research.

A comparison of overpressure magnitude for stress and fluid expansion mechanisms under a set of typical basin conditions is illustrated in Figure 7 using data for the northern Carnarvon Basin, northwest Australia. The figure illustrates that even for a relatively low sedimentation rate (100 m/m.y.) the magnitude of overpressure is significantly higher from disequilibrium compaction at 15 MPa (2200 psi) than from any of the fluid expansion mechanisms. Potential overpressure from hydraulic head, osmosis, and hydrocarbon buoyancy are not illustrated. The potential effect of chemical compaction cannot be quantified using existing data.

ACKNOWLEDGMENTS

We were part of the GeoPOP Phase 1 research group, funded by Agip, Amerada Hess, Amoco, Arco, Chevron, Conoco, Elf, Enterprise, Mobil, Norsk Hydro, Phillips Petroleum, Statoil, and Total. Thanks to Neil Goulty for identifying the need to use mean stress in pressure estimation and to Steve Larter for the calculation of volume change during maturation of kerogen. We wish to thank the GeoPOP research team at Durham, Newcastle, and Heriot-Watt universities and the sponsor representatives for their collective input and discussions. We are grateful for the comments of Rick Lahann and Glenn Bowers, which have led to many improvements in the clarity of this chapter.

REFERENCES CITED

Aplin, A. C., Y. Yang, and S. Hansen, 1995, Assessment of β, the compression coefficient of mudstones and its relationship with detailed lithology: Marine and Petroleum Geology, v. 12, p. 955–963.

Bachu, S., and J. R. Underschultz, 1993, Hydrogeology of formation waters, northeastern Alberta Basin: AAPG Bulletin, v. 77, p. 1745–1768.

Barker, C., 1972, Aquathermal pressuring—the role of temperature in development of abnormal pressures: AAPG Bulletin, v. 56, p. 2068–2071.

Bjorkum, P. A., and P. H. Nadeau, 1998, Temperature controlled porosity/permeability reduction, fluid migration, and petroleum exploration in sedimentary basins: Australian Petroleum Production and Exploration Association Journal, v. 38, no. 1, p. 453–464.

Bowers, G. L., 1994, Pore pressure estimation from velocity data: accounting for overpressure mechanisms besides undercompaction: International Association of Drilling Contractors/Society of Petroleum Engineers Drilling Conference, Society of Petroleum Engineers paper 27488, p. 515–530.

Bredehoeft, J. D., R. D. Djevanshir, and K. R. Belitz, 1988, Lateral fluid flow in a compacting sand-shale sequence, south Caspian Sea: AAPG Bulletin, v. 72, p. 416–424.

Bruce, C. H., 1984, Smectite dehydration—its relation to structural development and hydrocarbon accumulation in northern Gulf of Mexico basin: AAPG Bulletin, v. 68, p. 673–683.

Buhrig Jr., J. F., 1989, Geopressured Jurassic reservoirs in the Viking Graben: modelling and geological significance: Marine and Petroleum Geology, v. 6, p. 31–48.

Burrus, J., 1998, Overpressure models for clastic rocks, their relation to hydrocarbon expulsion: a critical reevaluation, *in* B. E. Law, G. F. Ulmishek, and V. I. Slavin, eds., Abnormal pressures in hydrocarbon environments: AAPG Memoir 70, p. 35–63.

Burrus, J. K., K. G. Osadetz, S. Wolf, B. Doligez, K. Visser, and D. Dearborn, 1996, A two-dimensional regional basin model of Williston basin hydrocarbon systems: AAPG Bulletin, v. 80, p. 265–291.

Capuano, R. M., 1993, Evidence of fluid flow in microfractures in geopressured shales: AAPG Bulletin, v. 77, p. 1303–1314.

Carr, A. C., 1991, A pressure dependent kinetic model for vitrinite reflectance, *in* D. Manning, ed., Organic geochemistry: 15th Meeting European Association of Organic Geochemists, p. 285–287.

Carr, A. C., 1999, A vitrinite reflectance kinetic model incorporating overpressure retardation: Marine and Petroleum Geology, v. 16, p. 355–377.

Colton-Bradley, V. A. C., 1987, Role of pressure in smectite dehydration—effects on geopressure and smectite to illite transition: AAPG Bulletin, v. 71, p. 1414–1427.

Dickinson, G., 1953, Geological aspects of abnormal reservoir pressures in Gulf Coast, Louisiana: AAPG Bulletin, v. 37, p. 410–432.

Fisher, A. T., and G. Zwart, 1996, Relation between permeability and effective stress along a plate-boundary fault, Barbados accretionary complex: Geology, v. 24, p. 307–310.

Goff, J. C., 1983, Hydrocarbon generation and migration from Jurassic source rocks in the E. Shetland Basin and Viking Graben of the northern North Sea: Journal of the Geological Society, v. 140, p. 445–474.

Goulty, N. R., 1998, Relationships between porosity and effective stress in shales: First Break, v. 16, p. 413–419.

Hanebeck, D., B. M. Krooss, and D. Leythaeuser, 1993, Experimental investigation of petroleum generation and migration under elevated pressure and temperature conditions (abs.): 1st Geofluids Conference, p. 51–53.

Harrold, T. W. D., R. E. Swarbrick, and N. R. Goulty, 1999, Pore pressure estimation from mudrock properties in Tertiary basins, southeast Asia: AAPG Bulletin, v. 83, p. 1057–1067.

Hart, B. S., P. B. Flemings, and A. Deshpande, 1995, Porosity and pressure: role of compaction disequilibrium in the development of geopressures in a Gulf Coast Pleistocene basin: Geology, v. 23, p. 45–48.

Henning, A., N. Yassir, T. Addis, A. Warrington, and S. Kravis, 1998, Pore pressure estimation in an active thrust region and its impact on exploration and drilling (abs.): American Association of Drilling Engineers Industry Forum on Pressure Regimes in Sedimentary Basins and their Prediction, unpaginated.

Heppard, P. D., H. S. Cander, and E. B. Eggertson, 1998, Abnormal pressure and the occurrence of hydrocarbons in offshore eastern Trinidad, West Indies, *in* B. E. Law, G. F. Ulmishek, and V. I. Slavin, eds., Abnormal pressures in hydrocarbon environments: AAPG Memoir 70, p. 215–246.

Hermanrud, C., L. Wensaas, G. M. G. Teige, E. Vik, H. M. Nordgard Bolas, and S. Hansen, 1998, Shale porosities from well logs on Haltenbanken (offshore mid-Norway) show no influence of overpressuring, *in* B. E. Law, G. F. Ulmishek, and V. I. Slavin, eds., Abnormal pressures in hydrocarbon environments: AAPG Memoir 70, p. 65–85.

Holm, G. M., 1998, Distribution and origin of overpressure in the Central Graben of the North Sea, *in* B. E. Law, G. F. Ulmishek, and V. I. Slavin, eds., Abnormal pressures in hydrocarbon environments: AAPG Memoir 70, p. 123–144.

Lahann, R., 2002, Impact of smectite diagenesis on compaction modeling and compaction equilibrium, *in* A. R. Huffman and G. L. Bowers, eds., Pressure regimes in sedimentary basins and their prediction: AAPG Memoir 76, p. 61–72.

Luo, X., and G. Vasseur, 1992, Contributions of compaction and aquathermal pressuring to geopressure and the influence of environmental conditions: AAPG Bulletin, v. 76, p. 1550–1559.

Mann, D. M., and A. S. Mackenzie, 1990, Prediction of pore fluid pressures in sedimentary basins: Marine and Petroleum Geology, v. 7, p. 55–65.

McTavish, R. A., 1998, The role of overpressure in the retardation of organic matter maturation: Journal of Petroleum Geology, v. 21, p. 153–186.

Meissner, F. F., 1978, Petroleum geology of the Bakken Formation, Williston basin, North Dakota and Montana, *in* 24th annual conference, Williston basin symposium: Montana Geological Society, p. 207–227.

Miller, T. W., 1995, New insights on natural hydraulic fractures induced by abnormally high pore pressures: AAPG Bulletin, v. 79, p. 1005–1018.

Mouchet, J. P., and A. Mitchell, 1989, Abnormal pressures while drilling: Manuels Techniques Elf Aquitaine, v. 2, 255 p.

Neuzil, C. E., 1994, How permeable are clays and shales?: Water Resources Research, v. 30, p. 145–150.

Neuzil, C. E., 2000, Osmotic generation of "anomalous" fluid pressures in geological environments: Nature, v. 403, p. 182–184.

Osborne, M. J., and R. E. Swarbrick, 1997, Mechanisms which generate overpressure in sedimentary basins: a reevaluation: AAPG Bulletin, v. 81, p. 1023–1041.

Osborne, M. J., and R. E. Swarbrick, 1999, Diagenesis in North Sea HPHT clastic reservoirs—consequences for porosity and overpressure prediction: Marine and Petroleum Geology, v. 16, p. 337–354.

Palciauskas, V. V., and P. A. Domenico, 1980, Microfracture development in compaction sediments: relation to hydrocarbon-maturation kinetics: AAPG Bulletin, v. 64, p. 927–937.

Pearson, M. J., and J. S. Small, 1988, Illite-smectite diagenesis and palaeotemperatures in northern North Sea Quaternary to Mesozoic shale sequences: Clay Minerals, v. 23, p. 111–132.

Swarbrick, R. E., and R. R. Hillis, 1999, The origin and influence of overpressure with reference to the North West Shelf, Australia: Australian Petroleum Production and Exploration Association Journal, v. 39, p. 64–72.

Swarbrick, R. E., and M. J. Osborne, 1996, The nature and diversity of pressure transition zones: Marine and Petroleum Geology, v. 14, p. 111–116.

Swarbrick, R. E., and M. J. Osborne, 1998, Mechanisms that generate abnormal pressures: an overview, *in* B. E. Law, G. F. Ulmishek, and U. I. Slavin, eds., Abnormal pressures in hydrocarbon environments: AAPG Memoir 70, p. 13–34.

Swarbrick, R. E., M. J. Osborne, and G. S. Yardley, 1998, The magnitude of overpressure from generating mechanisms under realistic basin conditions: American Association of Drilling Engineers Forum, 6 p.

Ungerer, P., E. Behar, and D. Discamps, 1983, Tentative calculation of the overall volume expansion of organic matter during hydrocarbon genesis from geochemistry data: implications for primary migration, *in* M. Bjoroy et al., eds., Advances in organic geochemistry: John Wiley, p. 129–135.

Yardley, G. S., and R. E. Swarbrick, 2000, Lateral transfer: a source of additional overpressure?: Marine and Petroleum Geology, v. 17, p. 523–538.

Yusoff, W. I., and R. E. Swarbrick, 1994, Thermal and pressure histories of the Malay Basin, offshore Malaysia (abs.): AAPG Bulletin, v. 78, p. 1171.

2

The Interrelationships between Overpressure Mechanisms and In-Situ Stresses

T. W. Miller
Knowledge Systems Inc., Stafford, Texas

C. H. Luk
Retired from ExxonMobil Upstream Research Company, Houston, Texas

D. L. Olgaard
ExxonMobil Upstream Research Company, Houston, Texas

ABSTRACT

In this chapter, we discuss how the two different excess pore-pressure–generating mechanisms that are primarily associated with burial affect stresses in different ways. We, and others before us, term these two mechanisms as either "compaction disequilibrium" or "source mechanisms." Rapid burial rates in association with low permeabilities are attributed to the former, whereas aquathermal expansion, smectite/illite diagenesis (or other diagenetic processes), kerogen maturation, and hydrocarbon cracking are examples of the latter.

The compaction disequilibrium mechanism is fundamentally different from the source mechanism. In the compaction disequilibrium case, pore-pressure increases are primarily a reaction of the fluid to pore-volume decreases that are a result of increased vertical loading (assuming minimal tectonic stresses). The magnitude of the pressure increase depends on the load increase and the relative magnitudes of the sediment pore-volume and pore fluid compressibilities. Regardless of the magnitude of the pore-pressure increase, mechanical equilibrium requires that the effective stresses of the sediment increase. As a result, both the horizontal and vertical (effective and total) stresses of a given sediment package increase as its burial depth increases, and pore-pressure increases due to burial are less than the increase in the overburden stress. In the source case, pore-pressure increases are a response to increases in the pore-fluid specific volume. In a source-dominated system, the sediment pore volume increases, and, consequently, the effective stresses decrease. In these cases, the pore pressures increase faster than the effective stresses decrease, and the horizontal and vertical total stresses increase. Because pore volumes increase, the increases in pore pressure can be larger than increases in overburden stress.

Miller, T. W., C. H. Luk, and D. L. Olgaard, 2002, The Interrelationships between Overpressure Mechanisms and In-Situ Stresses, in A. R. Huffman and G. L. Bowers, eds., Pressure regimes in sedimentary basins and their prediction: AAPG Memoir 76, p. 13–20.

INTRODUCTION

Although there are many mechanisms that cause overpressures, these mechanisms can be classified in general groups based on whether they are directly associated with burial and how they affect in-situ stresses. Compaction disequilibrium, aquathermal expansion, smectite/illite diagenesis (or other diagenetic processes), kerogen maturation, and hydrocarbon cracking are examples of overpressuring mechanisms associated with burial-depth increases. Shear failure (Yassir, 1989) and geometric factors such as large, tilted sand bodies (Flemings et al., 1998) are examples of other mechanisms that are not covered in this chapter.

This chapter focuses on those excess pore-pressure–generating mechanisms commonly associated with burial-depth increases and the fundamental differences between compaction disequilibrium and the other mechanisms. Of course, all of these mechanisms act simultaneously, so each mechanism contributes to excess pore pressures to some degree. When we identify a mechanism as the cause of overpressures, we mean that the particular mechanism is the dominant cause of overpressures. To identify the dominant mechanism, we should look at how the different mechanisms affect in-situ stresses and pore pressures.

Whether compaction disequilibrium or source mechanisms are the dominant cause of overpressures is ultimately due to the relative magnitudes of many factors such as burial rate, sediment permeability, mechanical behavior, and the thermal environment. These factors are compared via simple considerations of fluid flow and the sediment's mechanical behavior that are discussed in the following section. For simplicity, we assume one-dimensional fluid flow and deformation. Many of the conclusions are not strongly affected by this assumption. Some are, however, so the conclusions presented in this chapter should be carefully evaluated when one-dimensional assumptions are not appropriate.

THEORETICAL BACKGROUND

Fluid-Flow Considerations

The basic equation that describes fluid flow in a sediment undergoing burial provides key insights into how compaction disequilibrium and source mechanisms affect in-situ stresses. Following Luo and Vasseur (1992), using terms added to account for sediment-grain volume changes, the fluid-flow equation can be written in terms of the time rates of change in total stress and either the pore pressure or the effective stress. When the equation is written in terms of the pore pressure, p, and the total vertical stress, s_v, it is clear that the pore pressure increases with an increase in burial depth (s_v), inward Darcy flows, an increase in temperature, and with active fluid sources.

We assume compressive stresses and strains are positive and use the shorthand convention of a dot over the variable to mean the time differential of that variable.

$$\dot{p} = \left(\frac{1}{c_p + c_f - c_s}\right) \left\{c_p\dot{s}_v + \left[\frac{1}{\phi\rho}\nabla\cdot\frac{\rho k}{\mu}\nabla(p - \rho g z) + (\alpha_f - \alpha_s)\dot{T} + \dot{q}\right]\right\} \quad (1)$$

The variables T, q, ϕ, and z are temperature, fluid-specific volume source, porosity, and true vertical depth, respectively. Material constants c_p, c_s, c_f, k, ρ, μ, α_f, and α_s are, respectively, the one-dimensional compressibilities of the sediment pores and solid constituents, the fluid compressibility, the sediment permeability, the fluid density and viscosity, and the coefficients of thermal expansion of the fluid and solids; g is the acceleration of gravity. The pore compressibility, c_p, depends on ϕ, c_s, and the rock's one-dimensional bulk compressibility, c_b: $c_p = (c_b—c_s)/\phi$.

The first term in the curly brackets {} defines the rate of pressure increase due to burial-depth increases. The first term in the square brackets governs Darcy flow and is positive for fluid flows into the system. The second term in the square brackets describes aquathermal expansion, and the third term (in the square brackets) describes any other source such as smectite/illite diagenesis or kerogen maturation. Either of these latter terms in the square brackets can be thought of as a source mechanism.

Equation 1 demonstrates that both active source mechanisms and increasing burial depths tend to increase pore pressures. This is not the case with effective stresses, σ, which may tend to increase or decrease depending on the mechanism that causes overpressures. Because the basic fluid-flow and compaction equations involve porosity changes, we use the Terzaghi effective stress (Carroll, 1980), defined as

$$\sigma = s - p \quad (2)$$

where s is the total compressive stress.

Combining equations 1 and 2 gives the following relationship for the effective vertical stress in terms of the total vertical stress and the various fluid source terms:

$$\dot{\sigma}_v = \left(\frac{1}{c_p + c_f - c_s}\right) \left\{(c_f - c_s)\dot{s}_v - \left[\frac{1}{\phi\rho}\nabla\cdot\frac{\rho k}{\mu}\nabla(p - \rho g z) + (\alpha_f - \alpha_s)\dot{T} + \dot{q}\right]\right\} \quad (3)$$

Whether σ_v increases or decreases depends on the relative strengths of the source terms, the system's permeability, and the burial rate. Under one-dimensional vertical strain condition, the horizontal stresses respond directly to changes in the vertical effective stress and the pore pressure. As we show in the following section, both the magnitude of the horizontal stress and the relationship between horizontal and vertical stresses change depending on whether the effective stresses increase or decrease.

Stress-Strain Considerations

A sediment compacts inelastically (irreversibly) or elastically (reversibly) depending on its loading history. Figure 1 shows porosity vs. depth for virgin compaction (inelastic) and unloading/reloading (elastic) paths for normally pressured rocks. Porosity change follows the virgin compaction curve as long as the sediment's effective stress is increasing and at its maximum value. Porosity change follows the unloading/reloading curve, which is tied to the porosity at the sediment's maximum effective stress, whenever the effective stress is less than its previous maximum value. This latter response can be caused either by uplift and erosion or by active source terms.

The various loading behaviors can be modeled mathematically with a modified Cam-Clay model from critical state soil mechanics (Atkinson and Bransby, 1978). Here we modify the basic Cam-Clay log-linear plastic volume-change relationship to one based on a general Athy-type exponential compaction curve in which

$$\phi = \phi_0 \exp(-\lambda d) \quad (4)$$

where ϕ is the porosity; ϕ_0 is the porosity at the top of the sediment column, d is the depth, and λ is an empirical coefficient. The porosity actually depends on

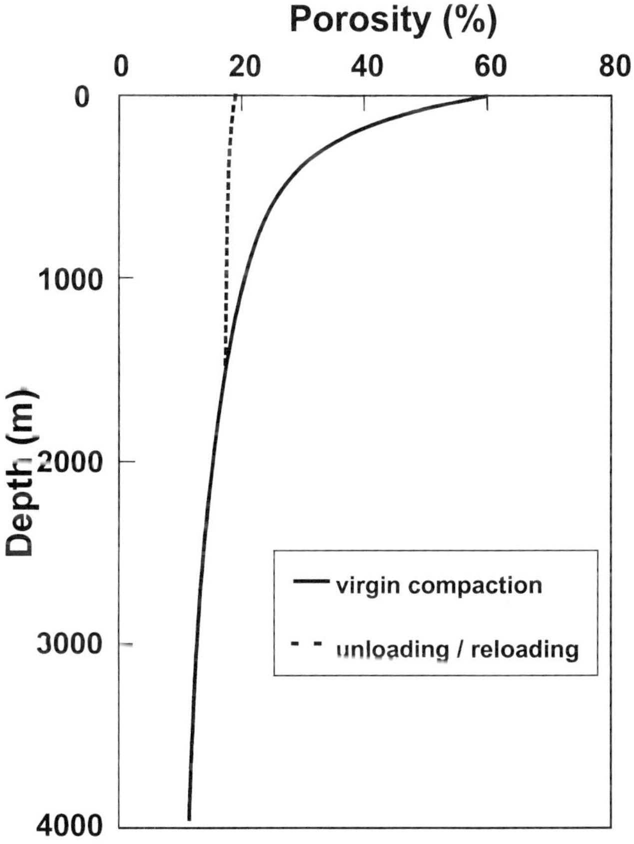

Figure 1. Schematic illustrating porosity vs. depth for a clastic sediment under two loading conditions. Porosity is reduced irreversibly during virgin compaction but rebounds/re-compacts reversibly during unloading/reloading.

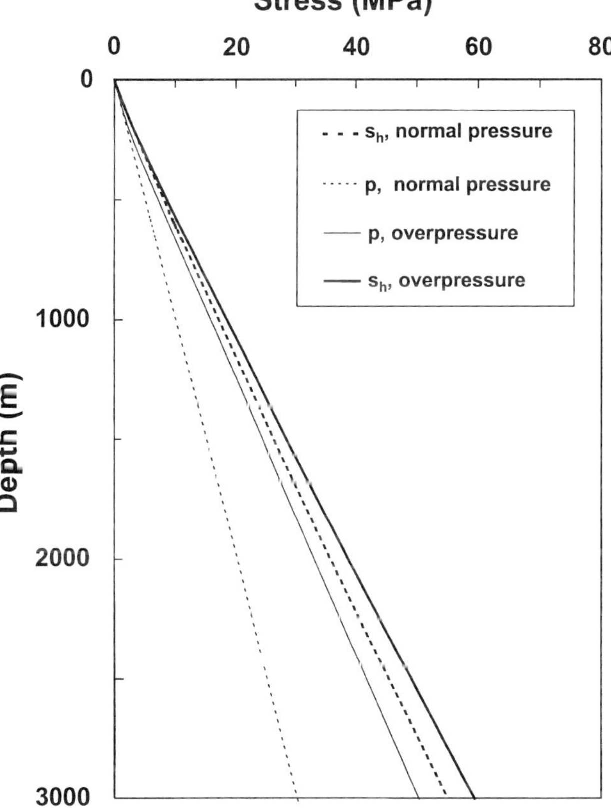

Figure 2. Overpressures caused by compaction disequilibrium (solid lines) increase both pore pressures, p, and total horizontal stresses, s_h. Normal pressures (dotted line) shown for reference.

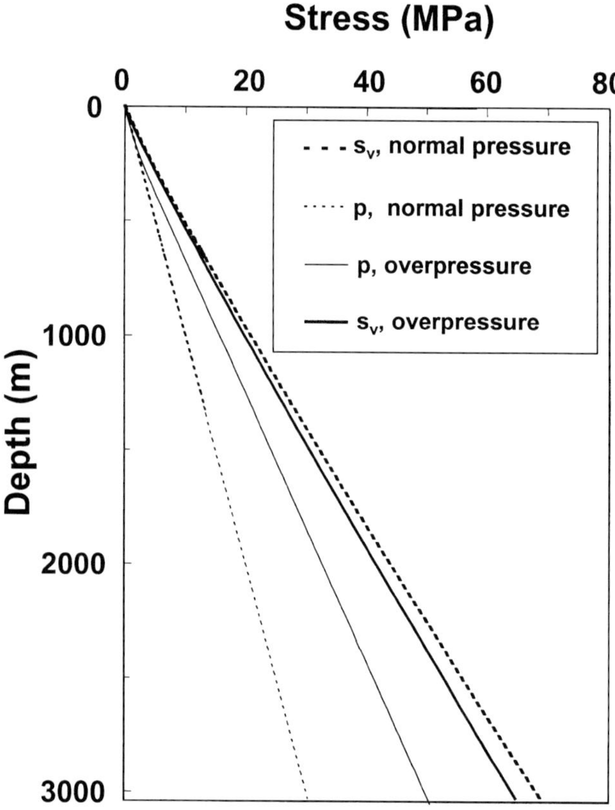

Figure 3. Overpressures caused by compaction disequilibrium decrease total vertical stresses, s_v. Other symbols the same as in Figure 2.

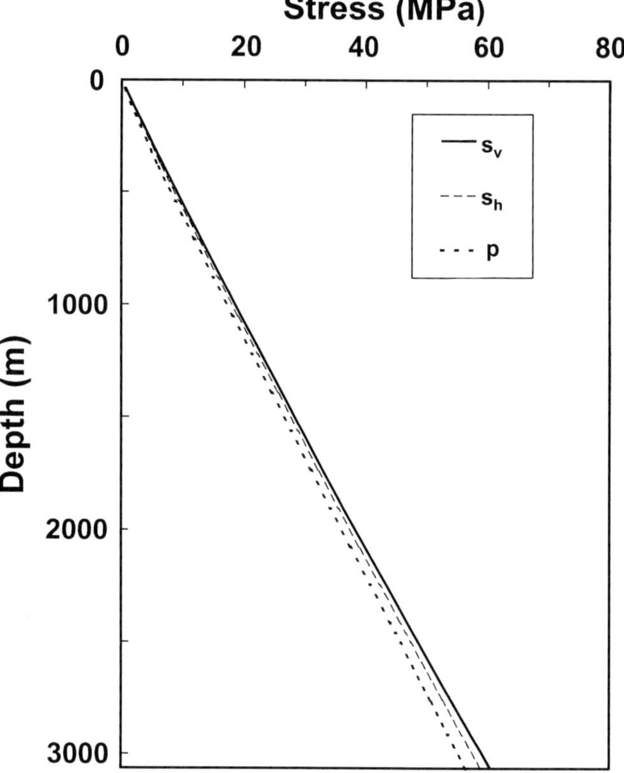

Figure 4. In sediments that are highly overpressured by compaction disequilibrium, $p \leq s_h \leq s_v$.

the vertical effective stress. An appropriate functional dependence between ϕ and σ_v can be found from such Athy relationships for normally pressured sediments. We call that function the compaction function, which is defined as

$$\phi = C(\sigma_v) \tag{5}$$

This compaction function applies whenever the sediment is at its maximum σ_v. In this case, the one-dimensional deformation assumption leads to

$$\dot{\sigma}_h \cong k_0 \dot{\sigma}_v \tag{6}$$

(Atkinson and Bransby, 1978). We assume equality in equation 6 because this condition ignores only a small thermoplastic effect. The parameter k_0 depends on the type of sediment. Experiments have shown (Karig and Hou, 1992; Vasseur et al., 1995) that k_0 remains constant over a wide range of stresses. Although it is not strictly a material property in that its value is a consequence of other material properties, it can be treated as one for uniaxial deformation conditions.

The parameter k_0 is not related to Poisson's ratio, ν, which should be considered as an elastic property only. In fact, the values of Poisson's ratio (ν^*) inferred from measured values of k_0 via the Eaton (1975) equation

$$\nu^* = \frac{k_0}{1 + k_0} \tag{7}$$

are approximately twice those measured in appropriate, drained stress-strain tests. For example, k_0 is around 0.70 for most shales and 0.55 for many unconsolidated sands. Thus the inferred Poisson's ratios for these materials would be 0.41 and 0.35, respectively, compared to reported elastic values of 0.25 and 0.20, respectively. Clearly misapplication of the apparent property into the incorrect governing equation could lead to large errors in calculated stresses.

The previous discussion points out the important differences between assuming elastic and inelastic mechanical responses. While the sediment is actively compacting, it behaves inelastically, but at stresses less than the maximum, sediments behave mechanically more like elastic materials. It follows (Miller, 1995) then that during unloading/reloading,

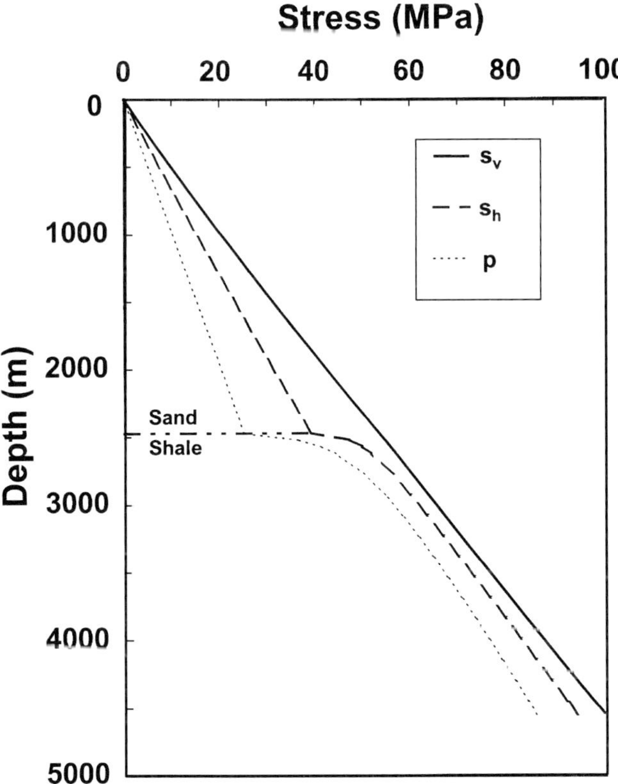

Figure 5. Total stresses, s_v and s_h, and pore pressures, p, in a sediment column of shale overlain by sand.

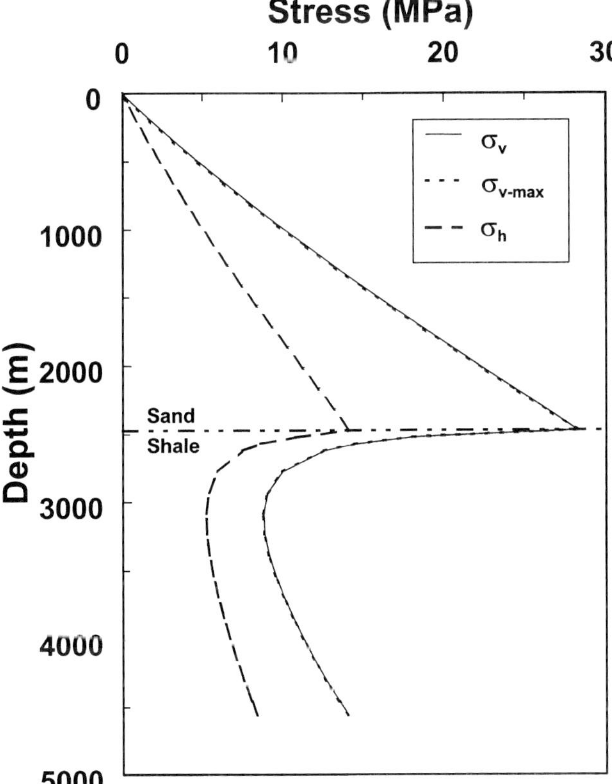

Figure 6. Stress reversals can be caused by changes in lithology without unloading ($\sigma_v = \sigma_{vmax}$).

$$\dot{\phi} = [(1-\phi)c_b - c_s]\dot{\sigma}_v \text{ and} \tag{8}$$

$$\dot{\sigma}_h = \frac{1}{1-\nu}\left[\upsilon\dot{\sigma}_v - \frac{c_s(1-2\upsilon)}{c_b}\left(\dot{p} - \frac{\alpha\dot{T}}{c_s}\right)\right] \tag{9}$$

The value of ν for this equation cannot be determined from stress data, and the inelastic bulk compressibility implicit in equation 5 is typically an order of magnitude greater than the equivalent elastic constant, c_b, in equations 8 and 9.

RESULTS

The following examples demonstrate some of the key differences in the distribution and magnitudes of pore pressures and in-situ stress conditions resulting from compaction disequilibrium and source-pressuring mechanisms. The results presented in this chapter were calculated using a one-dimensional simulator that calculates pressure, temperature, and stress histories of a sediment column. In addition to the basic equations described previously, this model couples fluid properties to temperature and pressure and includes a permeability vs. porosity algorithm that results in shale permeability predictions consistent with the average values given by Neuzil (1994).

Compaction Disequilibrium

Excess pore pressures resulting from compaction disequilibrium lead to porosities higher than, total vertical stresses less than, and total horizontal stresses greater than those found in equivalent normally pressured settings. Although the pressures can be high under rapid burial conditions, pore pressures and horizontal stresses cannot exceed the overburden stress.

Figures 2 and 3, respectively, show that overpressures increase the total horizontal stress, s_h, and decrease the overburden or total vertical stress, s_v. The total vertical stress is less than in the normally pressured case because the excess pressures inhibit porosity reduction, resulting in a less dense overburden. The total horizontal stress increases with fluid pressure because of a combination of several factors, but the most important is that lateral strains are constrained by adjacent rocks, which prevent free horizontal deformations. Consequently, horizontal loads (and total stresses) increase or decrease where lateral deformations are, respectively, inward or outward. Therefore,

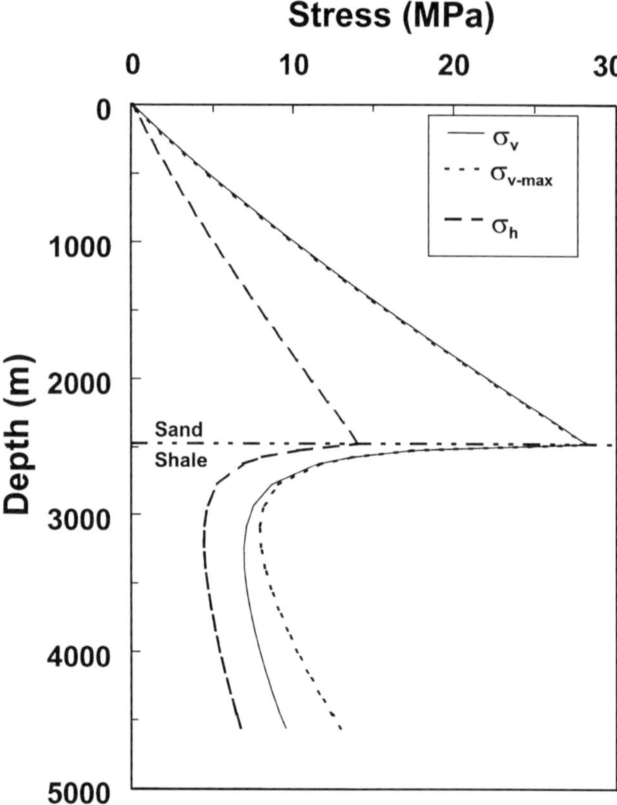

Figure 7. With a small fluid source in the shale, unloading ($\sigma_v < \sigma_{vmax}$) occurs because of the increase in pore pressure.

increased pore pressures, which tend to expand the rock, increase the horizontal total stresses.

In addition, because the vertical effective stress at any given location has always increased in time, σ_h can be calculated from equation 6 and hence is always less than σ_v.

Because the pore-pressure increases are caused by compression of the pores, mechanical equilibrium dictates that any increase in pore pressure resulting from burial-depth increases cannot exceed the increase in total vertical stress. This effect is illustrated in Figure 4. The stresses and pore pressure follow the same trend established in Figures 2 and 3, in which increasing pore pressures lead to decreased overburden stresses and increased horizontal stresses. Note that these simulations could be repeated with a higher burial rate (or lower permeability), and the horizontal and vertical stresses would approach the limiting condition of $p = s_h = s_v$.

Figure 5 is a plot of the pore pressures and vertical and horizontal total stresses of a sediment column that is predominantly shale in the lower half and predominantly sand in the upper half. In this example, both the sand and shale were buried at a constant rate of 1.5 km/Ma. Here the pore pressures and stresses in the sand are as expected for normally pressured sediments, and the elevated pore pressures and changed stresses are as expected in the shale. The general pore-pressure and stress conditions shown in Figure 5 are commonly mistaken for unloading because a plot of porosity vs. depth would show an increase in porosity below the sand/shale interface. Figure 6 is a plot of the current and maximum vertical and the current horizontal effective stresses vs. depth for the same sediment whose total stresses are shown in Figure 5. These effective-stress plots show that the current and maximum vertical stresses are the same, a result that indicates unloading has not occurred despite the effective stresses decreasing with depth.

Sources and Unloading

For unloading to occur, an active source, fluid inflows, or fluid expansion are required (equation 3). Although the relative rate of decrease in the horizontal and vertical effective stresses during unloading varies as a function of mechanical properties, source intensity, and temperature, the vertical effective stress tends to decrease faster than the horizontal effective stress (equations 8 and 9). Eventually, the horizontal total stresses may exceed the overburden stress. Furthermore, pore pressures can also increase, at least in principle if rock fracture would not occur, above the overburden stress. As a practical matter, leak-off tests in such an environment would generate horizontal fractures, and one can erroneously conclude that the horizontal stress would equal the vertical stress instead of exceeding it.

Figures 7 and 8 are plots of the current and maximum vertical and current horizontal effective stresses for the same sand-over-shale sediment package shown in Figures 5 and 6, but with a recently active source in the shale. For the small source case, shown in Figure 6, unloading has occurred because the current vertical effective stress is less than its maximum value, and the difference between the vertical and horizontal stresses is less than in the no-source case. The vertical effective stress, however, is still greater than the horizontal effective stress. For the larger source case, shown in Figure 8, the shale has unloaded further, and the vertical effective stress has decreased below the horizontal effective stress at depth, that is, horizontal total stresses exceed the overburden, and pore pressures almost equal the overburden, that is, $\sigma_v \approx 0$. High pore pressures are not by themselves conclusive evidence of any one overpressuring mechanism (Figure 9).

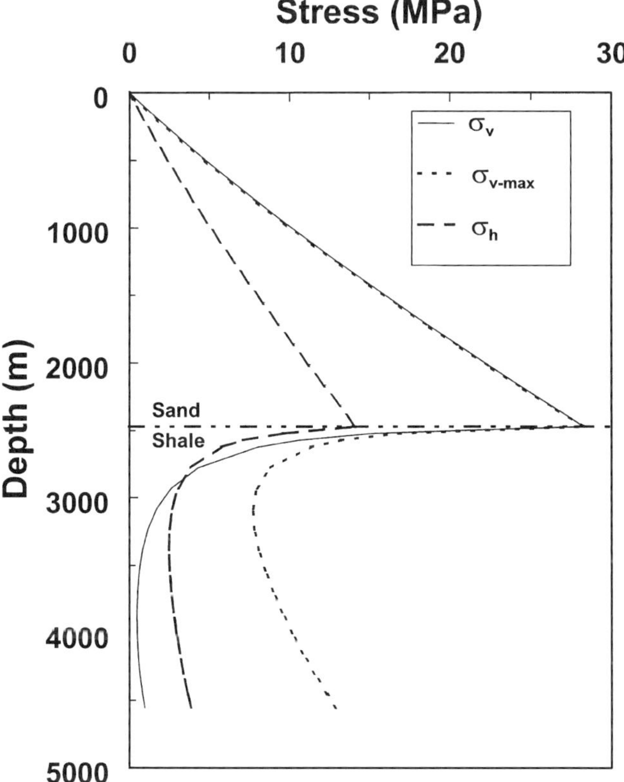

Figure 8. With a large fluid source in the shale, unloading occurs to such an extent that $\sigma_v < \sigma_h$ is possible.

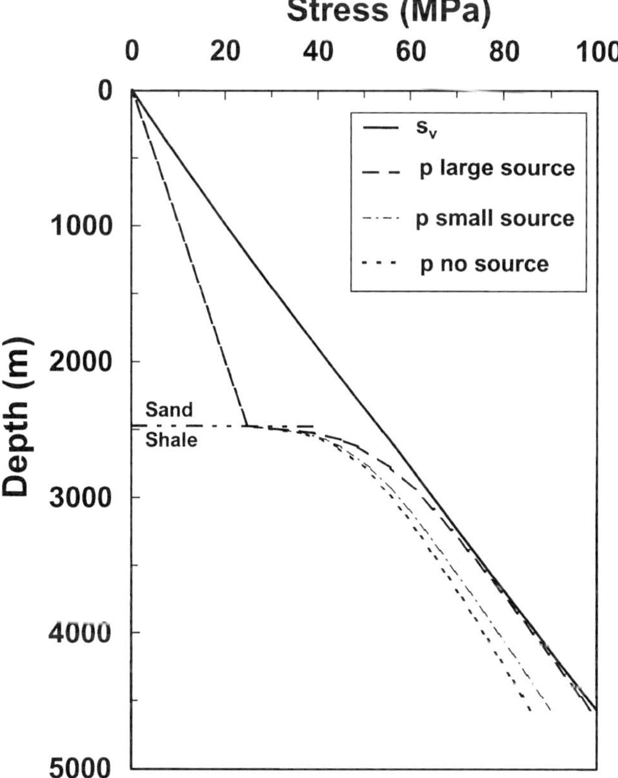

Figure 9. Pore pressure, *p*, vs. depth for no fluid source, small fluid source, and large fluid source magnitudes in the shale (see Figures 6–8).

DISCUSSION

The previous examples show that the compaction disequilibrium and source-dominated overpressuring mechanisms affect in-situ stresses in fundamentally different ways. In the case of compaction disequilibrium:

- The overburden stress (s_v) is less than in the normal pressure case
- Current stresses (σ_v, σ_h, s_v, and s_h) are always at their maximum value
- Horizontal stresses (σ_h and s_h) are less than the respective vertical stresses
- Pore pressures are always less than the overburden

In the case of fluid sources:

- The total overburden stress is virtually unchanged
- Current effective stresses are less than their maximum values
- Horizontal (effective and total) stresses can exceed their respective vertical stresses
- Pore pressures can exceed the overburden stress

REFERENCES CITED

Atkinson, J. H., and P. L. Bransby, 1978, The mechanics of soils: London, McGraw-Hill, 450 p.

Carroll, M. M., 1980, Mechanical response of fluid-saturated porous materials, *in* F. Rimrott and B. Tabarrok, eds., Theoretical and applied mechanics. Proceedings of 15th International Congress of Theoretical and Applied Mechanics, p. 251–262.

Eaton, B. A., 1975, The equation for geopressure prediction from well-logs: Society of Petroleum Engineers Paper 5544.

Flemings, P., B. B. Stump, T. Finkbeiner, and M. D. Zoback, 1998, Pressure differences between overpressured sands and bounding shales of the Eugene Island 330 field (offshore Louisiana, U.S.A.) with implications for fluid flow induced by sediment loading: Proceedings of the American Association of Drilling Engineers Industry Forum on Pressure Regimes in Sedimentary Basins and Their Prediction.

Karig, D., and G. Hou, 1992, High-stress consolidation experiments and their geologic implications: Journal of Geophysical Research, v. 97, no. B1, p. 289–300.

Luo, X., and G. Vasseur, 1992, Contributions of compaction and aquathermal pressuring to geopressure and the influence of environmental conditions: AAPG Bulletin, v. 76, no. 10, p. 1550–1559.

Miller, T. W., 1995, New insight on natural hydraulic fractures induced by abnormally high pore pressures: AAPG Bulletin, v. 79, no. 7, p. 1005–1018.

Neuzil, C. E., 1994, How permeable are clays and shales?: Water Resources Research, v. 30, no. 2, p. 145–150.

Vasseur, G., I. Djeran-Maigre, D. Grunberger, G. Rousset, D. Tessier, and B. Velde, 1995, Evolution of structural and physical parameters of clays during experimental compaction: Marine Petroleum Geology, v. 12, no. 8, p. 941–954.

Yassir, N. A., 1989, Undrained shear characteristics of clay at high total stress, *in* V. Maury and C. Fourmaintraux, eds., Rock at great depth: Proceedings of the International Symposium on Rock Mechanics, v. 2, p. 907–913.

3

The Primary Controls over Sediment Compaction

Phil Holbrook
Force Balanced Petrophysics, Houston, Texas

ABSTRACT

Mineralogic composition is the primary control over sediment compaction. Near-surface sediments and sedimentary rocks compact in proportion to the effective stress load applied to the grain matrix framework. The load borne by the grain framework is a volumetric force-balance relationship within and between solid particles. Mineral ionic bonds bear the load within particles and across direct grain-grain contacts. Electrostatic repulsive forces between negatively charged clay mineral particles also bears a part of the effective stress load.

A strong correlation exists between average particle size and initial porosity of natural marine sediments. Compaction resistance is also strongly correlated to average particle size. Clay mineral interparticle repulsive force explains both correlations. The relationship of particle size vs. compaction resistance is continuous from coarse sands to the finest particle-size clays. Graded bedding due to individual particle settling velocity sorting places mineral grains of similar size and mechanical properties together.

Five mineral-specific compaction functions were determined from in-situ petrophysical data. Average mineral ionic bond strength controls mineral hardness, solubility, and the plastic compaction intercept at zero porosity. Clay minerals have an additional interparticle electrostatic repulsive force that is inversely proportional to sedimentary clay particle size and directly proportional to clay surface area in a given rock volume. All these factors are simultaneously accounted for by two power-law compaction coefficients (σ_{max} and α).

Solidity (1.0 − porosity) is an end-of-plastic-compaction in-situ strain parameter. The (solidity = 1.0) intercept (σ_{max}) incorporates both elastic and plastic grain matrix strain into volumetric in-situ strain. The power-law compaction exponent (α) captures the interparticle and intraparticle compaction resistances mentioned previously with respect to (σ_{max}).

STRUCTURE, IONIC BOND STRENGTH, AND CHARGE DISTRIBUTION OF THE COMMON SEDIMENTARY MINERALS

Figure 1 shows the crystal lattice structures of the four most common sedimentary mineral types. The common nonclay minerals have a directionally neutral internal charge distribution. Positive and negative charges are equal on the angstrom scale within the mineral. Contacts between nonclay mineral particles are on the much larger micron scale and therefore have no net electrostatic charge. Adjacent nonclay mineral particles and those in direct contact have no net electrostatic repulsive forces between them.

Sedimentary clay minerals all possess a layered structure of dominantly silicon-centered tetrahedra and aluminum-centered octahedra as shown in Figure 1. The positively charged (+) aluminum, silicon, and

Holbrook, Phil, 2002, The Primary Controls over Sediment Compaction, in A. R. Huffman and G. L. Bowers, eds., Pressure regimes in sedimentary basins and their prediction: AAPG Memoir 76, p. 21–32.

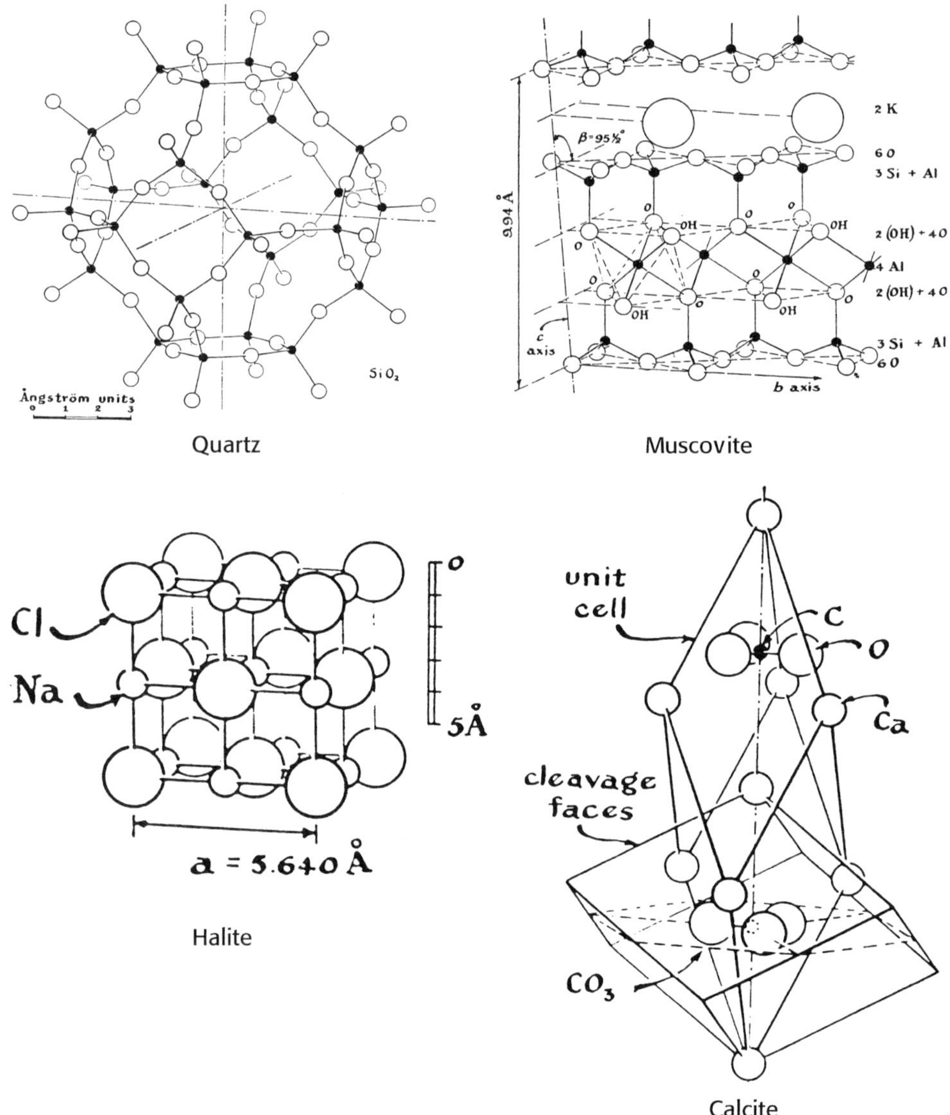

Figure 1. Crystal lattices of the four most common sedimentary mineral types. Quartz, muscovite (representative 2:1 clay), calcite, and halite (rock salt) are shown. All minerals have a net neutral internal change balance. The nonclay minerals cleave in all directions presenting net neutrally changed mineral grain surfaces. The clay minerals cleave preferentially perpendicular to the *c* axis presenting a negatively charged oxygen anion layer at the clay particle surface. Adapted from Berry and Mason (1959).

other high positive valence ions are sandwiched between negatively charged (−) oxygen layers.

Clays cleave preferentially normal to their *c* axis exposing large surfaces of negatively charged (−) oxygen anions. Sedimentary clay particles are thin platelets with negatively charged faces and positively charged edges. The negative to positive charge ratio over the entire surface of sedimentary clay mineral platelets is very large (Scott, 1963).

IONIC BOND FORCE BALANCE WITHIN MINERAL PARTICLES AND ACROSS NONCLAY MINERAL GRAIN CONTACTS

Nonclay minerals resist compaction through their mineral lattice and at direct grain contacts with other nonclay minerals. A given particle is in contact with its neighbors over a fractional area of that particle. Particle contact area limits are from zero in a particle fluid suspension to 1.0 where intergranular porosity is reduced to zero.

Force balance across the fractional contact area is the product of the number and strength of mineral ionic bonds. Harder minerals with stronger ionic bonds such as quartz can bear a given load over a smaller area than soft minerals like halite. Particle contact area varies in proportion to volumetric solidity. Both particle area and solid volume reach 1.0 at the sedimentary rock's upper plastic compaction limit (σ_{max}).

DIRECT NONCLAY MINERAL GRAIN CONTACT PRESSURE SOLUTION COMPACTION MECHANISM

Pressure solution (dissolution of grains at highly stressed intergranular contact points) is a major compaction mechanism on a geologic time scale. On a geologic time scale the mineral grain matrix tends to behave like an ideal plastic. Higher stress across neutrally charged grain-grain contacts is gradually equalized through pressure solution and generally local reprecipitation.

Mineral solubility in water can be related to hardness via an inverse power law function (Carmichael, 1982). All are fundamentally controlled by average mineral ionic bond strength.

Stress is everywhere proportional to strain in the earth's mechanical systems. The stress field is three-dimensional, and the coordination number of adjacent solid load-bearing grains is n-dimensional. The loading limb and unloading limb stress/strain relationships for porous granular solids are multidimensional, that is, power-law functions.

COMPACTION OF NONCLAY GRANULAR SEDIMENTS

Figure 2a shows a representative transition of neutrally charged spherical grains from initial grain contact to complete consolidation. The initial depositional porosity of rounded sedimentary grains is about 40%. The left border of the diagram represents the point of initial gravitational grain-grain contact. Zero effective stress exists at initial contact. Physically this represents the surface of the earth overlain by fluid or air. The rightmost edge of Figure 2a is the plastic compaction limit (solidity = 1.0 at σ_{max}).

Figure 2a represents solid volume conserved compaction of rounded neutrally charged mineral grains. As the grains are forced together, minerals are dissolved preferentially at the more highly stressed grain-

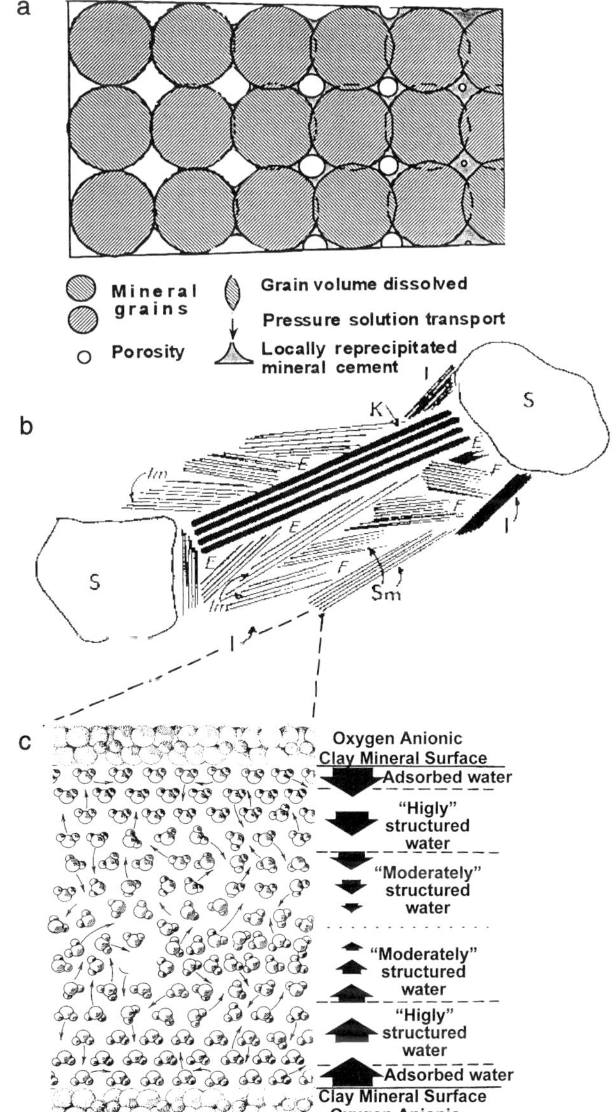

Figure 2. (a) Microscopic scale diagram showing the compaction of a spherical grain matrix from 40 to 0% porosity. Solid mass is conserved through pressure solution compaction. During compaction minerals are dissolved preferentially at grain-grain contacts and reprecipitated in the local pore space. The microscopic examination of quartz grainstones and intergranular cement look very much like Figure 2a as porosity is reduced from 40 to 0%. (b) Sedimentary particles and forms of water in a low-porosity clay soil. F = free or bulk water; E = external or intercluster water; Im = Interlammelar or intercluster water; S = silt; K = kaolinite; I = illite; Sm = smectite (modified from Hueckel, 1992). (c) Bound water occurs adjacent to negatively charged clay mineral surfaces. Two opposing negatively charged clay mineral oxygen anionic surfaces shown at the atomic scale. Negative charge density diminishes with distance from the oxygen anionic mineral surface. Negative charge density both orients the nearby water dipole molecules and repels other negatively charged clay mineral surfaces.

grain contacts and reprecipitated in the nearby pore space. Net pore volume is indicated in white in Figure 2a. Idealized pore shape changes progressively from an inverse sphere toward small isolated spheres where porosity reduction is accompanied by pressure solution followed by local reprecipitation.

Initial grain volume is not reduced from left to right in Figure 2a. The compaction portrayed is accomplished entirely by transporting mineral matter from higher stress grain-grain contacts into the available pore space. Natural quartz grains have different initial shapes. Mineral grains may be fractured, rehealed, and rearranged during compaction. If there is no net mineral volume added or subtracted, however, the volumetric strain of the grain framework is necessarily the same as the idealized case of Figure 2a.

Quartz cement has the same mineral lattice structure and the same load-bearing capacity as the quartz grain from which it was dissolved. Through pressure solution the neutrally charged grain matrix increases the number of mineral ionic bonds at each grain-grain contact to support the average effective stress load. The remainder of the total load is borne by pore-fluid pressure. Although an agglomerated grain matrix has some finite porosity, the total confining load is shared between (1) solid net neutrally charged mineral ionic bonds, (2), interparticle electrostatic repulsive forces, and (3) pore-fluid pressure.

SEDIMENTARY CLAY MINERAL PARTICLE SIZE VS. ELECTROSTATIC INTERPARTICLE REPULSIVE FORCE RELATIONSHIP

Most natural sediments and sedimentary rocks are a mixture of clay and nonclay minerals. Electrostatic interparticle repulsion is an important factor in the compaction of these natural sediments. Figure 2b shows a microscopic representation of a low-porosity clay dominated soil. The clay mineral packets shown (kaolinite, illite, and smectite) have similar particle-size/shape-aspect ratios. Clays are commonly sedimented with varying percentages of neutrally charged silt grains also shown. The net compaction resistance of a naturally deposited sedimentary mixture is the mineralogically weighted average of the individual grain compaction resistances (Holbrook et al., 1995).

Figure 2c is a molecular scale representation of the space between two adjacent clay mineral particles. The water dipoles are held tightly to the electrostatically charged clay mineral oxygen anionic surface as shown. Water dipoles oscillate more randomly as the negative charge density originating from the clay mineral surface falls to zero.

The different clay minerals shown in Figure 2b, c have important internal structural and compositional differences. These different minerals, however, have very similar micromechanical properties that are all related to particle size as indicated in Figure 3. An essentially 1:1 power-law relationship exists between clay mineral specific surface area and cation exchange capacity (CEC). Clay mineral particle size decreases as specific surface area and CEC increase, as shown in Figure 3.

Both clay and nonclay mineral particle size at initial sedimentation is determined by particle settling velocity. The clay mineral CEC–particle-size continuum shown in Figure 3 is an extension of the general Reynolds number particle-size settling velocity continuum for all particles that are settled gravitationally (Scott, 1963).

The dominant underlying control over clay mineral electrostatic repulsive forces and mechanical properties is also clay mineral particle size. This relationship is set gravitationally at the moment of initial particle settling agglomeration at the earth's surface. Natural marine sediments with longer water columns best demonstrate the particle size, initial porosity, and interparticle repulsive force relationships. (Liquid limit is the water content at which a clay starts behaving like a liquid; water content is the weight of water in the clay divided by the weight of the dry clay solids.)

MINERAL-DEPENDENT GRAIN SIZE–INTERPARTICLE REPULSIVE FORCE–INITIAL POROSITY RELATIONSHIPS

The zero effective stress compaction starting point for granular sediments can be determined from porosity–mineralogy–particle-size relationships in marine sediments at the sea floor (Shumway, 1960; Skempton, 1970). The approximately 10 cm below the surface line provides a sample that is close to zero effective stress. Figure 4 shows the interrelationships between observed initial depositional porosity, laboratory-measured liquid limit, and observed in-situ compactional stress/strain relationships. (Liquid limit is the water content at which a clay starts behaving like a liquid; water content is the weight of water in the clay divided by the weight of the dry clay solids.) Effective vertical stress is on the horizontal axis, and the two more common measures of strain, void ratio and porosity, are shown on the vertical axis. Liquid limit is strongly correlated to CEC (Nagaraj and Murthy, 1983). Therefore the effective stress compactional relationships shown in Figure 4 are also particle-size dependent.

Figure 5, adapted from Shumway (1960), shows the average sediment particle diameter (left-hand axis) plotted on the phi scale (right-hand axis) vs. surface depositional porosity. The conversion from particle diameter (mm) to phi scale value (not to be confused

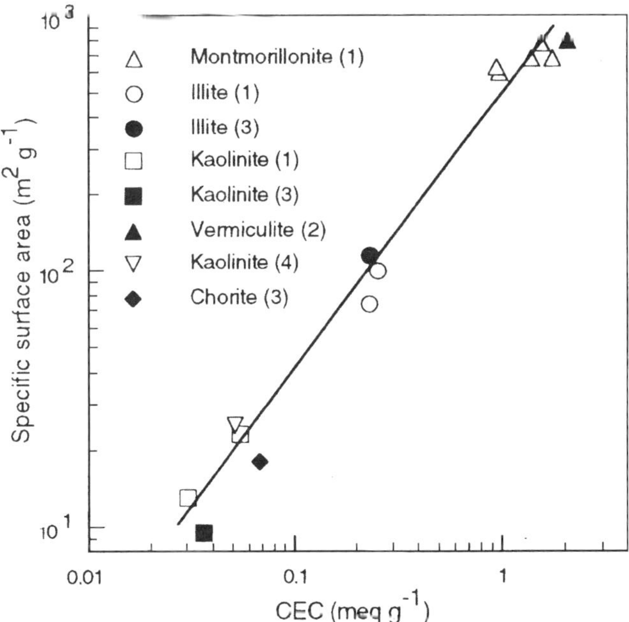

Figure 3. Specific surface area vs. CEC for various clay minerals. This strong correlation suggests that the surface charge density of a clay mineral portrayed in Figure 2c is essentially an areal constant. Given this, the dominant underlying control over clay mineral physical properties is most likely average clay mineral particle size. Taken from Revil et al. (1997), reprinted with permission from the Geological Society Publishing House. Numbers in parentheses refer to references, which are listed in Revil et al. (1977).

with porosity) is phi = −log(particle diameter)/log(2) (Bates and Jackson, 1987). The figure shows a systematic semi–log-linear relationship between almost pure clay mineral deposits with more than 80% depositional porosity and an average 2 μm particle size to an almost pure quartz sand deposit with an average particle size of more than 500 μm and a depositional porosity of only 40%. The highly correlated depositional porosity and particle-size distributions of the near-surface marine sediment are controlled by mineral particle settling velocity.

The grain density of quartz and clay are approximately equal, and the effective stress gravitational load in the uppermost meter of the sea floor is very low. The difference between 80% porosity for fine clay deposits and 40% porosity for medium quartz sand deposits is due to the repulsive force between individual clay particles. The interparticle repulsive force is weak but persistent with increasing effective stress compaction as indicated in Figure 4. The Figure 5 particle-size crossplot coincides with the 10 cm line in Figure 4.

Nagaraj and Murthy (1983) later explained Skempton's entire compactional relationship mechanically in terms of negatively charged mineral surface area. The CEC is essentially a direct measure of clay mineral negatively charged (−) oxygen anion surface area. The average electrostatic repulsive force between clay particles is also proportional to CEC. The compactional relationships of all clay minerals were normalized to the same curve based upon their average CEC.

Adjacent particles in a sedimentary layer have fallen through the same water medium at about the same speed. Clay particles in the sand size range have proportionally lower external surface area and therefore lower average interparticle repulsive force. Graded bedding places grains of like size and interparticle repulsive force adjacent to each other. Particle size and negative surface charge area work together in the marine environment to generate the strong interdependent relationship shown in Figures 3, 4, and 5.

GENERAL MINERALOGIC STRESS/STRAIN LOADING FOR GRANULAR SOLIDS

Here, voumentric strain is represented by the complement of porosity (1 − ϕ), termed "solidity." Solidity is assumed to follow a power law effective stress relation of the form proposed by Baldwin and Butler (1985). Figure 6 (adapted from Holbrook, 1999) shows the volumetric effective stress loading-limb compactional relationships for natural single mineral sedimentary deposits over the entire depth range of drilling interest. The loading-limb stress/strain relationships are global in nature, dependent principally upon mineralogic composition (Holbrook, 1995).

The effective stress compactional relationships were measured in situ considering overburden and pore-pressure force balance (Terzaghi, 1923). The volumetric effective stress theorem (Carroll, 1980) was also honored (Holbrook, 1999). These macromechanical stress/strain relationships are a composite of the micromechanical relationships (Figures 1–5) previously discussed. The composite of many microscale power-law relationships results in a macro–power-law relationship.

The four neutrally charged nonclay minerals—quartz, calcite, anhydrite, and halite—have subparallel power-law loading-limb effective stress/strain coefficients (α). These neutrally charged single-mineral stress/strain relationships are offset from each other in proportion to their plastic compaction intercepts (σ_{max}). The compaction intercept (σ_{max}) is positively related to mineral hardness and inversely related to mineral solubility (Table 1). All three are measures of the average interionic bond strength of these sedimentary minerals.

Naturally sedimented clay minerals also have power-law loading-limb effective stress/strain coeffi-

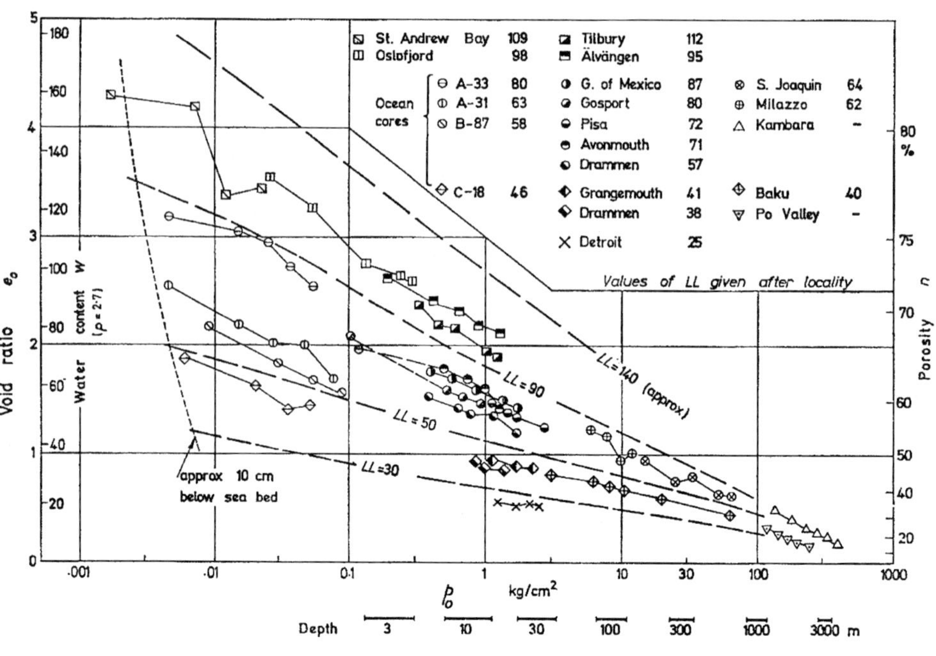

Figure 4. Relationships of in-situ compactional effective stress (kg/cm^2) to strain (void ratio) from many different basins worldwide. Compactional strain is proportional to the laboratory measured liquid limit (LL) of the reconstructed solid. The liquid limit is fundamentally controlled by clay mineralogy and average clay content and can be related to depositional porosity approximately 10 cm below the sea floor. Taken from Skempton (1970), reprinted with permission from the Geological Society Publishing House.

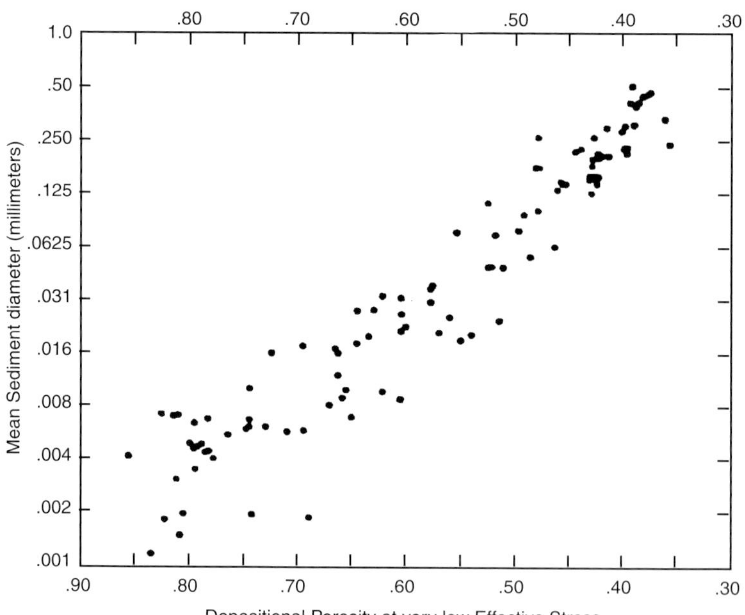

Figure 5. Mean sediment diameter vs. porosity for surface marine sediments taken from Shumway (1960); reprinted with permission from the Society of Exploration Geophysicists. Samples are from 10 cm below the sea floor. Depositional porosity ranges uniformly from 40% for (0.5 mm) quartz sands to 90% for a very fine (0.001 mm) sedimentary clay. Porosity at very low effective stress is a function of particle size and clay mineral interparticle repulsion.

cients (α) and a plastic compaction intercept (σ_{max}). As shown in Figure 6, the (σ_{max}) compaction intercept is well below granular quartz and above granular calcite. This corresponds to clay mineral ordinal rankings between quartz and calcite on the hardness and solubility physical property scales. This supports the conclusion that (σ_{max}) is a mineralogic compactional stress/strain physical property.

The average sedimentary claystone effective stress/strain coefficients (α) is distinctly different from the four neutrally charged minerals: quartz, calcite, anhydrite, and halite. So long as clay minerals have water-wet surfaces, the additional electrostatic repulsive forces between adjacent clay particles are effective. These electrostatic repulsive forces tend to oppose direct grain-to-grain contact, especially during initial compaction, which, in comparison to other sediment types, leads to the lower compaction coefficients (α) and the greater ductility observed in natural clay sediments and claystones.

This clay mineral repulsive force particle-size effect is shown by Skempton (1970) (Figure 4), Shumway

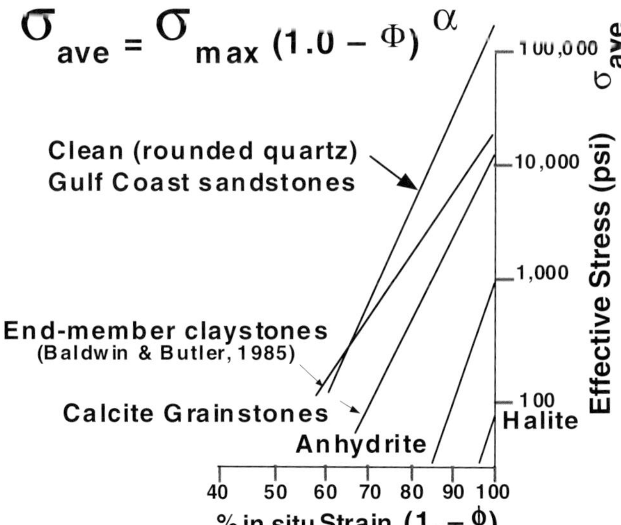

Figure 6. The first fundamental in-situ stress/strain relationship for the five most common single mineral sedimentary rocks. Average effective stress is borne by mineral ionic bonds and sedimentary clay mineral interparticle repulsive force. Average effective stress (σ_{ave}) is calculated directly from volumetric in-situ strain ($1.0 - \phi$) for single mineral and mixed mineral sedimentary rocks. The power-law effective stress/strain coefficients (σ_{max} and α) are sedimentary mineral grain properties that are independent of geologic age, depth, and pressure. The power-law compaction exponent (α) incorporates interparticle repulsion in proportion to clay volume. Adapted from Holbrook (1999).

(1960) (Figure 5), and rationalized by Nagaraj and Murthy (1983). This relative mineralogic compactional relationship is also observed in corresponding non–force-balanced depth vs. porosity relationships as shown subsequently.

THE GRANULAR QUARTZ VS. CLAYSTONE GRAVITATIONAL COMPACTION CROSSOVER

The porosities of gravitationally compacted granular quartz sediments and clay-rich sediments cross over each other in virtually all subsiding basins. Figure 7 shows average quartz sandstone and shale porosity vs. depth relationships from a well in the Gulf Coast. Above 2000 ft in Figure 7, clay-rich sediments have higher porosity than quartz sands. Below 2000 ft each of the two mineralogic end members continue along their own compaction gradients, and the curves diverge with increasing depth. Each mineralogic end member follows its own smooth continuous compaction vs. depth trend throughout the burial history shown.

Figure 6 has a corresponding mineralogic effective stress/strain crossover at about 300 psi and 35% porosity. With increasing effective stress the quartz grainstone and worldwide claystone compactional stress/strain relationships diverge as they do in Figure 7. Skempton's Figure 4 data also indicate a convergence in this same effective stress/strain region. Figures 4 and 6 show the effective stress data and force-balanced stress/strain relationships. Compactional porosity vs. depth data such as shown in Figure 7 correspond to the convergence and crossover shown on the effective stress functions of Figures 4 and 6.

This crossover occurs because of the claystone interparticle repulsive force. Quartz grainstones settle to the sea floor with about 40% porosity or less. Depending on particle size and related interparticle repulsive force, clay particles settle with an initial porosity of 50 to 95% (Lambe and Whitman, 1969). The individual quartz grains are much harder than clay minerals and have high compaction resistance. Claystone electrostatic interparticle repulsion and the minerals themselves are softer by comparison and therefore compact more easily. The primary controls over sediment compaction are these mineral-specific physical properties.

If effective stress were properly accounted for in Figure 7, the claystone points would plot on the worldwide claystone stress/strain power-law relationship whether they are overpressured or not. The same would be true of the quartz grainstone data points. The dashed line extensions of compaction trends in Figure 7 are entirely speculative. They do not consider the changing overburden load conditions nor do they consider the load borne by pore pressure. Plotting depth against porosity makes no mechanical sense, yet it is done frequently.

The power-law effective stress and strain axes of Figure 6 account for sediment compaction or lack of it in a mechanically sensible way. The compaction crossover, so-called normal compaction, and overpressured retracement are both related to average effective stress. Figure 6 explains both compaction features shown in Figure 7 as a simple mechanical system that is dependent on mineral physical properties.

LOADING LIMB COMPACTION VISUALIZATION

Figure 8 shows the three most common mineralogic end members as a stacked ternary diagram. The vertical axis of this diagram is effective stress on a logarithmic scale. The logarithmic scale linearizes the

Table 1. Power-Law Compaction Coefficients, Hardness, and Solubility for Naturally Sedimented Single-Mineral Grainstones and End-Member Claystones*

Mineral (or Rock)	σ_{max} (psi) Plastic Limit	α Compaction Exponent	Hardness (mhos)	Solubility (ppm)
Quartz sand	130,000	13.849	7.0	6
End-member claystone	18,461	9.348	3.0+	20
Calcite sand	12,000	13.637	3.0	120
Anhydrite	1585	20.646	2.5	3000
Halite	85	32.564	2.0	350,000

*Holbrook et al. (1995).

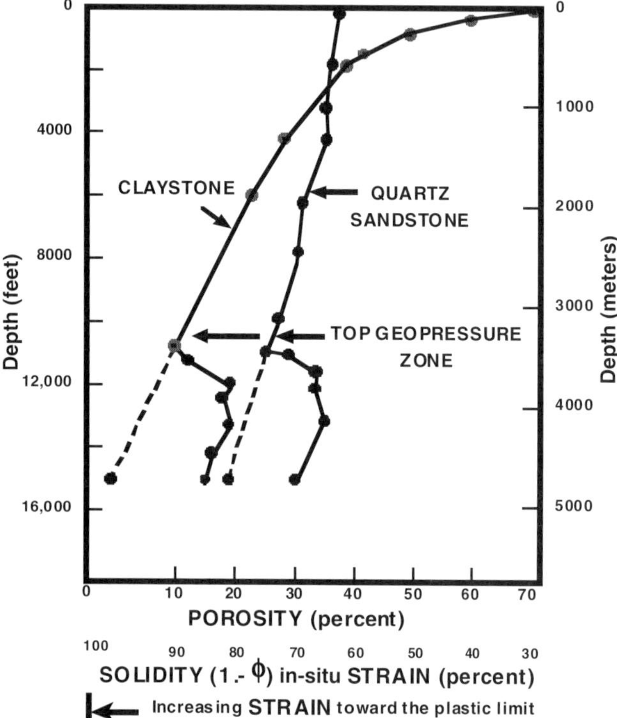

Figure 7. Quartz sand and shale porosity vs. depth compaction functions in the Gulf Coast. Modified from Stuart (1970). Shales have higher α plastic compressibility than do quartz sandstones. The two mineral-specific in-situ compaction curves cross at about 1000 ft.

power-law effective stress/strain relationship on the vertical scale. The horizontal quartz-calcite-clay plane is linear. Isoporosity lines are drawn on the surface of the two visible bimineral surfaces shown. The limestone-claystone continuum is on the left face of the diagram, and the quartz grainstone-claystone continuum is on the right.

The data supporting this diagram are about 300 continuous petrophysical logs from Normal Fault Regime Basins worldwide (Holbrook, 1996). In-situ–measured porosity, mineralogy, and effective stress were calibrated at each location. Porosity was estimated from an appropriately transformed density or resistivity log. Lithostratigraphic sequence type, limestone-claystone (L), or quartz grainstone-claystone (S) were assigned by an operator using local knowledge. Relative clay fraction (0 to 1.0) was assigned based upon a baseline-normalized gamma-ray log. Holbrook et al. (1995) describes these procedures.

The features and relative compactional relationships shown in Figure 8 were observed on all 300 petrophysical logs and are reasoned to be global in nature. The limestone-claystone face (L) of Figure 8 is characterized by a high-gamma-ray–high-porosity relationship. The quartz grainstone-claystone face (S) of Figure 8 is characterized by a high-gamma-ray–low-porosity relationship. These two lithostratigraphic sequence types produce parallel and hourglass patterns on petrophysical log suites as shown in Figure 9.

THE COMPACTION CROSSOVER IN THE QUARTZ GRAINSTONE-CLAYSTONE LITHOSTRATIGRAPHIC SEQUENCE

The isoporosity lines in Figure 8 are traces of isoporosity surfaces that pass through the ternary mixed mineralogy solid. This three mineral idealization is reasonably close to the gross mineralogic composition of many sedimentary rocks. Many of the patterns that are observed on petrophysical logs are explained by these mineralogic power-law linear compactional relationships.

The quartz grainstone-claystone compaction crossover occurs at about 300–500 psi on the right face of Figure 8. The 35% isoporosity line is parallel with effective stress on the quartz grainstone-claystone surface. This parallel stress/strain relationship is the physically representative equivalent of a compaction crossover.

Claystones compact more than quartz grainstones at effective stresses above 500 psi. This corresponds to the porosity divergence observed in the depth function shown in Figure 7. Any occurrence of calcite in a quartz

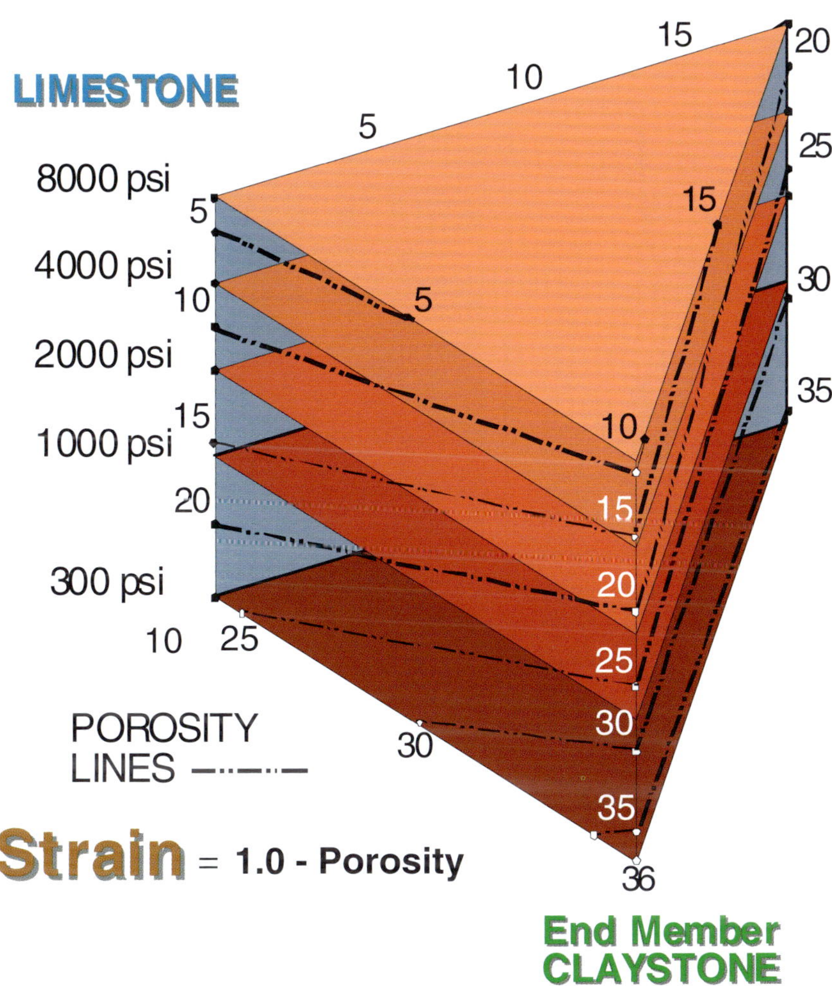

Figure 8. Ternary limestone-claystone-quartz grainstone mineralogic compaction resistance diagram. The vertical axis of this diagram is effective stress on a logarithmic scale. Isoporosity lines are drawn on the surface of the two lithostratigraphic continua shown. The limestone-claystone continuum is on the left face of the diagram, and the quartz grainstone-claystone continuum is on the right. Figure taken from Holbrook, 2001.

grainstone-claystone lithostratigraphic sequence tends to reduce rock porosity. Calcite has significantly lower load-bearing capacity than either quartz or sedimentary clays.

Granular Quartz–Calcite Mineralogic Mixtures

Thin isolated limestones commonly occur in dominantly quartz grainstone-claystone stratigraphic sequences. Where these thin limestones occur, they invariably have lower porosity than the neighboring quartz grainstones. This is because calcite (hardness 3) has much lower compaction resistance than quartz (hardness 7). The calcite-quartz mineralogic continuum is on the back face of the diagram shown in Figure 8, and the expected porosities are shown along the 8000 psi edge of the diagram.

Calcite cement in quartz grainstones invariably reduces porosity. This too is a natural consequence of the softer calcite's low compaction resistance. Whether the calcite mineral is a secondary deposit or not, the softer calcite would yield to the much harder quartz upon loading. Should a calcite grain be recrystallized or be transported from high to low stress locations by pressure solution, calcite would appear to be cement. The mechanical properties of the minerals present should be considered using the grain-cement nomenclature commonly used to describe relative compaction.

The calcite mineral lattice has the same load-bearing capacity whether it be recognized as grain or cement. The same is true of quartz. Minerals bear the average effective stress load through their crystalline lattice irrespective of their geometry. The whole load is borne by the whole rock that is composed of these minerals.

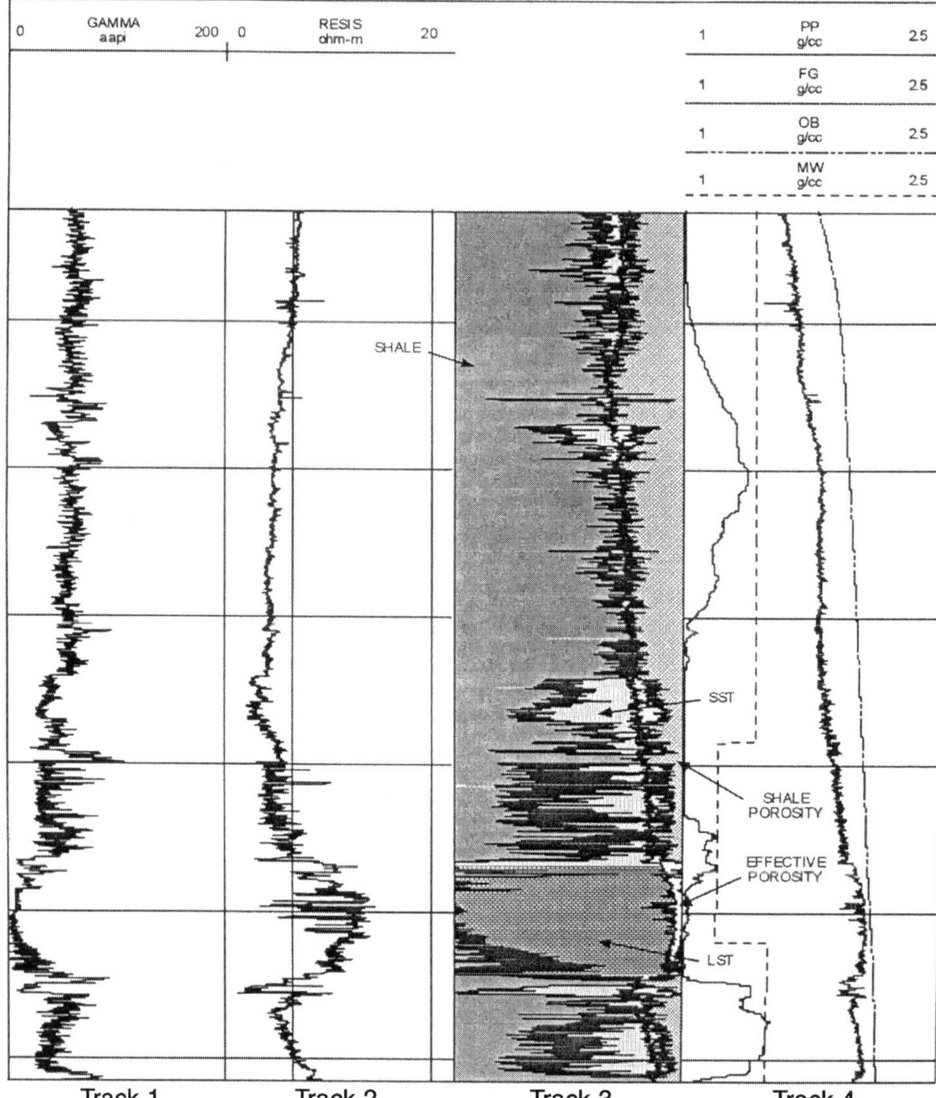

Figure 9. Composite wire-line and real-time log showing two lithostratigraphic sequence types. The quartz grainstone-claystone lithostratigraphic sequence has the opposite log pattern. The raw gamma-ray and raw resistivity logs in tracks 1 and 2 move parallel in quartz grainstone-claystone sequences. They move opposite in an hourglass pattern in limestone-claystone lithostratigraphic sequences. Figure taken from Holbrook, 1995.

Limestone-Claystone Lithostratigraphic Sequences

Calcite grainstones, chalk, and marl are rock types in the limestone-claystone lithostratigraphic continuum. Porosity is positively correlated to clay content in these sequences. Clay minerals are slightly harder than calcite so clay minerals have a slightly higher load-bearing capacity than calcite. Clay mineral interparticle repulsion, however, accounts for most of the increased porosity with respect to calcite. Much of the water in the marl pore space is probably electrostatically bound as shown in Figure 2c.

Pore pressure in limestone-claystone lithostratigraphic sequences can be significant. Loads borne by fluids are the other physical control of sedimentary rock compaction.

LOG EXAMPLE OF LITHOSTRATIGRAPHIC CONTROLS OVER COMPACTION OF SEDIMENTARY ROCKS

Figure 9 is a composite petrophysical log that shows both lithostratigraphic sequence types and their representative log patterns. Track 3 is a lithologic column showing fractional clay volume as an offset dash pattern. Quartz grainstone volume is shown by the dot pattern. Calcite volume is shown as a limestone block pattern. Porosity indicated as either bound water (gray shade) or free water (white) is the remainder of whole rock volume in track 3.

The raw gamma-ray and raw resistivity logs in tracks 1 and 2 move parallel in quartz grainstone-claystone lithostratigraphic sequences. They move opposite in an hourglass pattern in limestone-claystone lithostratigraphic sequences. This is the individual bed scale manifestation of the general power-law compactional relationships portrayed in Figures 6 and 8. Below the compaction crossover, quartz grainstones are more compaction resistant than claystones, and calcite is less compaction resistant than quartz and claystones. The parallel and hourglass log patterns observed on most petrophysical logs are related to these (Figures 6, 8) mineralogic compaction resistance relationships.

CONCLUSIONS

The compaction of sedimentary particles is explained in terms of mineral particle physical properties. The intraparticle load is borne within the mineral's lattice and across electrostatically neutral mineral grain contacts. Compaction of electrostatically neutral particles occurs at each of the n contacts with other particles. A power-law stress/strain relationship captures bulk compaction of n-coordinated particulate solids.

The repulsive electrostatic field between negatively charged clay mineral surfaces is also load bearing. The magnitude of this repulsive field is power-law related to clay particle surface area within a rock volume (Figure 3). Interparticle repulsive force is also a power-law function of distance between clay mineral surfaces (Figure 2c). The sum of these clay mineral power-law functions and the electrostatically neutral power-law function is a composite power-law function.

The net effects of interparticle and intraparticle load types on volumetric in-situ strain $(1.0 - \phi)$ are captured with two power-law compactional stress/strain coefficients (σ_{max} and α). The power-law compaction coefficients for the five most common sedimentary minerals were measured from in-situ strain after properly accounting for effective stress. Peak granular solid compaction resistance (σ_{max}) is positively correlated to mineral hardness and negatively correlated to mineral solubility. All three are power law relationships.

End-member sedimentary claystones have a significantly lower α value than the electrostatically neutral minerals. The unusually high 10–95% initial porosity of clay-rich sediments is power-law related to the log of average sediment particle size. Whole rock compaction is the volume-weighted average (σ_{max} and α) of its individual mineral specific stress/strain coefficients.

This general mineralogic (σ_{max} and α) sedimentary rock stress/strain compactional relationship has been tested in more than 300 wells in normal fault regime basins worldwide. Large-scale compaction trends (Figures 6, 7) are explained by this force-balanced stress/strain relationship. Observed sedimentary bed-scale compactional differences (Figures 4, 8, 9) are also explained by the same compactional relationship.

The entire load placed upon a sedimentary rock volume is borne by interparticle repulsion, intraparticle resistance, and pore-fluid pressure. Intraparticle compaction resistance is related to σ_{max} and hardness. Interparticle repulsion contributes additional compactional resistance to α in proportion to clay mineral surface area. Mineral ionic bond strength and clay mineral interparticle repulsion are believed to be primary controls over sediment compaction.

ACKNOWLEDGMENTS

Figure 9 is reprinted with permission of the Society of Professional Well Log Analysts.

REFERENCES CITED

Baldwin, B., and C. O. Butler, 1985, Compaction curves: AAPG Bulletin, v. 69, no. 4, p. 622–626.

Bates, R. L., and J. Jackson, eds., 1987, 3d ed., Glossary of geology: Falls Church, Virginia, American Geological Institute, 788 p.

Berry, L. G., and B. Mason, 1959, Mineralogy, concepts, descriptions, determinations: San Francisco and London, W. H. Freeman and Company, 630 p.

Carmichael, R. S., 1982, Handbook of physical properties of rocks: Boca Raton, CRC Press, 404 p.

Carroll, M. M., 1980, Compaction of dry or fluid-filled porous materials: Journal of Engineering Mechanics Division, Proceedings of the American Society of Civil Engineers, v. 106, no. EM5, p. 969–990.

Holbrook, P. W., 1995, The relationship between porosity, mineralogy and effective stress in granular sedimentary rocks: Society of Professional Well Log Analysts 36th Annual Logging Symposium, paper AA, 14 p.

Holbrook, P. W., 1996, The use of petrophysical data for well planning, drilling safety and efficiency: Society of Professional Well Log Analysts 37th Annual Logging Symposium, paper X, 14 p.

Holbrook, P. W., 1999, A simple closed-form force↔balanced solution for pore pressure, overburden and the principal effective stresses in the earth: Journal of Marine and Petroleum Geology, v. 16, p. 303–319.

Holbrook, P. W., 2001, Pore pressure through Earth mechanical systems: Houston, Texas, Force Balanced Press, 135 p.

Holbrook, P. W., D. A. Maggiori, and R. Hensley, 1995, Real-time pore pressure and fracture pressure determination in all sedimentary lithologies: Society of Petroleum Engineers Formation Evaluation, v. 10, n. 4, p. 215–222.

Hueckel, T. A., 1992, Water-mineral interaction in hygromechanics of clays exposed to environmental loads: a mixture-theory approach: Canadian Geotechnical Journal, v. 29, p. 1071–1086.

Lambe, T. W., and R. V. Whitman, 1969, Soil mechanics: New York, John Wiley & Sons, 553 p.

Nagaraj, T. S., and B. R. S. Murthy, 1983, Rationalization of Skempton's compressibility equation: Geophysique, v. 33, no. 4, p. 433–443.

Revil, A. P., A. Pezard, and M. Darot, 1997, Electrical conductivity, spontaneous potential and ionic diffusion in porous media, *in* M. A. Lovell and P. K. Harvey, eds., Developments in petrophysics: Geological Society Special Publication 122, p. 253–275.

Scott, R. F., 1963, Principles of soil mechanics: Reading, Massachusetts, Addison-Wesley, 550 p.

Shumway, G., 1960, Sound speed and absorption studies of marine sediments by resonance method—part II: Geophysics, v. 25, p. 659–682.

Skempton, A. W., 1970, The consolidation of clays by gravitational compaction: Quarterly Journal of the Geological Society of London, v. 125, pt. 3, p. 373–411.

Stuart, C. A., 1970, Geopressures: Proceedings of the Second Symposium on Abnormal Subsurface Pressure, 121 p.

Terzaghi, K. von, 1923, Die Berechnung der Durchässigkeitsziffer des Tones aus dem Verlauf der hydrodynamischen Spannungserscheinungen: Sitzungsberichte Akademie der Wissenschaften, Vienna, Mathematisch-Naturwissenschaftliche Klasse, abt. 2a, v. 132, p. 125–138.

4

Critical-Porosity Models

Jack Dvorkin
Stanford University,
Stanford, California

Amos Nur
Stanford University,
Stanford, California

ABSTRACT

Anomalous velocity and porosity are common indicators of abnormal pore pressure. Therefore, it is important to be able to link velocity to porosity and rock texture in a rational, first-principle–based manner. The critical-porosity concept allows for building such rock physics models. Critical porosity is the porosity above which the rock can exist only as a suspension. In sandstones the critical porosity is 36–40%, that is, the porosity of a random close pack of well-sorted rounded quartz grains. This pack is commonly the starting point for the formation of consolidated sandstones. Using this starting point for effective medium modeling, rational models can be built that relate velocity to porosity depending on rock texture and lithology.

INTRODUCTION AND CRITICAL-POROSITY CONCEPT

Porosity is one of the desired reservoir parameters that can be used, for example, for reserve estimation, reservoir simulation, and pore-pressure prediction. Derivation of porosity from such seismic observable properties as impedance or velocity requires a velocity-porosity relation. Such relations vary depending on lithology and rock texture. To appreciate the effect of texture on velocity, consider Figure 1, where P- and S-wave velocity are plotted vs. the total porosity for relatively clay-free gas-saturated sands at the differential pressure (confining minus pore pressure) of about 20 MPa.

All sandstone data points in Figure 1 represent rock that is mainly quartz with clay content not exceeding 10%. Yet, in the same porosity range, the P-wave velocity may span from 1.5 to more than 3 km/s, and the S-wave velocity from 1 to more than 2 km/s. One apparent reason for this large velocity difference between mineralogically similar samples is rock texture—the arrangement of the sand grains and pore-filling material in the pore space. In the sandstone samples from S. Strandenes (1991, unpublished data), the grains appear to be slightly cemented at their contacts, whereas the samples from Blangy (1992) are friable sands.

The velocity in the well-log data (Dvorkin et al., 1999b) is even smaller than that in the friable sands. These rocks are elastically similar to a handmade mixture of Ottawa sand and kaolinite where the small kaolinite particles fill the pore space without noticeably affecting the velocity.

We can create rational effective medium models to explain and predict the observed velocity-porosity behavior by examining the textural nature of sandstones. Consider Figure 2a where the compressional modulus (bulk density times the compressional-wave-velocity squared) of water-saturated clean sandstones and quartz marine sediment (suspensions) is plotted vs.

Dvorkin, Jack, and Amos Nur, 2002, Critical-Porosity Models, in A. R. Huffman and G. L. Bowers, eds., Pressure regimes in sedimentary basins and their prediction: AAPG Memoir 76, p. 33–41.

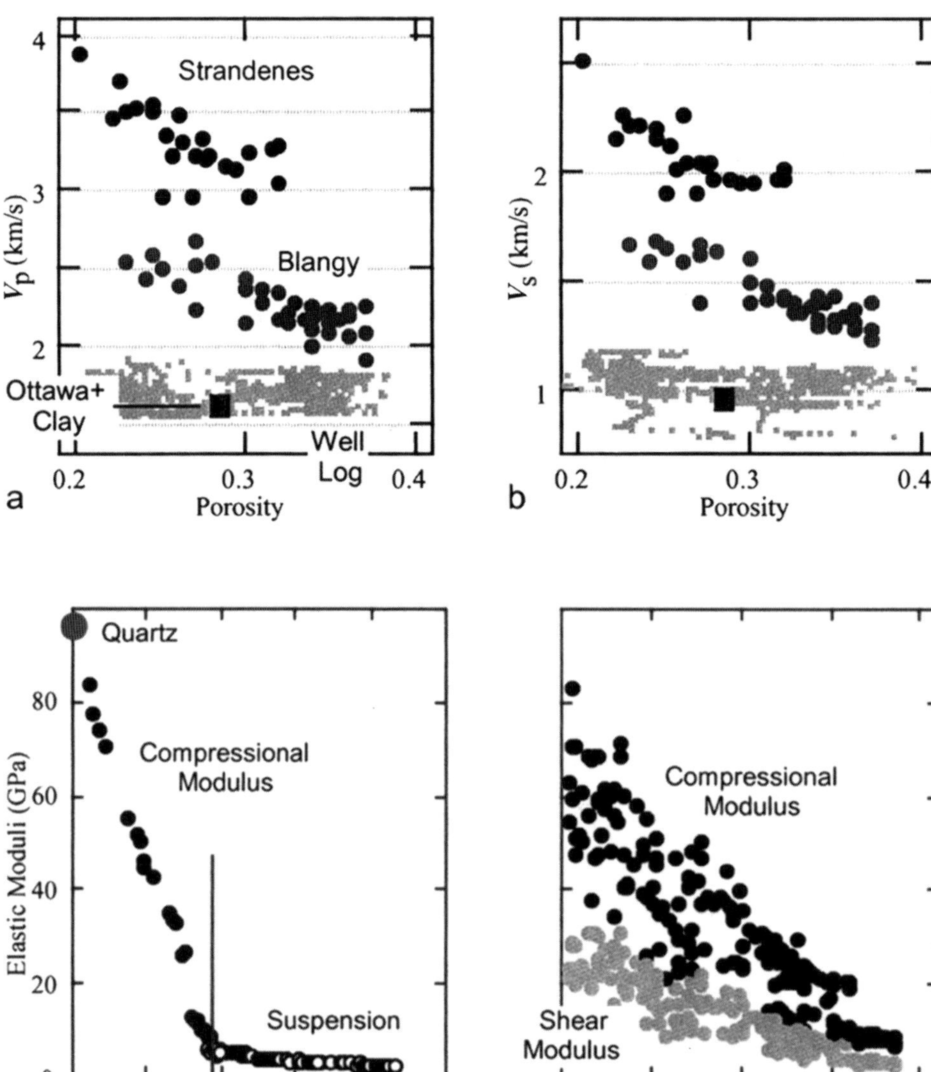

Figure 1. (a) P- (V_p) and (b) S-wave velocity (V_s) in rocks with gas at 20 MPa differential pressure. Circles represent laboratory data obtained on high-porosity fast (S. Strandenes, 1991, unpublished data) and slow (Blangy, 1992) sands; both data sets are from the North Sea. Gray symbols are from a Gulf Coast gas well. The filled square is for a handmade mixture of Ottawa sand and 10% kaolinite (Yin, 1993). Clay content for these data does not exceed 10%.

Figure 2. (a) Compressional modulus vs. porosity in clean water saturated sandstones and marine sediment. (b) Compressional and shear moduli of room-dry sandstones vs. porosity. The data used are discussed in Nur et al. (1998).

porosity. The porosity of 36–40% is the point where the modulus-porosity trend abruptly changes. In the lower porosity domain, the stiffness of the sandstone is determined by the framework of contacting quartz grains. In the higher porosity domain, the grains are not in contact anymore and are suspended in water. In this case, the stiffness of the sediment is determined by the pore fluid.

We call this threshold porosity "critical porosity" (Nur et al., 1998). The rocks where the solid phase is spatially continuous and dominates the stiffness of the rock have porosity that is smaller than the critical porosity. This fact is illustrated in Figure 2b where the compressional and shear moduli of many sandstone samples (room dry at 30–40 MPa differential pressure) are plotted vs. porosity.

The critical-porosity concept is valid not only for sandstones but also for other natural and artificial rocks. An example is given in Figure 3 where the compressional modulus is plotted vs. porosity for cracked igneous rocks and pumice (Nur et al., 1998). In the first case, the critical porosity is as small as 6%, whereas in the second case it reaches 70%. The reason is the peculiar microstructural topology of the rocks under examination. The igneous rocks are permeated by cracks that percolate and make the solid phase lose its spatial continuity at very small porosity. In the pumice, the honeycomb structure of the solid ensures its spatial continuity at high-porosity values. Nur et al. (1998) summarize the critical-porosity values for various rocks as shown in Table 1. In the following sections, we introduce the critical-concentration concept and de-

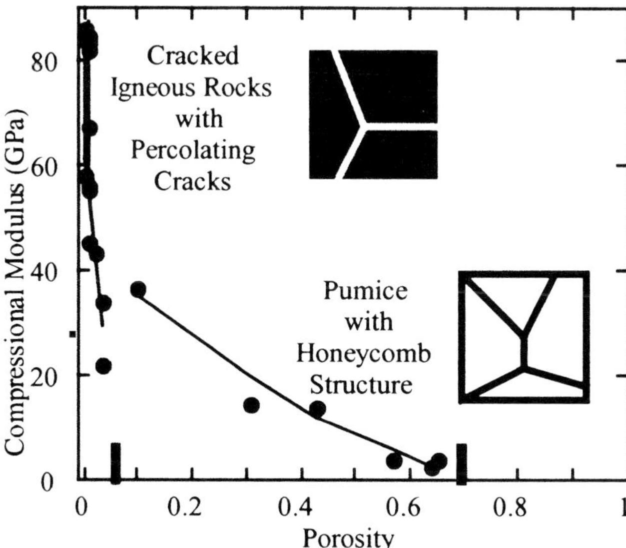

Figure 3. Compressional modulus vs. porosity in cracked igneous rocks and pumice. The data used are discussed in Nur et al. (1998).

Table 1. Critical-Porosity Values*

Material	Critical Porosity
Sandstones	40%
Limestones	40%
Dolomites	40%
Pumice	70%
Chalks	65%
Rock salt	40%
Cracked igneous rocks	5%
Oceanic basalts	20%
Sintered glass beads	40%
Glass foam	90%

*Data from Nur et al. (1998).

scribe several effective medium models that are based on the critical porosity concept.

CRITICAL-CONCENTRATION CONCEPT

The critical-porosity concept leads to the "critical-concentration" concept that Marion (1990) and Yin (1993) used to describe the properties of sands with shale. Consider the experimental data from Yin (1993) obtained on synthetic rocks made by mixing Ottawa sand and kaolinite at room-dry conditions. The volumetric clay content in the samples ranged from 0 to 100%.

The total porosity at 20 MPa differential pressure is plotted vs. the volumetric clay content in Figure 4a.

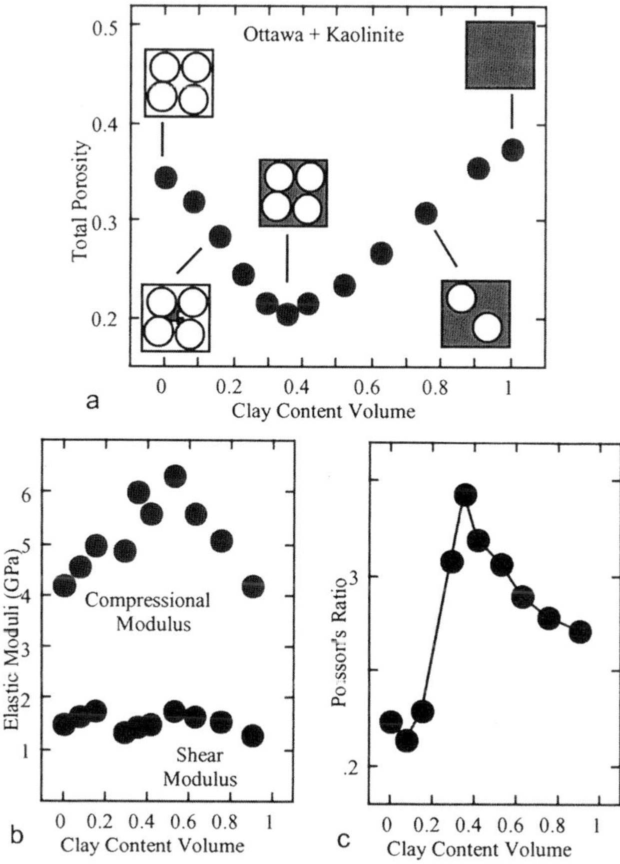

Figure 4. (a) Porosity, (b) elastic moduli, and (c) dynamic Poisson's ratio vs. volumetric clay content in Ottawa sand mixed with kaolinite at room-dry conditions and 20 MPa differential pressure (modified from Yin, 1993).

The two end members of the data set are the porosity of Ottawa sand at zero clay content and porosity of kaolinite at 100% clay content. The porosity of the mixture reaches its minimum at the point where the volumetric concentration of clay equals the porosity of Ottawa sand (which is close to the critical porosity for sandstones). This clay content is called "critical clay concentration."

The critical concentration affects not only the total porosity but also the dynamic (velocity-derived) elastic moduli of the mixture (Figure 4b). The elastic moduli of the mixture are maximum at the critical concentration and decrease as the clay content increases or decreases from the critical-concentration value. Dynamic dry Poisson's ratios behave in a similar way (Figure 4c).

The elastic properties of the synthetic mixture of Ottawa sand and kaolinite are plotted vs. the total porosity in Figure 5. The nonuniqueness of the elastic moduli, and, especially, Poisson's ratio in the crossplots is due to the grain-scale texture of the rock.

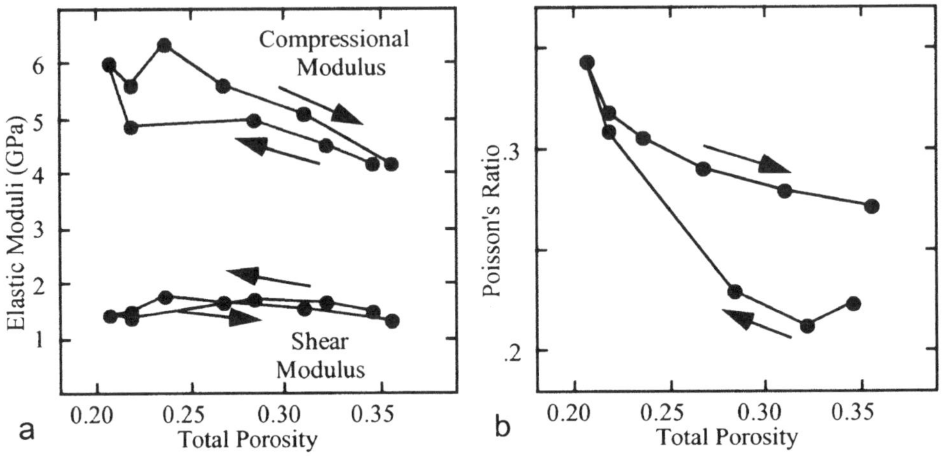

Figure 5. (a) Elastic moduli and (b) dynamic Poisson's ratio vs. total porosity in room-dry Ottawa sand mixed with kaolinite at room-dry conditions and 20 MPa differential pressure (modified from Yin, 1993). The arrows show increasing clay content.

This effect has to be considered when examining well-log data. In Figure 6a and b, we plot the bulk density and P-wave impedance vs. the gamma-ray values for a well in Colombia (M. Gutierrez, 1998, personal communication). Different trends are apparent for the low–gamma-ray and the high–gamma-ray branches. This effect depends on the rock's microstructure and results in nonuniqueness as the P-wave impedance is plotted vs. the bulk density and porosity (Figure 6c, d). Being aware of the physical reason underlying these nonunique crossplots allows the log analyst to separate the trends and arrive at accurate impedance-porosity transforms.

MODELS FOR HIGH-POROSITY SANDSTONES

The initial building point for effective medium models that describe high-porosity sandstones should be unconsolidated well-sorted sand, as proposed by the critical-porosity concept. In mathematical modeling, such sand is approximated by a dense pack of identical elastic spheres (Figure 7).

The contact-cement model (Dvorkin and Nur, 1996) assumes that porosity decreases from the initial critical-porosity value due to the uniform deposition of cement layers on the surface of the grains. This cement may be diagenetic quartz, calcite, or reactive clay (such as illite). The diagenetic cement dramatically increases the stiffness of the sand by reinforcing the grain contacts (Figure 8). The mathematical model is based on a rigorous contact-problem solution by Dvorkin et al. (1994).

In this model, the effective bulk (K_{dry}) and shear (G_{dry}) moduli of dry rock are

$$\begin{aligned} K_{dry} &= n(1-\phi_c)M_c S_n/6 \\ G_{dry} &= 3K_{dry}/5 + 3n(1-\phi_c)G_c S_\tau/20 \end{aligned} \tag{1}$$

where ϕ_c is critical porosity; K_c and G_c are the bulk and shear moduli of the cement material, respectively; $M_c = K_c + 4G_c/3$ is the compressional modulus of the cement; and n is the coordination number (average number of contacts per grain is 8–9). The variables S_n and S_τ are

$$S_n = A_n(\Lambda_n)\alpha^2 + B_n(\Lambda_n)\alpha + C_n(\Lambda_n)$$

$$A_n(\Lambda_n) = -0.024153\Lambda_n^{-1.3646}$$

$$B_n(\Lambda_n) = 0.20405\Lambda_n^{-0.89008}$$

$$C_n(\Lambda_n) = 0.00024649\Lambda_n^{-1.9864}$$

$$S_\tau = A_\tau(\Lambda_\tau, \nu_s)\alpha^2 + B_\tau(\Lambda_\tau, \nu_s)\alpha + C_\tau(\Lambda_\tau, \nu_s)$$

$$A_\tau(\Lambda_\tau, \nu_s) = -10^{-2}(2.26\nu_s^2 + 2.07\nu_s + 2.3)\Lambda_\tau^{0.079\nu_s^2 + 0.1754\nu_s - 1.342}$$

$$B_\tau(\Lambda_\tau, \nu_s) = (0.0573\nu_s^2 + 0.0937\nu_s + 0.202)\Lambda_\tau^{0.0274\nu_s^2 + 0.0529\nu_s - 0.8765}$$

$$C_\tau(\Lambda_\tau, \nu_s) = 10^{-4}(9.654\nu_s^2 + 4.945\nu_s + 3.1)\Lambda_\tau^{0.01867\nu_s^2 + 0.4011\nu_s - 1.8186}$$

$$\Lambda_n = 2G_c(1-\nu_s)(1-\nu_c)/[\pi G_s(1-2\nu_c)]$$

$$\Lambda_\tau = G_c/(\pi G_s)$$

$$\alpha = [(2/3)(\phi_c - \phi)/(1-\phi_c)]^{0.5}$$

$$\nu_c = 0.5(K_c/G_c - 2/3)/(K_c/G_c + 1/3)$$

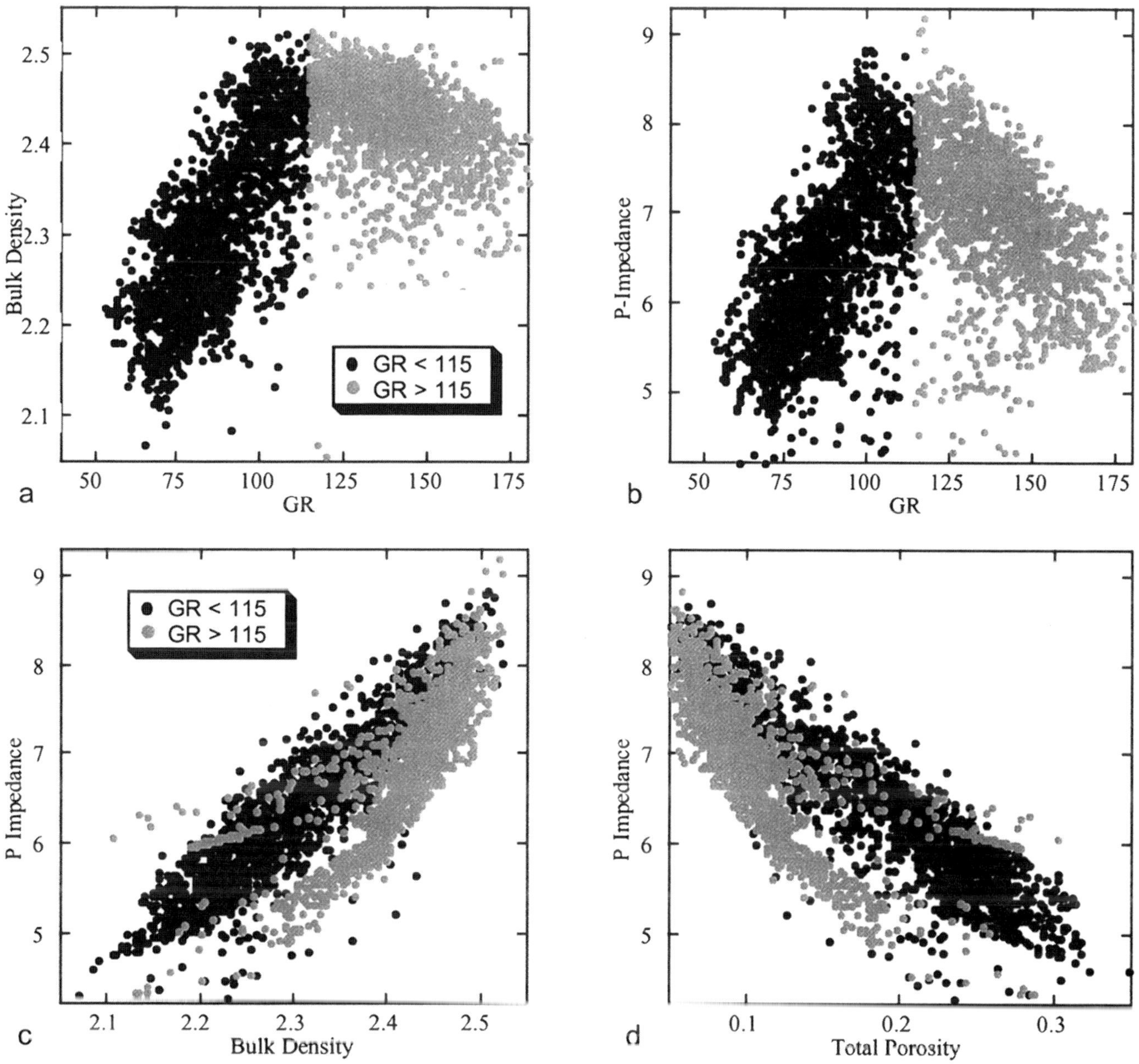

Figure 6. Well-log data. (a) Bulk density and (b) P impedance vs. gamma ray; P impedance vs. (c) bulk density and (c) total porosity.

$$\nu_s = 0.5(K_s/G_s - 2/3)/(K_s/G_s + 1/3)$$

where K_s and G_s are the bulk and shear moduli of the grain material. A detailed explanation of these equations and their derivation are given in Dvorkin and Nur (1996).

The contact-cement theory allows one to accurately model the velocity in fast high-porosity sands (Figure 9). One may find that the contact-cement model is appropriate for describing sands in high-energy depositional environments where the grains are well sorted and not covered by organic matter.

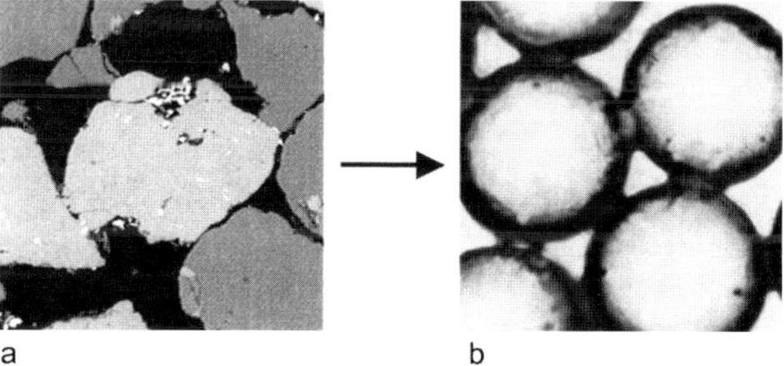

Figure 7. Approximating sand by a sphere pack. Microphotographs of (a) well-sorted sand and (b) a glass-bead pack.

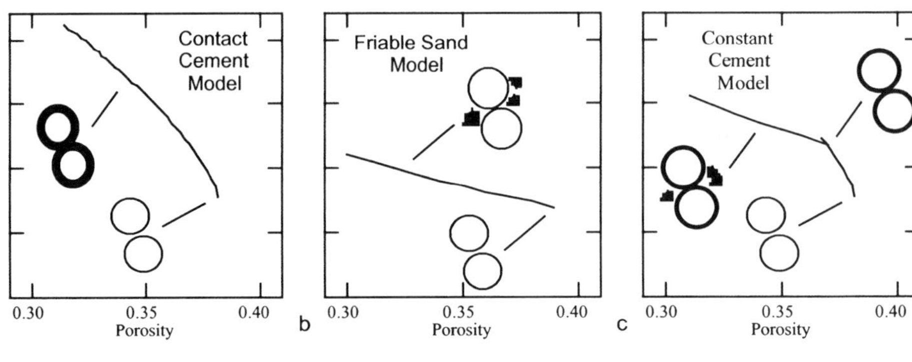

Figure 8. Schematic depiction of three effective-medium models for high-porosity sandstones and corresponding diagenetic transformations. (a) Contact-cement model, (b) friable sand model, and (c) constant-cement model.

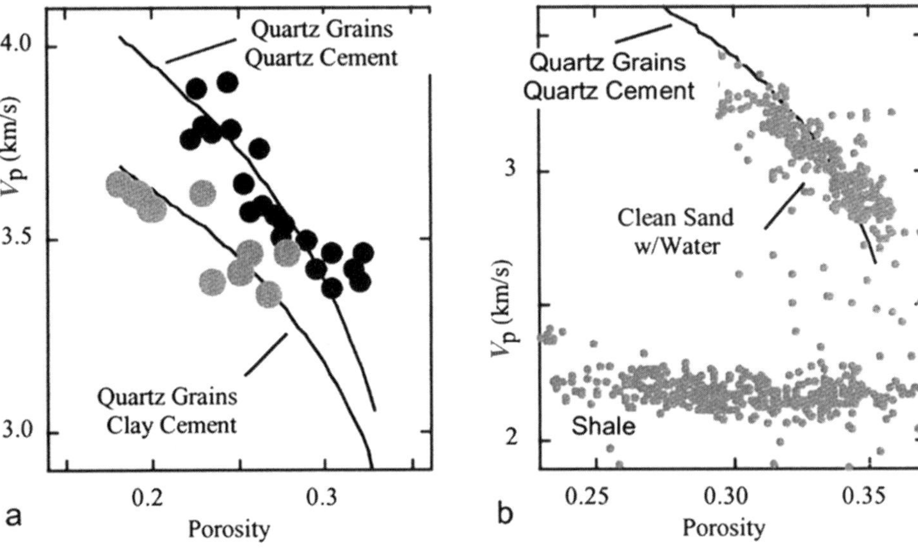

Figure 9. P-wave velocity vs. porosity. (a) Water-saturated-rock data based on laboratory measurements of fast high-porosity North Sea sandstones by S. Strandenes (1991, unpublished data). Solid black circles are for very clean samples. Solid gray circles are for samples with some clay. The curves are from the contact-cement model for pure quartz grains with quartz and clay cement. (b) Well-log data. The clean sand interval is saturated with water. The curve is from the contact-cement theory for pure quartz grains with quartz cement.

The friable sand model (Dvorkin and Nur, 1996) assumes that porosity decreases from the initial critical-porosity value due to the deposition of the solid matter away from the grain contacts. Such a diagenetic process of porosity reduction may correspond to deteriorating grain sorting. This noncontact additional solid matter weakly affects the stiffness of the rock (Figure 8b).

The theoretical effective-medium model connects two end points in the elastic-modulus-porosity plane. One end point is at critical porosity. The elastic moduli of the dry rock at that point are assumed to be the same as of an elastic sphere pack subject to confining pressure. These moduli are given by the Hertz-Mindlin (Mindlin, 1949) theory:

$$
\begin{aligned}
K_{HM} &= \left[\frac{n^2(1-\phi_c)^2G^2}{18\pi^2(1-\nu)^2}P\right]^{\frac{1}{3}} \\
G_{HM} &= \frac{5-4\nu}{5(2-\nu)}\left[\frac{3n^2(1-\phi_c)^2G^2}{2\pi^2(1-\nu)^2}P\right]^{\frac{1}{3}}
\end{aligned}
\tag{2}
$$

where K_{HM} and G_{HM} are the bulk and shear moduli at critical porosity ϕ_c, respectively; P is the differential pressure (total confining pressure minus pore pressure); G and ν are the bulk and shear moduli of the solid phase, and its Poisson's ratio, respectively; and n is the coordination number.

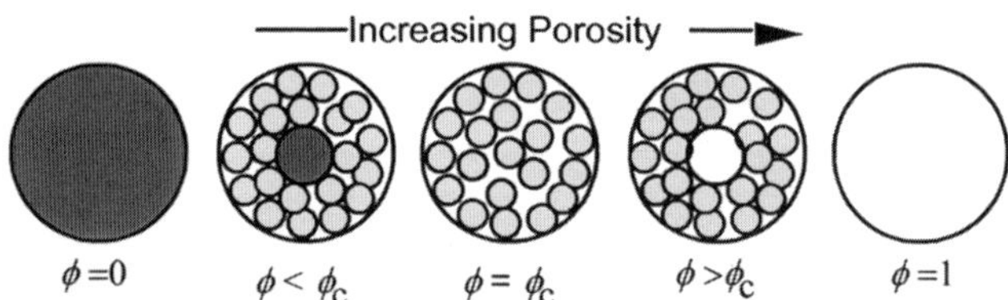

Figure 10. Hashin-Shtrikman arrangements of sphere pack, solid, and void.

The other end point is at zero porosity and has the bulk (K) and shear (G) moduli of the pure solid phase. These two points in the porosity-moduli plane are connected with the curves that have the algebraic expressions of the lower Hashin-Shtrikman (1963) bound (bulk and shear moduli) for the mixture of two components: the pure solid phase and the phase that is the sphere pack. The reasoning is that in unconsolidated sediment, the softest component (the sphere pack) en-

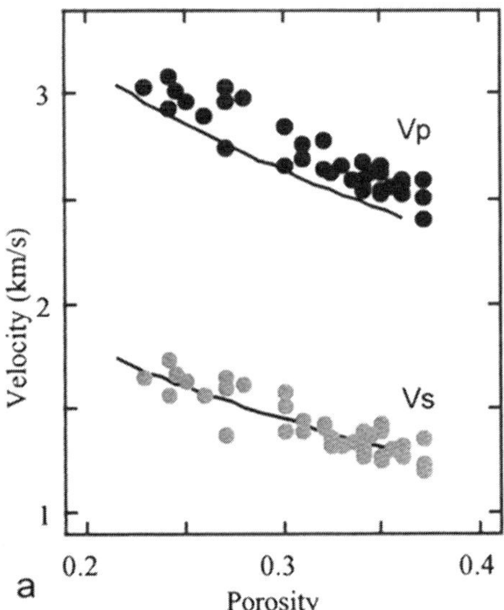

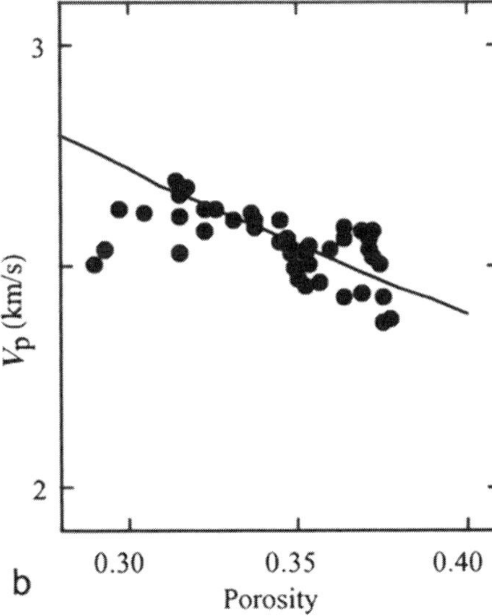

Figure 11. Velocity vs. porosity. (a) Water-saturated-rock data based on laboratory measurements of soft high-porosity North Sea sandstones by Blangy (1992). (b) Well-log data (Avseth et al., 1998) for oil-saturated pay zone. The curves are from the friable sand model.

velopes the stiffest component (the solid) in the Hashin-Shtrikman fashion (Figure 10).

At porosity ϕ the concentration of the pure solid phase (added to the sphere pack to decrease porosity) in the rock is $1 - \phi/\phi_c$ and that of the sphere-pack phase is ϕ/ϕ_c. Then the bulk (K_{dry}) and shear (G_{dry}) moduli of the dry frame are

$$K_{Dry} = \left[\frac{\phi/\phi_c}{K_{HM}+\frac{4}{3}G_{HM}} + \frac{1-\phi/\phi_c}{K+\frac{4}{3}G_{HM}}\right]^{-1} - \frac{4}{3}G_{HM}$$

$$G_{Dry} = \left[\frac{\phi/\phi_c}{G_{HM}+z} + \frac{1-\phi/\phi_c}{G+z}\right]^{-1} - z \quad (3)$$

$$z = \frac{G_{HM}}{6}\left(\frac{9K_{HM}+8G_{HM}}{K_{HM}+2G_{HM}}\right)$$

The friable sand model allows one to accurately predict velocity in soft high-porosity sands (Figure 11). This model is appropriate for describing sands where contact-cement deposition was inhibited by organic matter deposited on the grain surface.

The constant-cement model (Avseth et al., 1998) assumes that the initial porosity reduction from critical porosity is due to the contact-cement deposition. At some high porosity, this diagenetic process stops, and after that porosity reduces because of the deposition of the solid phase away from the grain contacts as in the friable sand model (Figure 8c). This model is mathematically analogous to the friable sand model except that the high-porosity end-point bulk and shear moduli (K_b and G_b, respectively) are calculated at some cemented porosity ϕ_b from the contact-cement model. Then the dry-rock bulk and shear moduli are

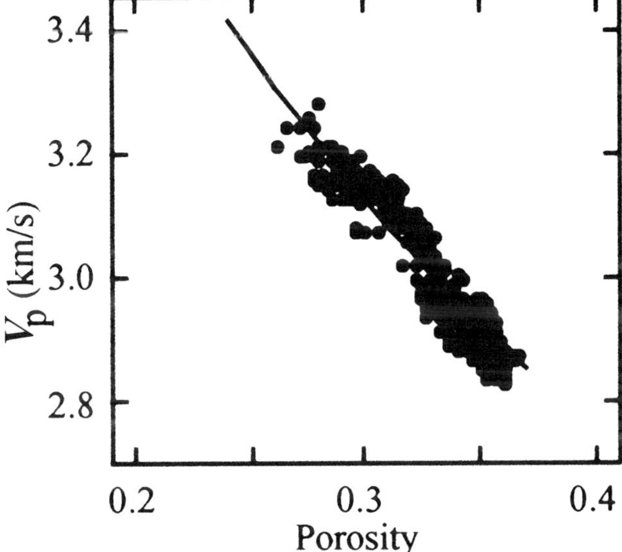

Figure 12. Velocity vs. porosity. Well-log data (Avseth et al., 1998) for oil-saturated pay zone. The curve is from the constant-cement model.

$$K_{dry} = \left(\frac{\phi/\phi_b}{K_b+4G_b/3} + \frac{1-\phi/\phi_b}{K_s+4G_b/3}\right)^{-1} - 4G_b/3$$

$$G_{dry} = \left(\frac{\phi/\phi_b}{G_b+z} + \frac{1-\phi/\phi_b}{G_s+z}\right)^{-1} - z \quad (4)$$

$$z = \frac{G_b}{6}\left(\frac{9K_b+8G_b}{K_b+2G_b}\right)$$

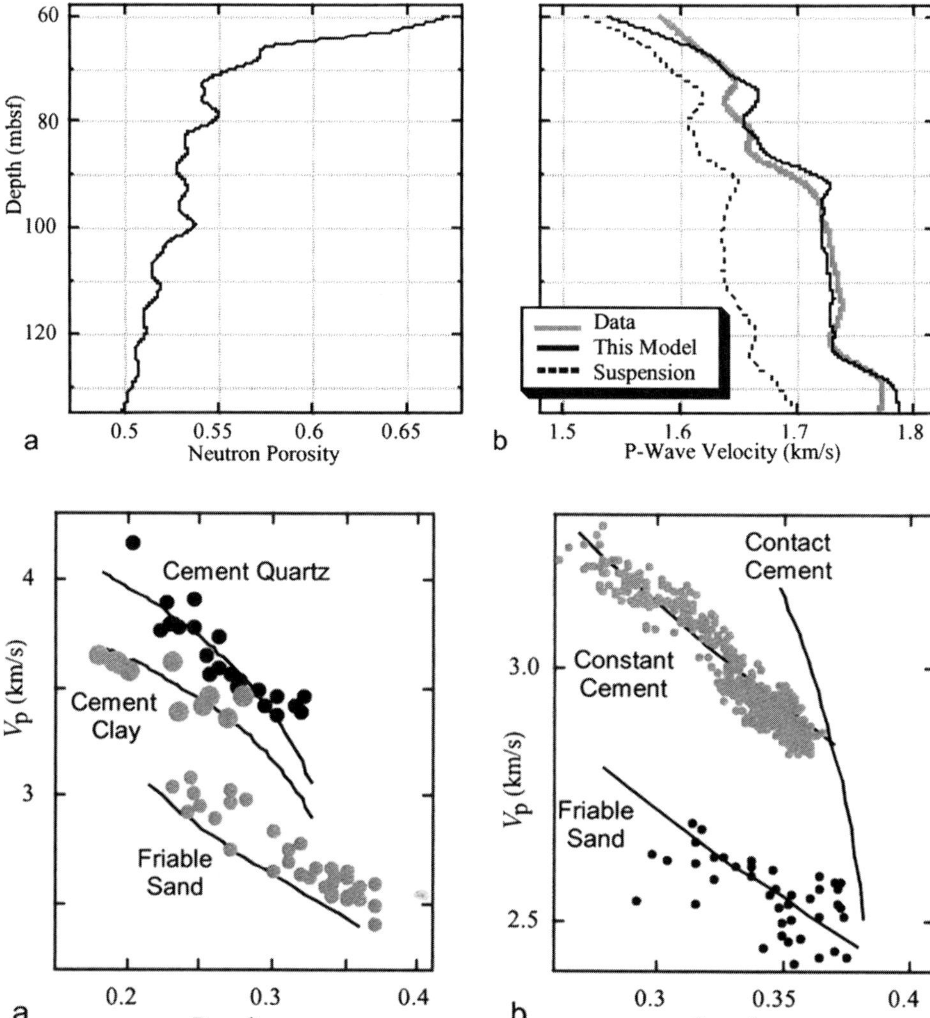

Figure 13. Deep Sea Drilling Project (DSDP) Well 974. (a) Neutron porosity vs. depth; mbsf = meters below sea floor. (b) Velocity vs. depth: data, our model, and suspension model. All curves are smoothed.

Figure 14. Velocity vs. porosity. Theoretical curves superimposed on data allow one to identify the rock type. (a) Data from Figures 9a and 11a. (b) Data from Figures 11b and 12.

An example of applying this model to well-log data is given in Figure 12.

MODELS FOR UNCONSOLIDATED MARINE SEDIMENT

This model (Dvorkin et al., 1999a) is analogous to the friable sand model but covers the porosity range above critical porosity. One end point is the critical porosity where the elastic moduli of the sphere pack are given by equation 2. To arrive at higher porosity, we add empty voids to the sphere pack (Figure 10). In this case the voids are placed inside the pack in the Hashin-Shtrikman fashion. Now the pack is the stiffest component, so we have to use the upper Hashin-Shtrikman limit.

At porosity $\phi > \phi_c$, the concentration of the void phase is $(\phi - \phi_c)/(1 - \phi_c)$ and that of the sphere-pack phase is $(1 - \phi)/(1 - \phi_c)$. Then the effective dry-rock frame bulk and shear moduli are

$$K_{Dry} = \left[\frac{(1-\phi)/(1-\phi_c)}{K_{HM}+\frac{4}{3}G_{HM}} + \frac{(\phi-\phi_c)/(1-\phi_c)}{\frac{4}{3}G_{HM}}\right]^{-1} - \frac{4}{3}G_{HM}$$

$$G_{Dry} = \left[\frac{(1-\phi)/(1-\phi_c)}{G_{HM}+z} + \frac{(\phi-\phi_c)/(1-\phi_c)}{z}\right]^{-1} - z \quad (5)$$

$$z = \frac{G_{HM}}{6}\left(\frac{9K_{HM}+8G_{HM}}{K_{HM}+2G_{HM}}\right)$$

The saturated-rock elastic moduli can be calculated using Gassmann's (1951) equation.

An example of applying this model to log data is given in Figure 13 (Dvorkin et al., 1999a). A good agreement between the model and the data is apparent. At the same time, the commonly used suspension model fails to correctly mimic the data. This model's departure from the data increases with depth, which is due to the effect of confining pressure that adds stiffness to the dry frame of the sediment, thus making the suspension model inadequate.

CONCLUSION

The critical-porosity and critical-concentration concepts allow the geophysicist to better understand the diversity of well-log and core-elastic data. Effective-medium models built on the basis of the critical-porosity concept can accurately model data. By superimposing theoretical model curves on velocity-porosity and elastic-moduli-porosity crossplots, one may mathematically diagnose rock, that is, determine the texture of the sediment (e.g., contact-cemented vs. friable). Rock-physics diagnostic can be conducted with velocity, impedance, or elastic moduli. Examples of diagnostic rock curves are given in Figure 14. Such diagnostic curves have implications for fluid detection (Avseth et al., 1998) and strength and permeability estimation (Dvorkin and Brevik, 1999). Moreover, the rock physics diagnostic may be crucial for pore-pressure prediction from velocity because at the same pressure, velocity may vary depending on rock texture (Dvorkin et al., 1999c). As a result, texture-related velocity variations may be mistakenly attributed to pore-pressure anomalies.

ACKNOWLEDGMENTS

This work was supported by the Stanford Rock Physics Laboratory.

REFERENCES CITED

Avseth, P., J. Dvorkin, G. Mavko, and J. Rykkje, 1998, Diagnosing high-porosity sands for reservoir characterization using sonic and seismic: Society of Exploration Geophysicists 66th Annual Meeting, Expanded Abstracts, p. 1024–1025.

Blangy, J. P., 1992, Integrated seismic lithologic interpretation: the petrophysical basis: Ph.D. thesis, Stanford University, Stanford, California, 357 p.

Dvorkin, J., and I. Brevik, 1999, Diagnosing high-porosity sandstones: strength and permeability from porosity and velocity: Geophysics, v. 64, p. 795–799.

Dvorkin, J., and A. Nur, 1996, Elasticity of high-porosity sandstones: theory for two North Sea datasets: Geophysics, v. 61, p. 1363–1370.

Dvorkin, J., A. Nur, and H. Yin, 1994, Effective properties of cemented granular materials: Mechanics of Materials, v. 18, p. 351–366.

Dvorkin, J., M. Prasad, A. Sakai, and D. Lavoie, 1999a, Elasticity of marine sediments: Geophysical Research Letters, v. 26, p. 1781–1784.

Dvorkin, J., D. Moos, J. Packwood, and A. Nur, 1999b, Identifying patchy saturation from well logs: Geophysics, v. 64, p. 1–5.

Dvorkin, J., G. Mavko, and A. Nur, 1999c, Overpressure detection from compressional- and shear-wave data: Geophysical Research Letters, v. 26, p. 3417–3420.

Gassman, F., 1951, Elasticity of porous media—Uber die elastizitat poroser medien: Vierteljahrsschrift der Naturforschenden Gesselschaft, v. 96, p. 1–23.

Hashin, Z., and S. Shtrikman, 1963, A variational approach to the elastic behavior of multiphase materials: Journal of Mechanics and Physics of Solids, v. 11, p. 127–140.

Marion, D., 1990, Acoustical, mechanical, and transport properties of sediments and granular materials: Ph.D. thesis, Stanford University, Stanford, California, 136 p.

Mindlin, R. D., 1949, Compliance of elastic bodies in contact: Journal of Applied Mechanics, v. 16, p. 259–268.

Nur, A., G. Mavko, J. Dvorkin, and D. Galmudi, 1998, Critical porosity: a key to relating physical properties to porosity in rocks: The Leading Edge, v. 17, p. 357–362.

Yin, H., 1993, Acoustic velocity and attenuation of rocks: isotropy, intrinsic anisotropy, and stress induced anisotropy: Ph.D. thesis, Stanford University, Stanford, California, 227 p.

5

The Role of Shale Pore Structure on the Sensitivity of Wire-Line Logs to Overpressure

G. L. Bowers
Applied Mechanics Technologies, Houston, Texas

T. John Katsube
Geological Survey of Canada, Ottawa, Ontario, Canada

ABSTRACT

Petrophysical characteristics of shales have been analyzed to improve our understanding of wire-line log response to overpressure. Bulk density and neutron porosity logs sometimes mask pore-pressure increases that are clearly evident on sonic and resistivity logs. This may be because sonic and resistivity logs respond to transport properties, whereas neutron and density logs reflect bulk properties.

Results of this study indicate that shale pore structure can be characterized by a storage-connecting pore system, with connecting pore sizes on the order of 2–20 nm. Laboratory compaction tests indicate that connecting pores are mechanically more flexible compared to storage pores and are likely to have lower aspect ratios. Consequently, connecting pores are likely to undergo more elastic rebound (widening) compared to storage pores as a result of effective stress reductions caused by overpressure.

Essentially, these results provide evidence that suggest sonic and resistivity logs respond to transport properties, whereas neutron and density logs reflect bulk properties, as previously proposed. This suggests that, whereas overpressure resulting from compaction disequilibrium may be detected by all four logs, fluid expansion overpressure may be best detected by sonic and resistivity logs.

INTRODUCTION

Pore-pressure increases are sometimes masked by the density log (Carstens and Dypvik, 1981; Grauls and Cassignol, 1993; Hermanrud et al., 1998). Fortunately, from the standpoint of overpressure detection, sonic velocity and resistivity commonly respond to pore-pressure changes where the density log does not. In fact, a drop in sonic velocity and electrical resistivity without a comparable decrease in density is commonly an indication of severe overpressure. Figure 1 shows an example of this from the Gulf of Mexico (GOM).

Hermanrud et al. (1998) investigated the relative sensitivity of sonic, resistivity, density, and neutron porosity logs to overpressure in the Haltenbanken area of offshore Norway. They were able to track wire-line data from the same shale across 28 wells in which the pore pressure of the shale ranged from normal to relatively high overpressure (up to 1.8 g/cm^3 [15.0 lb/gal] equivalent mud weight).

They found no significant differences between the responses for normal pressure and overpressure from density and neutron porosity logs. Sonic velocities and

Bowers, G. L., and T. John Katsube, 2002, The Role of Shale Pore Structure on the Sensitivity of Wire-Line Logs to Overpressure, in A. R. Huffman and G. L. Bowers, eds., Pressure regimes in sedimentary basins and their prediction: AAPG Memoir 76, p. 43–60.

resistivities in the higher pressure wells (>1.4 g/cm^3 [11.7 lb/gal] equivalent mud weight), however, were substantially below the trends for lower pressures (Figure 2). Because the data were collected from similar depths over an area in which the shale shows little lateral variation, Hermanrud et al. (1998) ruled out lithologic/diagenetic differences as important factors in these log responses.

Hermanrud et al. (1998) recognized that sonic and resistivity logs both respond to changes in transport properties, whereas neutron and density logs reflect bulk properties. This led them to suggest that the sonic and resistivity logs were sensing textural changes induced by overpressure. Sonic velocity decreases were thought to result from a reduction in intergranular contact stresses, whereas resistivity decreases were attributed to enhanced fluid connectivity, possibly due to microfracturing.

Hermanrud et al. (1998) also explored possible causes of the overpressure at Haltenbanken but were unable to draw any definite conclusions. Actually, their explanation for the pressure sensitivity of the sonic and resistivity logs limits the possible overpressure causes to internal, sourcelike mechanisms. Stress-related mech-

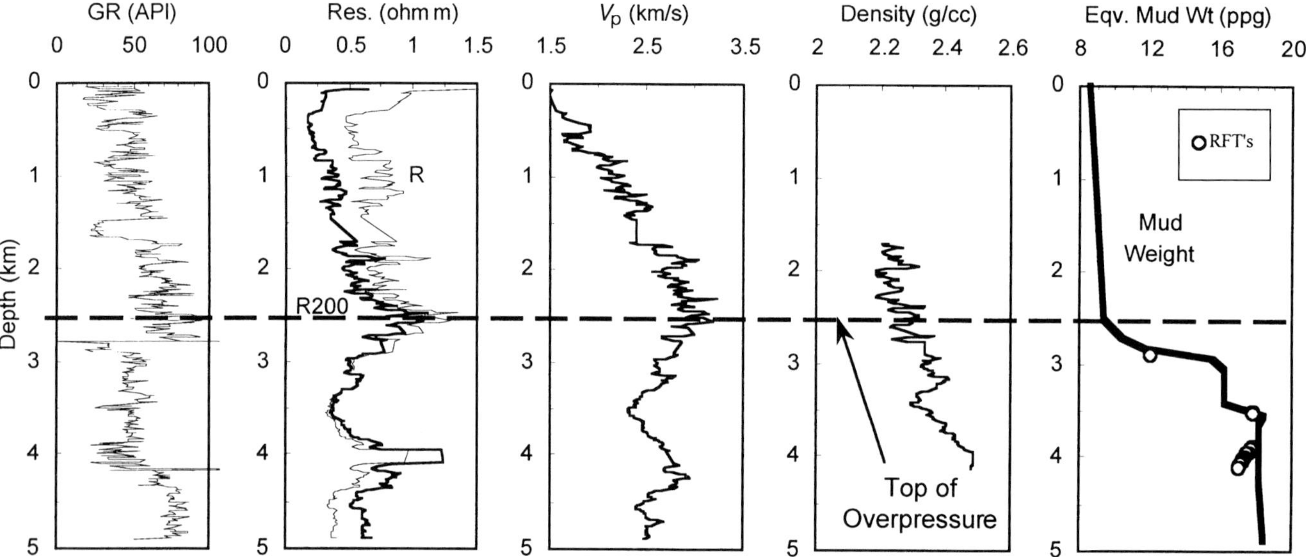

Figure 1. Wire-line data from an overpressured well in which sonic velocity and resistivity show a greater response to the onset of overpressure than bulk density data. The curve labeled "R" is the raw resistivity data; the curve labeled "R200" is the resistivity data normalized to a common temperature of 200°F (93°C).

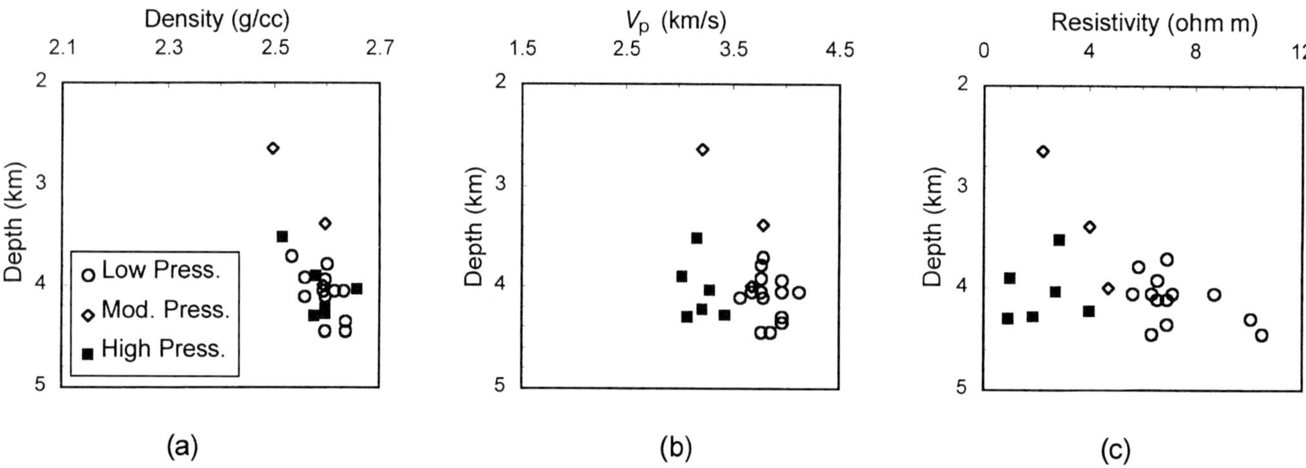

Figure 2. Wire-line density, sonic velocity, and resistivity for the Not formation in Haltenbanken, offshore middle Norway (Hermanrud et al., 1998). "Low Press.," "Moderate Press.," and "High Press." imply overpressures of less than 7.5 MPa, 7.5–20 MPa, and more than 20 MPa, respectively.

anisms (undercompaction, tectonic compression, tectonic extension) are not viable. Undercompaction cannot cause stress reductions, and neither can tectonic compression, unless it is accompanied by uplift and erosion, which appears not to be the case in the Haltenbanken area (Hermanrud et al., 1998). Tectonic extension can cause stress reductions and fracturing, but not overpressure.

Most internal pressure sources can be grouped into the category called fluid expansion mechanisms (Bowers, 1995), which includes lateral transfer, aquathermal pressuring, hydrocarbon generation, and the expulsion/expansion of bound water during clay diagenesis. Load transfer from smectite grains to pore water during clay diagenesis (Lahann, 1998) can also be viewed as an internal pressure mechanism because it can generate overpressure without a change in the in-situ total stress state. Precisely which mechanisms are involved at Haltenbanken remains an open question.

Figure 4 of Hermanrud et al. (1998) indicates that the pore pressures in their study area are all well below the overburden stress curve. Consequently, the log responses are unlikely to be due to pressure-induced microfracturing, because as discussed by Miller (1995), at incipient fracturing conditions, the pore pressure should be close to, and in principle, possibly even greater than the overburden stress. Pore pressures near the overburden stress would also be required to keep previously formed microfractures from closing. The sensitivity of transport properties to internal pressure sources, however, can be explained without invoking microfracturing. Such effects could also result from the widening of preexisting void spaces.

Toksoz et al. (1976) and Cheng and Toksoz (1979) analyzed the pressure sensitivity of porous media by modeling pores as oblate spheroids. Their relations suggest that pores with aspect ratios (minimum dimension over maximum dimension) in the range of 0.001 to 0.10 should be most responsive to internal pressuring. Pores with aspect ratios (α) greater than 0.10 (possibly representative of vugs and intergranular pores) are too rigid. And cracklike pores with very low aspect ratios are too flexible; they close at low stress levels and require pressures near the fracturing level to reopen. On the basis of these, we divide the pores into three types: elastically rigid ($\alpha > 0.1$), elastically flexible ($\alpha = 0.001$–0.1), and collapsing, cracklike pores ($\alpha < 0.001$).

Evidence (Katsube et al., 1992) exists that rock pore structures are a combination of relatively large, high aspect ratio storage pores linked together by a network of smaller, lower aspect ratio connecting pores, with transport properties controlled by the connecting pores. Recent studies (Katsube et al., 1999b) further indicate that connecting pores have a much higher sensitivity to laboratory stress changes than do storage pores, particularly in well-consolidated rocks. The oblate spheroid pore model (Toksoz et al., 1976; Cheng and Toksoz, 1979) suggests that storage pores tend to have higher aspect ratios than connecting pores.

Therefore, we propose that differences between the sensitivity of transport properties and bulk properties to overpressure can be explained by this storage pore/connecting pore model. With bulk properties (density, porosity), the contributions of storage pores and connecting pores are weighted equally. With transport properties, the two pore types are in series, which implies connecting pores are a more dominant factor. Storage pores are more resistant to volumetric changes than connecting pores. By the same token, volumetric losses in storage pores tend to be permanent, whereas the more flexible connecting pores are capable of elastic rebound.

These results suggest that bulk and transport properties should be equally responsive to overpressure caused by undercompaction because pressuring simply slows down or arrests the compaction process. Mismatches between the response of bulk and transport properties to pressure changes occur where overpressure is internally generated. The excess pressure widens the flexible connecting pores without significantly altering the more rigid storage pores.

In this chapter, first we present petrophysical evidence that supports the storage/connecting pore model. Second, we present relevant pore-structure theory (equations) and experimental and analytical procedures, including some new analytical techniques, used to obtain the connecting/storage porosity vs. pressure data that was used in this study. Finally, we discuss the shale storage/connecting pore characteristics and their implication on overpressure effect to log responses.

BASIC PORE-STRUCTURE CONCEPTS AND MODELS

Pore-Structure Models

Effective porosity of a rock represents the pore-space of all interconnected pores, excluding isolated pores. Total porosity represents all pore space in a rock, including isolated pores. Effective porosity is comprised of a three-dimensional network of void spaces that can vary in size and shape from sheetlike cracks to spherical chambers. Bulk density and effective porosity only depend on net pore volume. Transport properties such as permeability, resistivity, and sonic velocity, however, are sensitive to pore sizes, shapes, and how the pores are interconnected. These properties are grouped into a category called "pore structure."

Pore-size distributions are investigated through mercury porosimetry (Washburn, 1921). These tests also provide information on how the pore structure is divided between pores that control flow and pores that do not (Wardlaw and Taylor, 1976). Katsube and Williamson (1994) call the first group connecting pores and the second group storage pores. In the literature, connecting and storage pores are commonly referred to as "throats" and "pores," respectively.

Storage pores consist of relatively large, interior void spaces that can only be accessed through smaller connecting pores, or blind pores that branch off from the interconnected pore network (Katsube and Mareschal, 1993). The flow of fluid and electrical current through a rock is assumed to be controlled by the network of connecting pores (see Figure 3a). The sheetlike connecting pores in Figure 3 are simplified representations of connecting pores with aspect ratios of less than 0.001 to about 0.1. The volume fraction of storage pores (ϕ_S) plus connecting pores (ϕ_c) must equal the effective porosity (ϕ_E):

$$\phi_E = \phi_S + \phi_C \tag{1}$$

Permeability and electrical resistivity measurements provide information on the connecting porosity, connecting pore size, and tortuosity. Katsube and Williamson (1994) interpret these data using a pore-structure model in which connecting pores are idealized as a series of parallel, possibly saw-toothed–shaped cracks of uniform width (see Figure 3b). Storage pores are assumed to have no impact on flow properties; they are treated as nodes along the connecting pore network.

Mercury Injection/Withdrawal Porosimetry Tests

Pore-size distributions are determined from mercury porosimetry injection tests (Washburn, 1921). Mercury is forced into a rock sample in a series of pressure increments, and the volume injected during each step is recorded (Figure 4a). The latest pore size invaded during each injection step is estimated from the equation (Washburn, 1921):

$$d = (4\ \gamma\ \cos\theta)\ /\ (P_{Hg} - P_{AIRO}) \tag{2}$$

where d is the equivalent diameter of the pore, P_{Hg} is the mercury pressure of the injection step, γ is the sur-

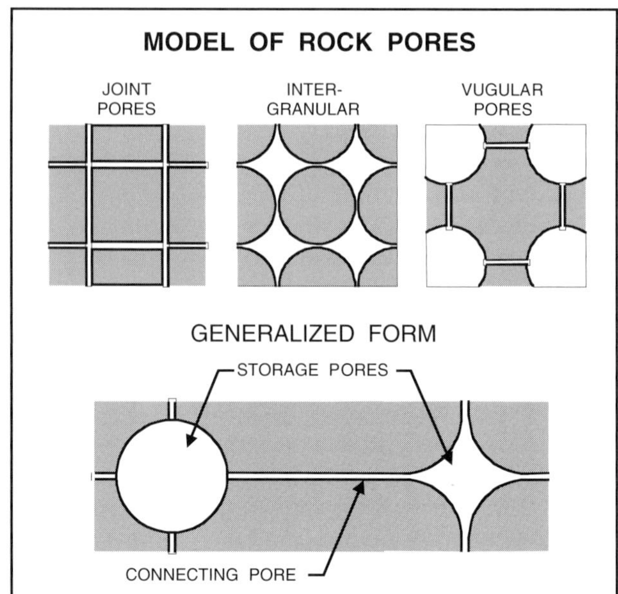

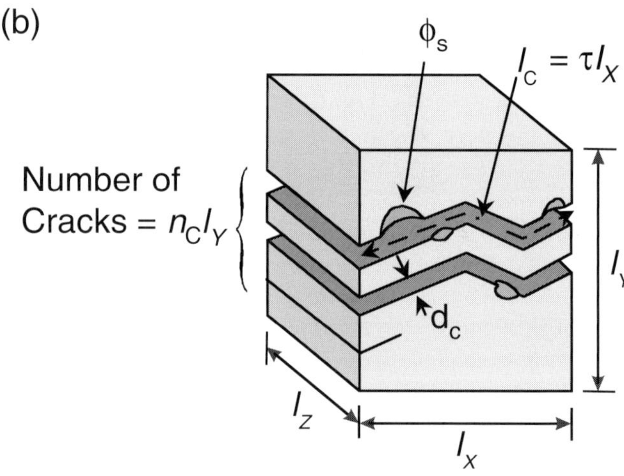

$$\phi_c = \frac{(n_C l_Y) d_C l_C l_Z}{(l_X l_Y l_Z)} = n_C d_C \tau \qquad \phi_E = \phi_S + \phi_C$$

$$F = \frac{\tau^2}{\phi_c}$$

$$S_c = \frac{2(n_C l_Y) l_C l_Z}{(l_X l_Y l_Z)} = 2 n_C \tau \qquad k = \frac{\phi_c^{\,3}}{3\tau^2 S_c^{\,2}}$$

Figure 3. Some pore-structure models used for characterizing shales (Katsube and Williamson, 1994, 1998): (a) storage-connecting pore model, and (b) tortuous connecting and storage pore model (modified from Katsube and Kamineni, 1983).

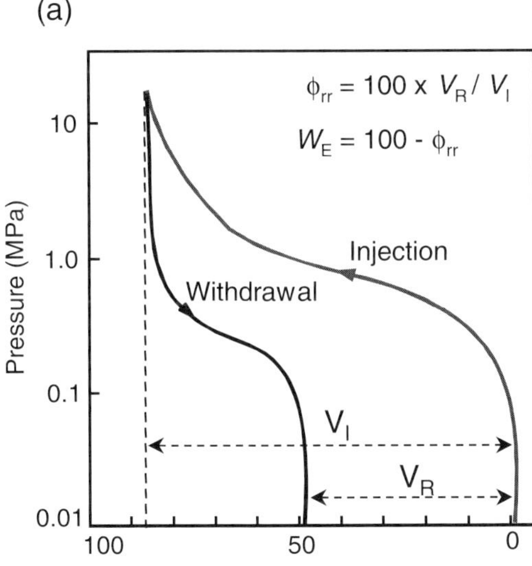

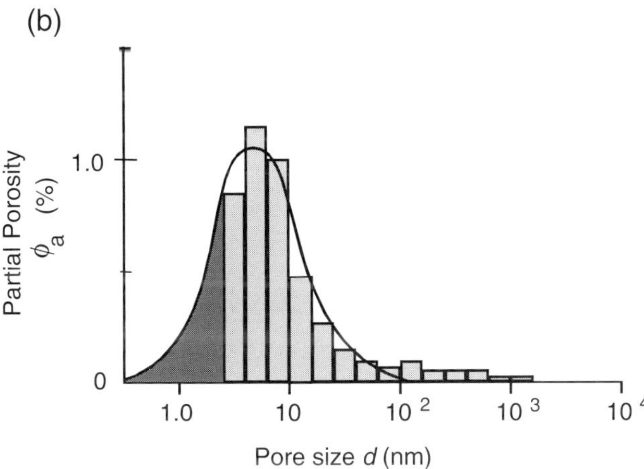

Figure 4. Diagram modified from (a) Wardlaw and Taylor (1976) describing the mercury intrusion and extrusion curves and explaining the definitions and methods for determining withdrawal efficiency (W_E) and storage porosity ratio (ϕ_{rr}), and (b) typical pore-size distribution for tight shales (Katsube and Williamson, 1994, 1998). V_I and V_R are the total mercury intrusion volume and total residual volume, respectively.

face tension of the air/mercury meniscus, θ is the angle the meniscus makes with the pore wall, and P_{AIR0} is the air pressure at the start of the test. The total volume of pores with a diameter d is equated to the volume of mercury injected during that step. The net result is a pore-size distribution plot (Figure 4b).

After the maximum injection pressure is reached, additional pore-structure information can be obtained by monitoring the volume of mercury expelled as the confining pressure is reduced (Figure 4a), termed withdrawal (Wardlaw and Taylor, 1976). We interpret the volume of mercury that is not recovered to represent the volume of storage pores and the volume of mercury that is recovered to represent the volume of connecting pores. The fractional part of the effective porosity comprised of storage pores is referred to as the residual porosity ratio (ϕ_{rr}), where

$$\phi_{rr} = \phi_S / \phi_E \tag{3}$$

The ϕ_{rr} values for shales are generally in the range of 0.4–0.8 (e.g., Katsube et al., 1999b), a range of values that provides proof of the existence of the two types of pores: storage and connecting pores. Further details of the principles of this method are described in Appendix 1.

Permeability/Formation Factor

The interpretation of permeability/formation factor data begins with the one-dimensional flow model shown in Figure 3b. As discussed previously, storage pores are assumed to have little impact on either electrical or fluid flow properties. Connecting pores are represented by a set of parallel cracks of uniform width d_C (Katsube and Williamson, 1994). The number of cracks per unit distance perpendicular to the direction of flow is termed the flow-path density η_C. Tortuosity (τ) is the ratio of the length of the actual flow path followed through a porous medium divided by the length of the assumed straight-line flow path. With the flow model in Figure 3b, tortuosity can be considered as giving the cracks a saw-toothed shape.

The actual network of connecting pores in a natural rock can be a very complex, three-dimensional structure. To make the analysis tractable, however, we model the connecting pore network as three mutually perpendicular sets of cracks (Figure 5), with identical tortuosities, crack widths, and flow-path densities (Katsube et al., 1991). Therefore, the total connecting porosity ϕ_C, and S_C, the connecting pore-surface area per unit bulk volume, are (Figure 3b):

$$\phi_C = n_C d_C \tau \tag{4}$$

$$S_C = 2 n_C \tau \tag{5}$$

where

$$n_C = 3\eta_C \tag{6}$$

is the total number of cracks in all three dimensions per unit length.

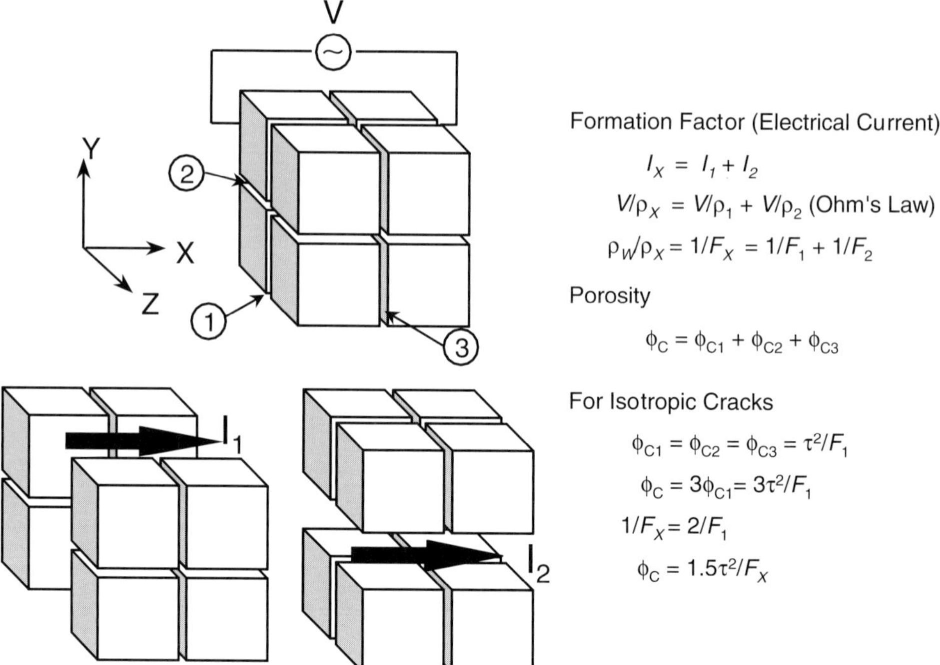

Figure 5. Three-dimensional extension of the tortuous connecting and storage pore model.

This model leads to the following relationships for ϕ_C, ϕ_E, ϕ_S, and d_C as a function of τ, permeability (k), and formation factor (F):

$$\phi_C = b_1\tau^2/F \quad (7)$$

$$\phi_E = \phi_S + b_1\tau^2/F \quad (8)$$

$$d_C = \sqrt{(12Fk)} \quad (9)$$

Appendix 2 shows the derivation of equations 7–9. The parameter b_1 equals

$$b_1 = 3/N \quad (10)$$

where N depends on the amount of coupling between mutually perpendicular crack sets. The planar crack model shown in Figure 5 corresponds to $N = 2$, $b_1 = 1.5$ (Katsube et al., 1991). A pore network with more out-of-plane branching than the model in Figure 5 would have N closer to 3 and b_1 near 1. Values for b_1 are chosen based on the assumed configuration of the connecting pore network.

Of the seven parameters in equations 7–9, ϕ_E, F, and k are measured, and the rest are derived or interpreted. As illustrated in Figure 6, the product $b_1\tau^2$ is obtained by linear regression of ϕ_E vs. $1/F$ data obtained at multiple effective confining pressures (P_e) (Katsube, 2000) using equation 8. The ϕ_C and ϕ_S values are then derived by inserting the F and F and ϕ_E values for each pressure step into equations 7 and 8, respectively, using the $b_1\tau^2$ value. The connecting pore width (d_C) is derived by inserting the F and k values, for each pressure step, into equation 9. Tortuosity (τ) can be computed from $b_1\tau^2$ using an assumed value for b_1.

Linear $\phi_E - 1/F$ trends indicate that ϕ_S and τ are both, on average, constant (see equation 8). Nonlinear $\phi_E - 1/F$ trends imply that ϕ_S and/or τ change with effective pressure. To interpret nonlinear trends, such as the data in Figure 7a, the following relations for ϕ_S and τ are substituted into equation 8:

$$\phi_S = \phi_{S0} \exp(-\varepsilon P_e) \quad (11)$$

$$\tau = \tau_0 \exp(\omega P_e) \quad (12)$$

Equations 11 and 12 follow from equation 8, and the observations of Rubey and Hubbert (1959) and Katsube (2000), respectively, that ϕ_E and F can be approximated as exponential functions of effective stress. A nonlinear curve fitting tool in Excel ("Solver") is used to find b_1, ϕ_{S0}, ε, τ_0, and ω values that best fit the $\phi_E - 1/F$ data. Tortuosity is then analytically calculated from equation 12, whereas ϕ_S and ϕ_C are computed at each pressure step from equations 7, 8, and 12. Figure 7 illustrates the process. It should be noted that this approach is new and still under evaluation. For the laboratory data presented subsequently, only the sample of sea floor mud (VSF-1) required a nonlinear analysis.

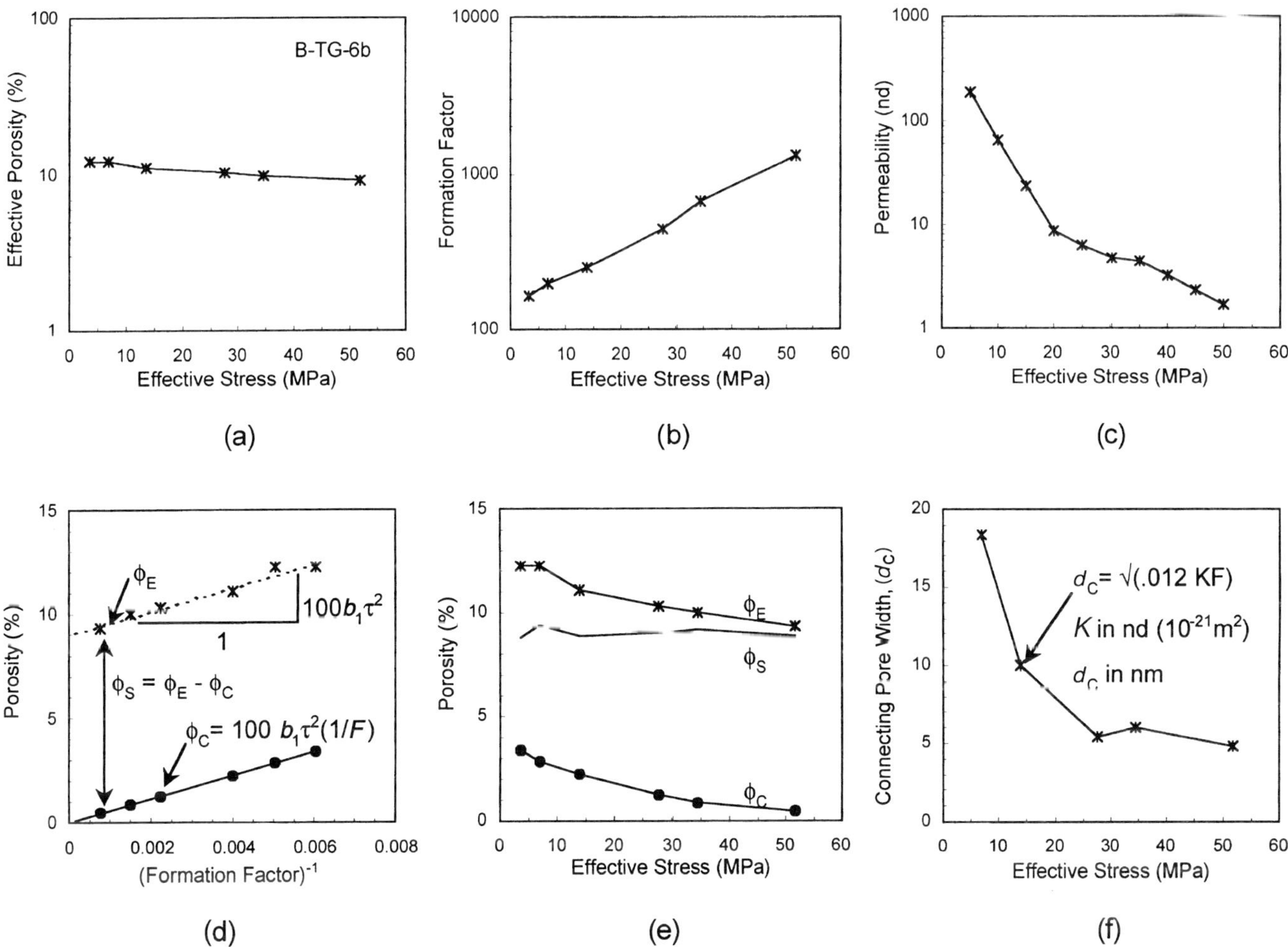

Figure 6. Pore-structure analyses for multiple confining pressure tests: (a) effective porosity ϕ_E, (b) formation factor F, and (c) permeability k are measured at multiple effective stresses P_e; (d) tortuosity τ is determined from the slope of a crossplot of ϕ_E vs. $1/F$; b_1 is a constant, typically assumed equal to 1.5; (d, e) connecting porosity $\phi_C = 100\ b_1\tau^2/F$; storage porosity $\phi_S = \phi_E - \phi_C$; (f) connecting pore width d_c is calculated from F and k. The original data source is Katsube et al. (1996a).

EXPERIMENTAL APPROACH

Petrophysical Data and Samples

Physical data used in this study have been previously published (e.g., Katsube, 2000). The samples used to produce these data were commonly obtained from 1-in. plugs taken from 4-in. (101.6 mm) split-core samples from various wells in North America. A typical sampling procedure that was used is documented in the literature (Katsube et al., 1991). Several disc specimens, 0.5–1.5 cm in thickness, were cut from each of these plugs for the permeability and formation factor measurements. Cuttings, disks, or partial disk specimens from the same samples were used for the other measurements (e.g., porosity and shale texture including scanning electron microscopy and x-ray diffraction).

Pore-Size Distributions

Pore-size distributions were calculated from mercury porosimetry injection tests using equation 2. Values typically used for the surface tension γ and contact angle θ are 0.48 N/m and 140°, respectively (Katsube et al., 1991). Injection pressures covered a range of 0.14–420 MPa, with equilibration times of 45 s for each high-pressure step and 10 s for low-pressure steps.

The maximum injection pressure of 420 MPa that was used should theoretically be capable of invading a minimum pore size of 3–4 nm, so the populations of smaller nanopores may be missed. Storage pores shielded from mercury invasion by smaller connecting pores have their volumes inadvertently included in the population of the pore size that controls invasion.

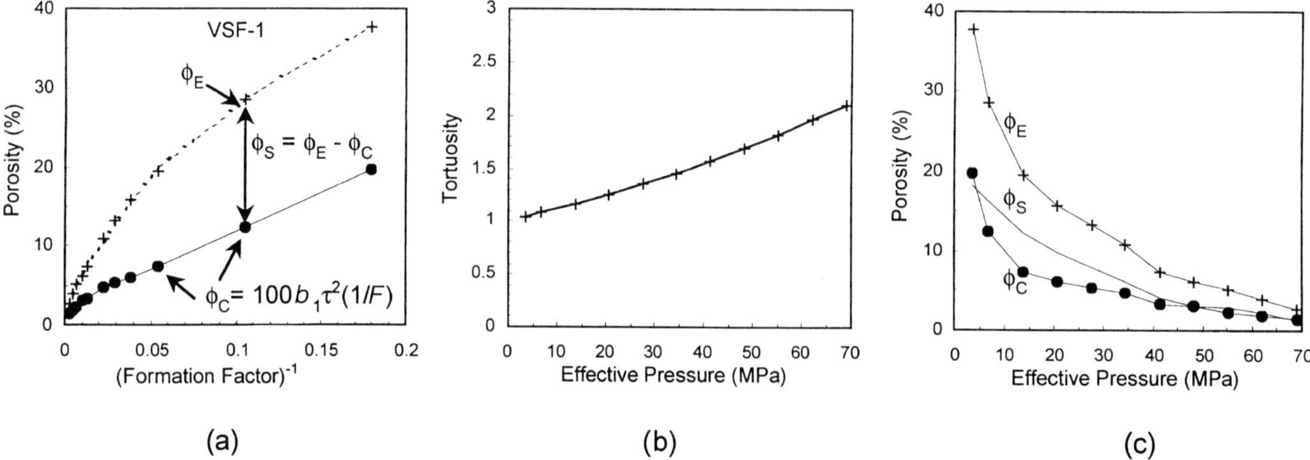

Figure 7. Storage porosity/connecting porosity analysis for nonlinear $\phi_E - 1/F$ trends; ϕ_S and τ are assumed to be exponential functions of effective stress, associated parameters are found by nonlinear curve fit of $\phi_E - 1/F$ algorithm. Tortuosity is then analytically calculated and used to find ϕ_C and ϕ_S from $1/F$ and ϕ_E data. Original data source is Katsube (2000).

Effective Porosity/Residual Porosity Ratio

Immersion, helium porosimetry, and mercury-injection porosimetry methods have been used to determine effective porosity (ϕ_E) (e.g., Katsube et al., 1992) using cuttings (5–10 g) from core samples. As previously discussed, the residual porosity ratio (ϕ_{rr}) is determined from mercury porosimetry withdrawal tests by taking the ratio of the total mercury residual volume (V_R), after extrusion, over the total intrusion volume (V_I) (Figure 4) (Wardlaw and Taylor, 1976).

Effective porosities were also measured at multiple confining pressures under drained (constant pore pressure) conditions. Thin disc specimens of 0.5–1.5 cm in thickness were used to minimize pore-pressure equilibration times. Initial porosities were found from mercury immersion bulk volume and helium porosimeter grain volume measurements. The pore volume change caused by each confining pressure increment was equated to the volume of brine squeezed out of the sample, with each load step held until the flow of brine stopped. Details of the testing and sampling procedures can be found in the original references (e.g., Loman et al., 1993).

Formation Factor

The multiple confining pressure tests typically included 1000 Hz bulk electrical resistivity measurements. Formation factors computed directly from these data include pore-surface conductivity effects, whereas equations 7–9 require true formation factor (F) data free of pore-surface conductivity effects. Such effects are routinely eliminated using a well-documented procedure (e.g., Patnode and Wyllie, 1950; Katsube et al., 1991), but this technique is difficult to apply to tests run at multiple confining pressures.

Therefore, pore-surface conductivity effects were removed through a new analytical technique developed by Katsube (1999). The technique is based on the assumption that F can be represented by an exponential function of effective pressure (P_e):

$$F = F_0 \exp(\beta P_e) \qquad (13)$$

where F_0 is the true formation factor at atmospheric pressure, and β is a coefficient. This expression is justified by the fact that permeability (k) is generally an exponential function of P_e (Katsube and Coyner, 1994) and that both k and F are determined by the connecting pore configuration. Further details of this procedure are described in Appendix 3.

Permeability

Permeability at atmospheric pressure (k_0) is commonly extrapolated from measurements made at higher pressures (Katsube and Coyner, 1994) by the transient pulse technique (Brace et al., 1968), using the following equation (Katsube et al., 1991):

$$k_0 = k \exp(\nu P_e) \qquad (14)$$

where k_0 and k are the fluid permeabilities at atmospheric pressure and at given effective pressures P_e, respectively, and ν is a coefficient.

PORE-STRUCTURE DATA

Pore-Structure Evolution during Burial

Figure 8 illustrates pore-structure evolution during burial under normal pressure conditions using mercury porosimetry data from three wells in the Beaufort-Mackenzie basin. On the left are plotted effective porosities (ϕ_E) and mean pore sizes (d_{Hg}) vs. depth. To the right are pore-size (d) distributions for samples obtained from the depths marked by shaded horizontal lines.

Pore-size distributions are unimodal at all depths but have larger porosities and pore sizes at shallower depths. Effective porosities decrease from 30% at 952 m to 7% at 2460 m. The mean pore size starts out in the vicinity of 230 nm at 952 m, abruptly drops to 32 nm by 1640 m, and then remains in the 20–40 nm range from there on. The mode (peak value) of each pore-size distribution changes more with depth than the mean, decreasing from roughly 800 nm at 952 m to approximately 8 nm at 2460 m. As previously mentioned, however, the mode could be biased by smaller pores that shield larger pores from mercury invasion.

Storage porosities (ϕ_s), and residual porosity ratios (ϕ_{rr}) were also determined for these samples. Residual porosity ratios show no apparent depth dependence. Values for most samples lie between 0.45 and 0.60; the

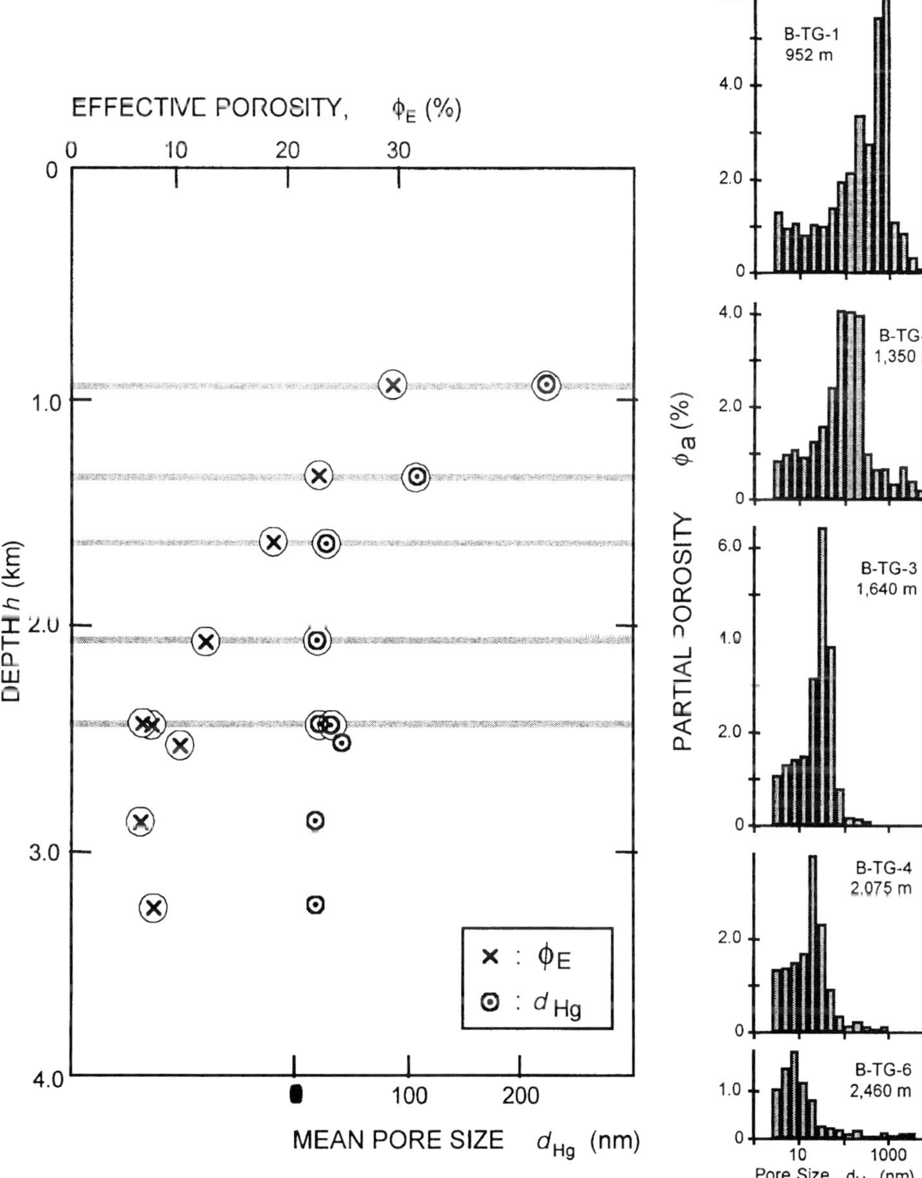

Figure 8. Mercury porosimetry data: effective porosity (ϕ_{Hg}), mean pore size (d_{Hg}), and pore-size distribution (d) as function of depth for shale samples from three wells in the Beaufort-Mackenzie basin (Northern Canada) (Katsube and Issler, 1993).

exception is the sample at 2460 m, for which $\phi_{rr} = 0.77$. Table 1 summarizes these results and also lists porosities obtained from helium porosimetry measurements. Helium porosities tend to be higher than mercury injection values because helium can penetrate smaller pore sizes than mercury.

In-Situ Porosity-Effective Stress Trends

Figure 9a shows possible fits of porosity-effective stress data derived from the helium porosity measurements in Table 1. Effective stresses were estimated by subtracting a 0.0104 MPa/m (0.46 psi/ft) normal pressure profile from an overburden stress curve calculated from the helium porosities in Table 1 and from shallow core density data from the GOM shelf. Porosities derived from the GOM shallow core data are also plotted in Figure 9a.

For low to intermediate effective stresses, we used the following form of a porosity-effective stress relation proposed by Butterfield (1979) for soils and by Baldwin and Butler (1985) for shales:

$$\phi_E = 1 - \mu\sigma^{\lambda} \tag{15}$$

where σ is the vertical effective stress, and μ and λ are coefficients. For intermediate to high stresses, we switched to a generalized version of the Rubey and Hubbert equation (1959):

$$\phi_E = \phi_{MIN} + \phi_0 \exp(-\zeta\sigma) \tag{16}$$

where ϕ_{MIN}, ϕ_0, and ζ are all coefficients.

Results of using two versions of equation 16 are shown (Figure 9); one with $\phi_{MIN} = 0$ and a second with $\phi_{MIN} = 5\%$. The second case represents the Katsube and Williamson (1994) hypothesis that at porosities somewhere between 5 and 10%, the dominant cause of porosity loss in shales changes from mechanical compaction to diagenetic/chemical processes. Lahann (1998) also advocates a nonzero minimum porosity, with ϕ_{MIN} lying somewhere between 3 and 10%, for rocks of smectite and illite composition.

Table 1. Mercury and Helium Porosimetry Data from the Beaufort Mackenzie Basin

Sample	Depth (m)	d_{mean} (nm)	d_{mode} (nm)	ϕ_E (%)	ϕ_C (%)	ϕ_S (%)	ϕ_{rr}	ϕ_{He}
B-TG-1	952	229	800	29.5	13.3	16.2	0.55	29.5
B-TG-2	1350	115	100	22.8	8.9	13.9	0.61	24.5
B-TG-3	1640	32	32	18.6	10.2	8.4	0.45	18.4
B-TG-4	2075	26	20	12.6	5.9	6.7	0.53	17.3
B-TG-6	2460	36	8	6.92	1.62	5.3	0.77	10.6

*d_{mean} = mean of the pore-size distribution; d_{mode} = mode (peak value) of the pore-size distribution; ϕ_{Ehg} = effective porosity determined from mercury injection; ϕ_C = connecting porosity = volume of mercury recovered/sample bulk volume; ϕ_S = storage porosity = $\phi_E - \phi_C$; ϕ_{rr} = residual porosity ratio = ϕ_S/ϕ_E; and ϕ_{Ehe} = effective porosity determined from helium injection. Data from Issler and Katsube, 1994.

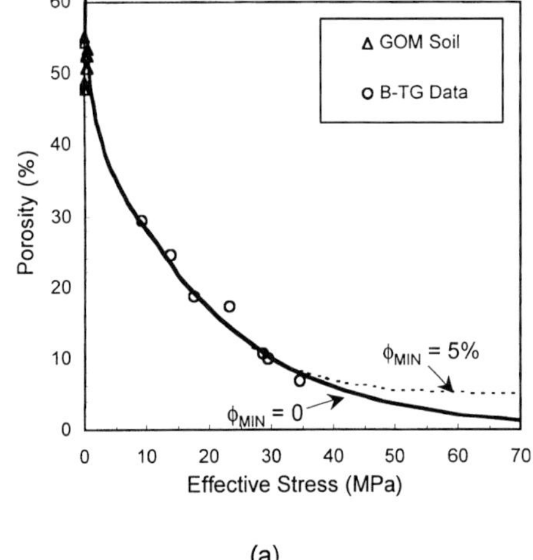

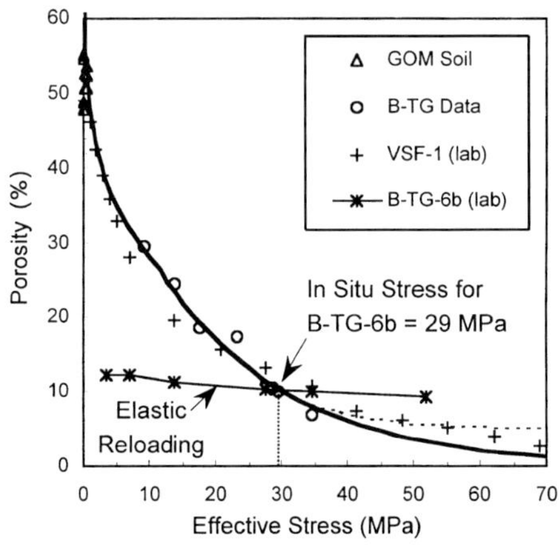

Figure 9. (a) Porosity-effective stress trends developed from helium porosimetry measurements for Beaufort-Mackenzie shales in Figure 7; possible low-stress trends represented by GOM soil data; (b) effective stress trends in (a) compared with laboratory compaction tests for a sea floor mud (VSF-1) and a Beaufort-Mackenzie shale (B-TG-6b) (Katsube et al., 1996a). Behavior of sea floor mud is consistent with in-situ profiles; the B-TG-6b sample shows no evidence of inelastic deformation even where loaded well beyond its estimated in-situ stress state.

Laboratory Compaction Behavior vs. In-Situ Trends

Figure 9b compares the estimated in-situ porosity-effective stress trends in Figure 9a with two laboratory compaction tests. Sample B-TG-6b is the Beaufort-Mackenzie sample from 2460 m whose pore-size distribution is plotted in Figure 8; VSF-1 is a sea floor mud sample from offshore Nova Scotia (Katsube et al., 1996b). The B-TG-6b data follow a much flatter compaction trend than either the in-situ profiles or the sea floor mud.

To a certain extent, this difference is expected. Compaction, especially in shales, is predominately an inelastic process. As a result, only a small amount of elastic rebound occurs when the effective stress acting on a formation is reduced. During reloading, the deformation remains elastic until the past maximum effective stress is exceeded. This means that inelastic deformation cannot occur in the laboratory until the effective confining pressure exceeds the maximum in-situ stress. Sample B-TG-6b has an estimated in-situ vertical effective stress of 29 MPa = 4200 psi, whereas VSF-1 is essentially being compacted for the first time. Therefore, it is not surprising that VSF-1 would undergo substantially more inelastic deformation at low pressures in the lab than B-TG-6b.

This does not explain, however, why the B-TG-6b sample appears to continue deforming elastically after its in-situ effective stress is exceeded. Such an effect could be caused by cementation, but no cement was found in this sample. Another possible explanation could be the several orders of magnitude difference between laboratory and geologic loading rates. Laboratory tests also do not account for effects such as smectite dehydration (Bird, 1984; Colten-Bradley, 1987), clay diagenesis (Lahann, 1998), and pressure solution (Houseknecht, 1987). Exactly why these discrepancies occur remains an open question.

Regardless of whether a sample has undergone elastic rebound during its burial history, it certainly will have experienced elastic rebound after coring. Therefore, at effective confining pressures below the maximum in-situ stress state, and possibly at higher levels, laboratory compaction data reflect only elastic behavior. This, however, is precisely the deformation regime we are interested in, because our focus is pore-structure response during elastic rebound.

We make the assumption that the effective stress path followed by a pore-structure parameter during elastic reloading in the laboratory is indicative of the path that would be tracked during elastic rebound induced by overpressure. We realize that laboratory samples may contain artificial fractures caused by coring, drying, and handling. These effects, however, can commonly be identified, and they provide additional information on the response of bulk and transport properties to the opening and closing of cracklike void spaces.

Laboratory Compaction Data

Numerous data have been published (Coyner et al., 1993; Loman et al., 1993; Katsube and Coyner, 1994; Katsube et al., 1996a, b, 1999b) on the pressure characteristics of effective porosity (ϕ_E), formation factor (F), and permeability (k) for shales. Compaction test results for samples B-TG-6b, VSF-1, and four additional samples are presented here. Samples VSF-1, V-7, V-8, and V-9 are from offshore eastern Canada, whereas samples B-TG-6b and EJA-2 are from northern Canada.

Table 2 lists in-situ depths and the available x-ray diffraction mineralogy data for these samples. The

Table 2. Mineral content (XRD wt. %) of Laboratory Compaction Test Samples*

Sample	Depth (m)	ϕ_E** (%)	Total Clay	Qz	Fld	Ca	Dl	Sd	Py	Kl	Ch	Il/Mc	Il/Sm
VSF-1†	7.6	37.8											
EJA-2	896.4	33.3	25	70	5	0	0	0	0	5	3	15	2
B-TG-6b††	2460	12.2	50	38	1	0.3	0.9	6.3	0.2	7	3	-	40
V-8	5270	7.7	20	73	5	Tr	0	0	2	0	10	7	3
V-7	5270	7.2	24	70	4	Tr	0	tr	2	0	7	12	5
V-9	5550					10	2	0	4		4	10	3

*Data taken from Katsube et al. (1996b, 1999b), Block and Issler (1996), and Katsube and Williamson (1994).

Qz = quartz; Dl = dolomite; Kl = kaolinite; Mc = mica; Fld = feldspar; Sd = siderite; Ch = chlorite; Sm = smectite; Ca = calcite; Py = pyrite; Il = illite.

**Effective porosity measured after the first load step, which corresponded to an effective confining pressure of 3.5 MPa = 508 psi for samples VSF-1, EJA-2, B-TG-6b, V-8, V-7; 4.4 MPa = 638 psi for sample V-9.

†No XRD data available.

††No XRD data available; values shown are for a sample from 2530 m in the same well.

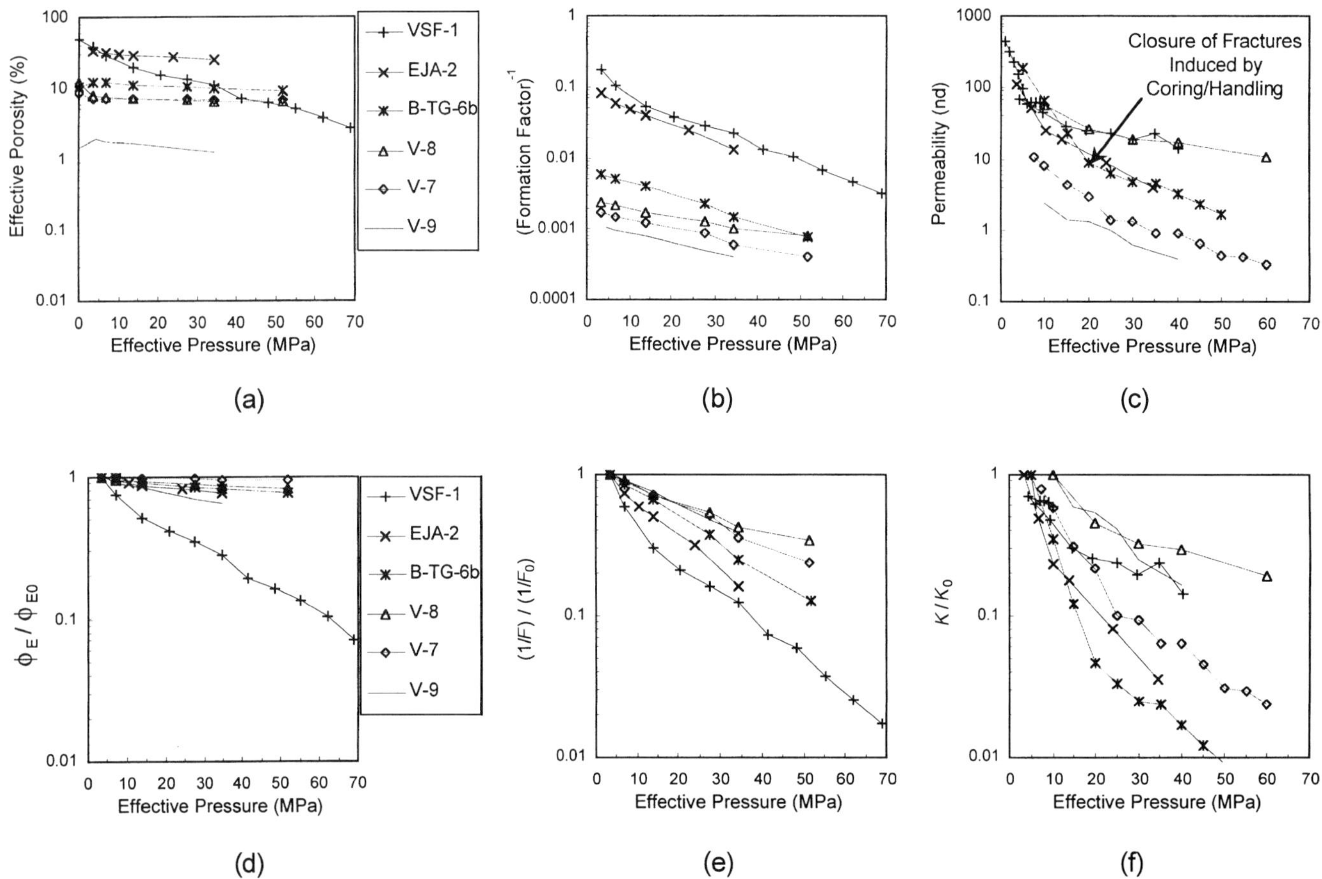

Figure 10. Laboratory compaction data: (a) effective porosity (ϕ_E), (b) inverse formation factor (1/*F*), and (c) permeability (*k*) as a function of effective stress (P_e); (d, e, f) data normalized to the measurement recorded after the first the load step. Original data source is Katsube (2000).

mineralogy data shown for sample B-TG-6b are actually from a sample obtained 70 m deeper in the same well. By GOM shale standards, most of these samples have very low clay contents. Sample B-TG-6b has the largest clay content (50%); the other samples contain less than 25% clay.

All tests were run at drained conditions, with each confining pressure increment held constant until the pore pressure and sample had stabilized. During the ϕ_E and F tests the pore pressure was vented to the atmosphere, whereas base pore pressures of 5–10 MPa (725–1450 psi) were used for the permeability tests. True formation factors (F) were determined from published apparent formation factor (F_a) data (Loman et al., 1993; Katsube et al., 1996a) using the technique described in Appendix 3 (Katsube, 1999).

The sensitivity of effective porosity (ϕ_E), formation factor (plotted as $1/F$), and permeability (k) to effective pressure (P_e) changes are displayed in Figure 10. Figure 10a, b, c presents the data in their original form, while Figure 10d, e, f replots the data normalized to the measurement made after the first load step. Only the sea floor mud sample (VSF-1) displays a significant change in effective porosity (ϕ_E) with increased P_e; the other samples appear to undergo primarily elastic deformation.

The $1/F$ values for all samples show considerable change with increased effective stress (Figure 10b, e), and the permeability data are even more responsive (Figure 10c, f). All of the permeability curves undergo a reduction in slope somewhere between 6 and 25 MPa (870–7250 psi), which could reflect closure of fractures induced by coring and handling. The three samples with the steepest permeability curves in the low stress range (VSF-1, EJA-2, B-TG-6b) are the only samples that have not experienced diagenetic cementation.

The ϕ_E vs. $1/F$ trends for samples VSF-1, EJA-2, and B-TG-6b, and the storage porosity (ϕ_S) and connecting porosity (ϕ_C) estimates derived from these data are displayed in Figure 11a, b, c. Similar plots for samples B-TG-6b, V-8, V-7, and V-9 are presented in Figure 11d, e, f. The connecting porosities all decrease with in-

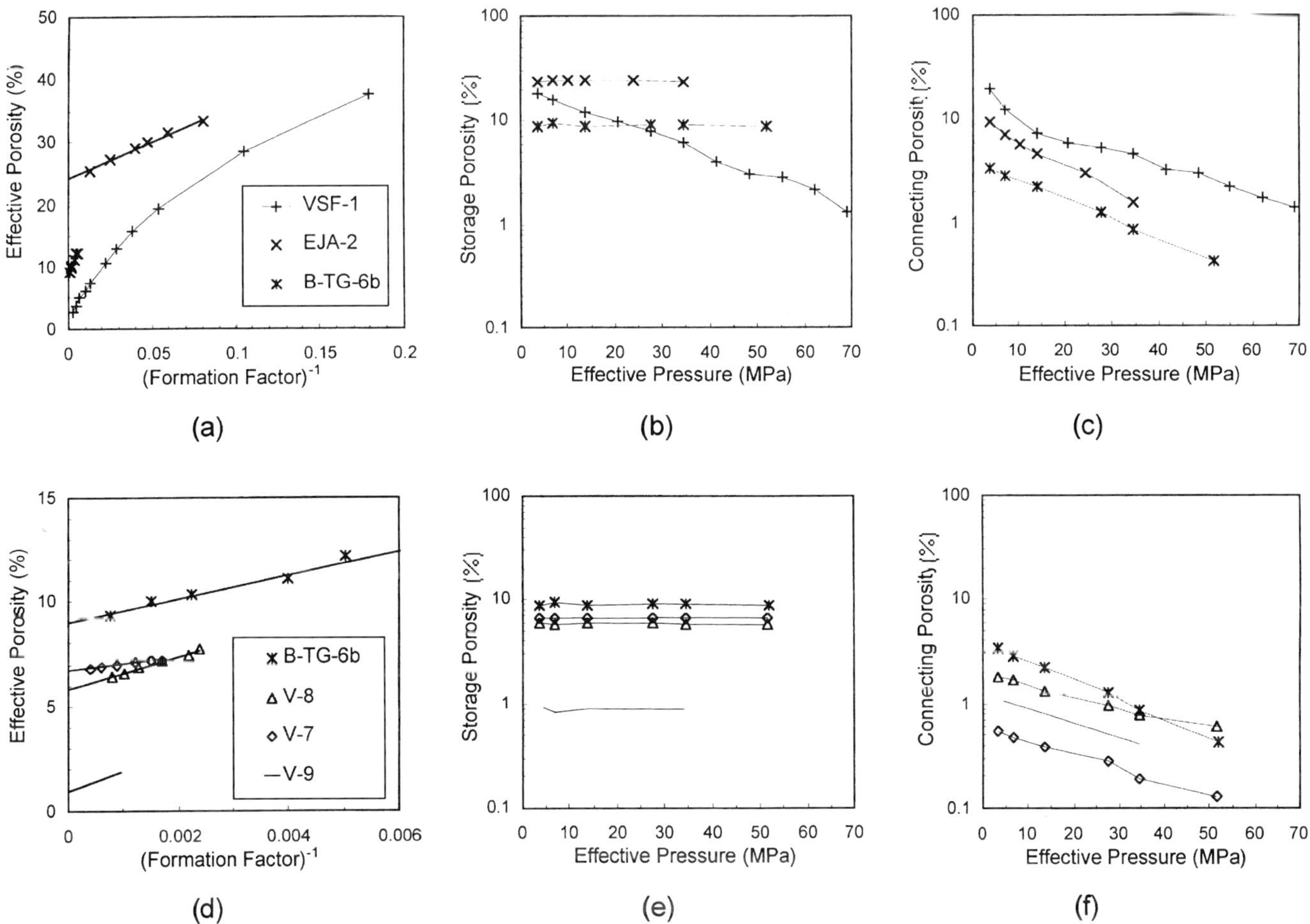

Figure 11. Storage porosity (ϕ_S)/connecting porosity (ϕ_C) analyses: (a, d) ϕ_E vs. b_1/F; (b, e) ϕ_S vs. P_e; and (c, f) ϕ_C vs. P_e. Linear $\phi_E - b_1/F$ trends imply constant tortuosity, and on average, constant storage porosity; the curved $\phi_E - b_1/F$ trend for sample VSF-1 suggests tortuosity and storage porosity may both change with effective stress.

creasing effective stress, whereas the storage porosities for every sample except VSF-1 (the sea floor mud) show little, if any, change.

Figure 12a, b, c shows the input data and results for the connecting pore width (d_C) analyses for samples VSF-1, EJA-2, and B-TG-6b; corresponding data for samples B-TG-6b, V-8, V-7, and V-9 are presented in Figure 12d, e, f. The d_C values range considerably, as would be expected, because these samples range from coarse-grained framework supported sandy shales (V-8) to finer grained framework-supported silty shales (V-7, EJA-2) or matrix-supported shales (V-9, B-TG-6b) (Katsube and Williamson, 1994; Katsube et al., 1996a, 1999a, b). The coarse-grained sandy shale (V-8) shows the largest d_C values. As with the permeability data (Figure 12e), the d_C curves for all samples in Figure 12f undergo a slope reduction after the first 6–25 MPa of loading, which, again, we believe reflect sample damage effects.

Surprising is that the samples from the shallowest burial depths, VSF-1 (sea floor mud) and EJA-2 (896.4 m), show the smallest d_C values (Figure 12c). These two samples have not experienced any diagenetic cementation, their clay contents show no indications of being higher than the rest (Table 2), and their silt content is high (>50–60 wt. % (Katsube et al., 1996b; Katsube et al., 1999b). The location of the clay particles within the texture of the rocks from shallower burial depth (VSF-1 and EJA-2) may be different (Katsube et al., 1999b) compared to those of the deeply buried rocks that have undergone diagenesis and have pore surfaces coated with authogenic clays (Katsube and Williamson, 1994). The textures of these samples are now being investigated using a scanning electron microscope.

In Table 3, the values of ϕ_E, ϕ_S, ϕ_C, and d_C in Figures 10–12 are compared with those determined by mercury porosimetry. The τ and b_1 values for the $\phi_E - 1/F$ data are also listed. All of the mercury porosimetry-based

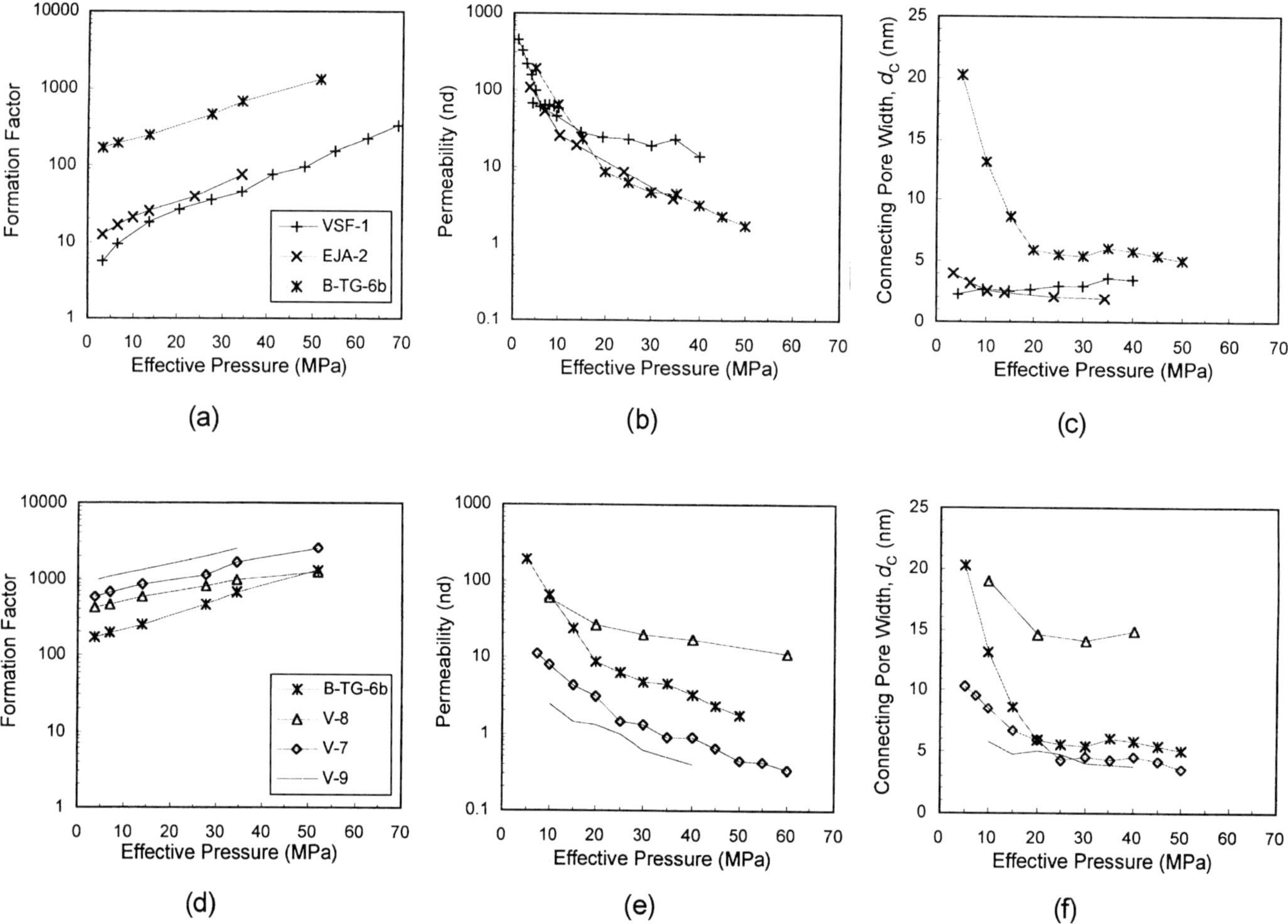

Figure 12. Connecting pore width (d_C) analyses: (a, d) F vs. P_e; (b, e) k vs. P_e; and (c, f) d_C vs. P_e.

data in Table 3, except d_{CHg}, are measured values; the d_{CHg} values represent the modes of calculated pore-size distributions. With the compaction tests, only ϕ_E is measured; the remaining data are calculated from F, k, and ϕ_E. Having two independent methods to determine the same pore-structure parameter provides credibility to the results.

With the exception of sample EJA-2, the two sets of ϕ_S and ϕ_C values show the same consistency as the ϕ_E measurements. The reason for the relatively large discrepancy between the two EJA-2 storage porosity estimates is still being explored. The two sets of flow-path size estimates agree to within the same order of magnitude, which is the level of accuracy expected. The laboratory compaction data indicate that shale connecting pore sizes are in the range of 2–20 nm, whereas the mercury porosimetry tests produce values in the range of 4–8 nm.

DISCUSSION

Pressure Sensitivity of Bulk and Transport Properties

The formation factor and permeability data in Figure 10 both show a much stronger response to elastic reloading than the effective porosity measurements. This suggests that elastic rebound induced by internal overpressuring would cause a greater change in transport properties (at least resistivity and permeability) than bulk properties. Such behavior is consistent with the wire-line log responses observed by Hermanrud et al. (1998) at Haltenbanken (see Figure 2).

Storage and Connecting Pore Properties

The $\phi_S - P_e$ and $\phi_C - P_e$ data for sample VSF-1 (Figure 11b, c) demonstrate that both storage and con-

Table 3. Pore-Structure Parameters Computed from Mercury Porosimetry (Subscript "Hg") vs. Range of Values Determined from Laboratory Compaction Tests (Subscript "Pe")*

	Effective Porosity (ϕ_E)		Storage Porosity (ϕ_S)		Connecting Porosity (ϕ_C)		Flow Path Size (d_C)		Tortuosity	
Sample	ϕ_{EHg} (%)	ϕ_{EPe} (%)	ϕ_{SHg} (%)	ϕ_{SPe} (%)	ϕ_{CHg} (%)	ϕ_{CPe} (%)	d_{CHg} (nm)	d_{CPe} (nm)	(τ)	b_1
VSF-1	–	37.8–2.7	–	18–1.3	–	20–1.4	–	2–4	1–2.1	1
EJA-2	29.5**	33.3–24.5	12	24	17.5	9–1.4	5; 250†	4–2	1	1.2
B-TG-6b	7.8	12.2–9.3	5.3	9	2.5	3–0.4	8	20–5	1.9	1.5
V-8	9.1	7.7–6.4	3.9	6	5.2	1.8–0.2	7	19–14	2.3	1.5
V-7	5.9	7.2–6.8	3.2	7	2.7	0.5–0.1	6	10–4	1.5	1.5
V-9	2.3	2.0–1.3	0.4	1	1.3	1.8–0.1	6	6–4	2.6	1.5

*Tortuosity and the b_1 parameter are only obtained from compaction tests. The basic data used for these determinations were taken from other publications (Katsube et al., 1996 a, b, 1999b, 2000).
**Effective porosity measured by helium porosimetry.
†Sample exhibited bimodal pore-size distribution.

necting pores can undergo significant volume decreases during first-time compaction. The data for the other samples in Figure 11, however, indicate that porosity changes during elastic reloading are almost entirely due to connecting porosity. Consequently, it appears that volume reductions in storage pores tend to be permanent, whereas connecting pores are more flexible and capable of elastic rebound. The oblate spheroid pore model (Toksoz et al., 1976; Cheng and Toksoz, 1979) would further suggest that connecting pores generally have lower aspect ratios than storage pores, possibly something like the shapes drawn in Figure 3a.

Wire-Line Log Responses to Overpressure

Because connecting pores appear to be mechanically flexible, they are likely to open up or widen when the effective stress decreases due to fluid expansion. Storage pores, however, are expected to remain essentially unchanged. Compaction disequilibrium cannot cause effective stress reductions, so this pressure mechanism is not capable of causing either the connecting or the storage pores to undergo widening.

This study has provided evidence that the contributions of storage pores and connecting pores to bulk properties (density, porosity) are weighted equally. With transport properties, however, the two pore types are in a series relationship, implying the effect of connecting pores is dominant. Bulk and transport properties should be equally responsive to overpressure caused by compaction disequilibrium, which can only slow down or arrest the compaction process. Mismatches between the response of bulk and transport properties to pressure changes occur where overpressure is internally generated, such as by fluid expansion. The excess pressure widens connecting pores without significantly altering storage pores.

CONCLUSIONS

Results of this study indicate that shale pore structure can be characterized by a storage-connecting pore system, with connecting pore sizes on the order of 2–20 nm. Laboratory compaction tests show connecting porosity (ϕ_C) decreasing during elastic reloading, whereas storage porosity (ϕ_S) remains essentially constant. This suggests that connecting pores are mechanically more flexible than storage pores and are therefore likely to have lower aspect ratios. Consequently, connecting pores would be expected to undergo more elastic rebound (widening) than storage pores where fluid expansion mechanisms cause effective stress reductions.

This study shows evidence that the contributions of storage pores and connecting pores to bulk properties (density, porosity) are weighted equally. With transport properties, however, the two pore types are in a series relationship, implying that the effect of connecting pores is dominant. Bulk and transport properties should be equally responsive to overpressure caused by compaction disequilibrium because this pressure mechanism can only slow down or arrest the compaction process. Mismatches between the response of bulk and transport properties to pressure changes occur where overpressure is internally generated, such as by fluid expansion. The excess pressure widens connecting pores without significantly altering storage pores.

ACKNOWLEDGMENTS

We express our sincere thanks to S. Connell (Geological Survey of Canada) for drafting and preparing many of the diagrams contained in this chapter.

APPENDIX 1: MERCURY POROSIMETRY TESTS

Because mercury is a nonwetting liquid, mercury invasion is resisted by menisci that form at the entrances to pores. Surface tension enables a meniscus to support a pressure differential, termed the capillary pressure (P_c), which from equilibrium considerations, is equal to

$$P_c = P_{Hg} - P_{AIR} = (\gamma \cos\theta) \times C/A \quad (17)$$

where P_{Hg} is the pressure of the mercury, P_{AIR} is the pressure of the air inside the pore, γ is the meniscus' surface tension, θ is the angle the meniscus makes with the pore wall, and C and A are the pore's perimeter and cross sectional area, respectively. Mercury injection data are commonly interpreted by modeling the pores as cylindrical capillary tubes. From equation 17, the pressure P_{Hgi} required to invade a specific cylindrical pore with diameter d_i is

$$P_{Hgi} = P_{AIR0} + (4 \gamma \cos\theta)/d_i \quad (18)$$

where P_{AIR0} is the air pressure at the start of the test. Through equation 18, the latest pore size invaded during each injection step can be estimated as follows:

$$d_i = (4 \gamma \cos\theta)/(P_{Hgi} - P_{AIR0}) \quad (19)$$

The total volume of pores with that diameter is assumed to equal the volume of mercury that is injected during that step. The net result is a pore-size distribution plot, like the one shown in Figure 4b. A limitation of this technique is that large pores shielded from invasion by smaller pores are included in the volume estimates for the smaller pores (Wardlaw and Taylor, 1976).

Mercury expulsion during withdrawal results from the expansion of trapped air; the air expands to keep equation 17 satisfied as the confining pressure is reduced. A pore becomes mercury free when its air pressure returns to P_{AIR0}, and equation 18 is once again satisfied. Consequently, mercury recovery successively occurs from smallest to largest pores.

For a system of parallel, constant diameter capillary tubes, all of the mercury that is injected would be recovered. Large interior pores that can only be accessed through smaller pores, however, may not be able to empty (Wardlaw and Taylor, 1976). If mercury continuity is broken, the smaller pores expel their mercury first and leave behind menisci where they join a larger pore. By the time this occurs, the pressure of the mercury is too low to cause reinvasion of the smaller pores, so the mercury in the large pore becomes trapped.

The volume of mercury that is not recovered, including all types of trapped mercury, is interpreted to represent the volume of storage pores. The fractional part of the effective porosity, referred to as residual porosity ratio (ϕ_{rr})

$$\phi_{rr} = \phi_S/\phi_E \quad (20)$$

is related to withdrawal efficiency (W_E) in the literature (Wardlaw and Taylor, 1976) by

$$W_E = 1 - \phi_{rr} \quad (21)$$

APPENDIX 2: PERMEABILITY/FORMATION FACTOR RELATIONS

Wyllie and Spangler (1952) adapted the general solution for fluid flow in a pipe to derive the following version of the Kozeny-Carman equation for permeability (k) as a function of porosity (ϕ), pore-surface area per unit bulk volume (S), and tortuosity (τ):

$$k = \phi^3/(S^2 \tau^2 c_0) \quad (22)$$

The parameter c_0 is a shape factor that generally falls somewhere in the range of 2.5–3.5 (Wyllie and Spangler, 1952); Walsh and Brace (1984) assumed $c_0 = 3.0$.

Walsh and Brace (1984) followed a similar approach to obtain a relation for formation factor (F) as a function of τ and ϕ

$$F = \tau^2/\phi \quad (23)$$

In this model, a porous medium has a higher resistivity than a volume of water of the same dimensions because the electrical flow path followed through pores (1) is longer, (2) has a smaller cross sectional area, and (3) is generally oriented at an angle from the assumed direction of flow.

Equations 22 and 23 are adapted to our pore-structure model by replacing ϕ with connecting porosity ϕ_C. Likewise, S is taken as the surface area of the connecting pores per unit bulk volume (S_C). Equations 4 and 5 give the relationships for ϕ_C and S_C as a function of connecting pore width (d_C), tortuosity (τ), and three-dimensional crack density (n_C).

Equations 22 and 23 are assumed to define the porosity-permeability and porosity–formation factor relationships for one set of cracks. To account for three-dimensional effects, a constant b_1 is introduced into the formation factor relationship:

$$F = b_1\tau^2/\phi_C \quad (24)$$

where b_1 is defined in equation 10. The b_1 term accounts for the fact that all three crack sets contribute to porosity, whereas less than three may contribute to flow along a principal flow direction (X, Y, and Z axes in Figure 5). By similar logic, b_1 is also required in the denominator of the permeability relation, so equation 22 becomes the following:

$$k = [\phi^3/(S_C{}^2\tau^2 c_0)]/b_1 \quad (25)$$

With $c_0 = 3.0$, equations 4, 5, and 25 yield the following relation for permeability:

$$k = \phi_C d_C^2/(12b_1\tau^2) \quad (26)$$

Finally, through equations 4, 5, 24, and 26, all of the key connecting pore parameters (d_C, ϕ_C, n_C, S_C) can be expressed solely in terms of F, k, and τ:

$$\phi_C = b_1\tau^2/F \quad (27)$$

$$d_C = \sqrt{(12Fk)} \quad (28)$$

$$S_C = 2\, b_1\tau^2/\sqrt{(12F^3k)} \quad (29)$$

$$n_C = b_1\tau\ /\sqrt{(12F^3k)} \quad (30)$$

APPENDIX 3: FORMATION FACTOR DETERMINATION

In principle, the formation factor, F, is determined from the bulk-rock electrical resistivity, ρ_r, and the electrical resistivity, ρ_W, of the pore fluid that saturates the rock using the Archie (1942) equation:

$$\rho_r = F_{a/\rho_W} \quad (31)$$

where F_a is the apparent formation factor. Because F_a is not always equal to F due to conductive layers existing on the pore surfaces, equation 31 is replaced with the Patnode and Wyllie (1950) equation:

$$1/\rho_r = 1/(F\rho_W) + 1/\rho_c \quad (32)$$

where ρ_c is the pore-surface resistivity. A value of F free from surface conduction effects can be obtained by (1) measuring ρ_r for several pore-fluid solutions having different values of ρ_W, (2) crossplotting $1/\rho_r$ as a function of $1/\rho_W$, and (3) performing a linear regression on the results. The slope equals $1/F$, whereas the intercept corresponds to $1/\rho_c$.

The true formation factor (F) has been successfully determined for various types of rocks (e.g., Katsube and Walsh, 1987), including shales (Katsube et al., 1991), using this method. With low-permeability shales, however, it can be extremely time consuming and costly to repeat this technique at multiple pressures. To overcome this problem, a new method proposed by Katsube (1999) assumes F to be an exponential function of the effective pressure P_e:

$$F = F_0\exp(\beta P_e) \quad (33)$$

where F_0 is the true formation factor at atmospheric pressure, and β is a coefficient. This is based on the observation that permeability (k) appears to be generally an exponential function of P_e (Katsube and Coyner, 1994) and that both k and F are determined by the connecting pore configuration. Equations 32 and 33 suggest that with the correct ρ_c value, a semi-log plot of ($1/\rho_r - 1/\rho_c$) vs. P_e should follow a straight line, that is,

$$1/\rho_r - 1/\rho_c = \exp(-\beta P_e)/(\rho_W F_0) \quad (34)$$

Therefore, an iterative procedure is used to find a pore-surface resistivity that does the best job of aligning ($1/\rho_r - 1/\rho_c$) vs. P_e data along a semi-log straight-line trend. This ρ_c value and the ρ_W of the pore fluid used during the tests are then substituted into equation 32 to find the value of F.

REFERENCES CITED

Archie, G. E., 1942, The electrical resistivity log as an aid in determining some reservoir characteristics: Transactions of the American Institute of Mining, Metallurgical and Petroleum Engineers, v. 146, p. 54–67.

Baldwin, B., and C. O. Butler, 1985, Compaction curves: AAPG Bulletin, v. 69, no. 4, p. 622–626.

Bird, P., 1984, Hydration-phase diagrams and friction of montmorillonite under laboratory and geologic conditions, with implications for shale compaction, slope stability, and strength of fault gouge: Tectonophysics, v. 107, p. 235–260.

Bowers, G. L., 1995, Pore pressure estimation from velocity data: accounting for overpressure mechanisms besides undercompaction: Society of Petroleum Engineers Drilling and Completion, June, p. 89–95.

Brace, W. F., J. B. Walsh, and W. T. Frangos, 1968, Permeability of granite under high pressure: Journal of Geophysical Research, v. 73, p. 2225–2236.

Butterfield, R., 1979, A natural compression law for soils (an advance on e-logp): Geotechnique, v. 29, no. 4, p. 469–480.

Carstens, C., and H. Dypvik, 1981, Abnormal formation pressure and shale porosity: AAPG Bulletin, v. 65, p. 344–350.

Cheng, H. C., and M. N. Toksoz, 1979, Inversion of seismic velocities for the pore aspect ratio spectrum of rock: Journal of Geophysical Research, v. 84, no. 813, p. 7533–7543.

Colten-Bradley, V. A., 1987, Role of pressure in smectite dehydration-effects on geopressure and smectite-illite transformation: AAPG Bulletin, v. 71, p. 1414–1427.

Coyner, K., T. J. Katsube, M. E. Best, and M. Williamson, 1993, Gas and water permeability of tight shales from the Venture gas field offshore Nova Scotia: Geological Survey of Canada, Current Research, no. 1993-D, p. 129–136.

Grauls, D., and C. Cassignol, 1993, Identification of a zone of fluid pressure-induced fractures from log and seismic data—a case history: First Break, v. 11, no. 2, p. 59–68.

Hermanrud, C., L. Wensaas, G. M. G. Teige, H. M. Nordgard Bolas, S. Hansen and E. Vik, 1998, Shale porosities from well logs on Haltenbanken (offshore mid-Norway) show no influence of overpressuring, *in* B. E. Law, G. F. Ulmishek, and V. I. Slavin, eds., Abnormal pressures in hydrocarbon environments: AAPG Memoir 70, p. 65–85.

Houseknecht, D. W., 1987, Assessing the relative importance of compaction processes and cementation to reduction of porosity in sandstones: AAPG Bulletin, v. 71, p. 633–642.

Issler, D. R., and T. J. Katsube, 1994, Effective porosity of shale samples from the Beaufort-MacKenzie basin: Geological Survey of Canada, Current Research, no. 1994-1B, p. 19–26.

Katsube, T. J., 1999, True formation factor determination by non-linear curve fitting: Geological Survey of Canada, Current Research, no. 1999-D, p. 27–34.

Katsube, T. J., 2000, Shale permeability and pore-structure evolution characteristics: Geological Survey of Canada, Current Research, no. 2000-E15, 9 p.

Katsube, T. J., and K. Coyner, 1994, Determination of permeability(k)-compaction relationship from interpretation of k-stress data for shales from eastern and northern Canada: Geological Survey of Canada, Current Research, no. 1994-D, p. 169–177.

Katsube, T. J., and D. R. Issler, 1993, Pore-size distribution of shales from the Beaufort-MacKenzie basin, northern Canada: Geological Survey of Canada, Current Research, no. 1993-E, p. 123–132.

Katsube, T. J., and D. C. Kamineni, 1983, Effect of alteration on pore structure of crystalline rocks: core samples from Atikokan, Ontario: Canadian Mineralogist, v. 21, p. 637–646.

Katsube, T. J., and M. Mareschal, 1993, Petrophysical model of deep electrical conductors: graphite lining as a source and its disconnection due to uplift: Journal of Geophysical Research, v. 98, no. B5, p. 8019–8030.

Katsube, T. J., and J. B. Walsh, 1987, Effective aperture for fluid flow in microcracks: International Journal of Rock Mechanics and Mining Sciences and Geomechanics Abstracts, v. 24, p. 175–183.

Katsube, T. J., and M. A. Williamson, 1994, Effects of diagenesis on shale nano-pore structure and implications for sealing capacity: Clay Minerals, v. 29, p. 451–461.

Katsube, T. J., and M. A. Williamson, 1998, Shale petrophysical characteristics: permeability history of subsiding shales, *in* J. Schiber, W. Zimmerle, and P. S. Sethi, eds., Shales and mudstones II: Stattgard, Germany, E. Schweizerbart'sche Verlagslouchhandlung, p. 69–91.

Katsube, T. J., M. E. Best, and B. S. Mudford, 1991, Petrophysical characteristics of shales from the Scotian Shelf: Geophysics, v. 56, p. 1681–1688.

Katsube, T. J., M. Williamson, and M. E. Best, 1992, Shale pore structure evolution and its effect on permeability, *in* Thirty-third annual symposium of the Society of Professional Well Log Analysts (SPWLA), symposium v. 3: Society of Core Analysts Preprints, Paper SCA-9214, p. 1–22.

Katsube, T. J., D. R. Issler, and K. Coyner, 1996a, Petrophysical characteristics of shales from the Beaufort-MacKenzie basin, northern Canada: permeability, formation factor and porosity versus pressure: Geological Survey of Canada, Current Research, no. 1996-B, p. 45–50.

Katsube, T. J., G. N. Boitnott, P. J. Lindsay, and M. Williamson, 1996b, Pore structure evolution of compacting muds from the sea floor offshore Nova Scotia: Geological Survey of Canada, Current Research, no. 1996-D: p. 17–26.

Katsube, T. J., J. Bloch, and W. C. Cox, 1999a, The effect of diagenetic alteration on shale pore-structure and its implications for abnormal pressures and geophysical signatures, *in* A. Mitchell and D. Grauls, eds., Overpressure in petroleum exploration—Proc. Workshop: Bulletin Centre Recherche Elf Exploration and Production, Memoir 22, p. 49–54.

Katsube, T. J., S. R. Dallimore, T. Uchida, K. A. Jenner, T. S. Collett, and S. Connell, 1999b, Petrophysical environment of sediments hosting gas-hydrate, JAPEX/JNOC/GSC Mallik 2L-38 gas hydrate research well, *in* S. R. Dallimore, T. Uchida, and T. S. Collett, eds., Scientific results from JAPEX/JNOC/GSC Mallik 2L-38 gas hydrate research well, Mackenzie Delta, North West Territories, Canada: Geological Survey of Canada Bulletin 544, p. 109–124.

Lahann, R., 1998, Impact of smectite diagenesis of compaction profiles and compaction equilibrium: American Association of Drilling Engineers Industry Forum: Pressure Regimes in Sedimentary Basins and Their Prediction, 6 p.

Loman, J. M., T. J. Katsube, J. M. Correia, and M. A. Williamson, 1993, Effect of compaction on porosity and formation factor for tight shales from the Scotian Shelf, offshore Nova Scotia: Geological Survey of Canada, Current Research, no. 1993-E, p. 331–335.

Miller, T. W., 1995, New insights on natural hydraulic fractures induced by abnormally high pore pressures: AAPG Bulletin, v. 79, p. 1005–1018.

Patnode, H. W., and M. R. J. Wyllie, 1950, The presence of conductive solids in reservoir rocks as a factor in electric log interpretation: Transactions of the American Institute of Mining, Metallurgical and Petroleum Engineers, v. 189, p. 47–52.

Rubey, W. W., and M. K. Hubbert, 1959, Role of fluid pressure in mechanics of overthrust faulting, part II: Bulletin of the Geological Survey of America, v. 70, p. 167–206.

Toksoz, M. N., C. H. Cheng, and A. Timur, 1976, Velocities of seismic waves in porous rocks: Geophysics, v. 41, no. 4, p. 621–645.

Walsh, J. B., and W. F. Brace, 1984, The effect of pressure on porosity and the transport properties of rocks: Journal of Geophysical Research, v. 89, p. 9425–9431.

Wardlaw, N. C., and R. P. Taylor, 1976, Mercury capillary pressure curves and the interpretation of pore structure and capillary behavior in reservoir rocks: Bulletin of Canadian Petroleum Geology, v. 24, p. 225–262.

Washburn, E. W., 1921, Note on a method of determining the distribution of pore sizes in a porous material: Proceedings of the National Academy of Science, v. 7, p. 115–116.

Wyllie, M. R., and M. B. Spangler, 1952, Application of electrical resistivity measurements to problems of fluid flow in porous media: AAPG Bulletin, v. 36, p. 359–403.

6

Impact of Smectite Diagenesis on Compaction Modeling and Compaction Equilibrium

Richard Lahann
Conoco Exploration Production Technology, Houston, Texas

ABSTRACT

Shale compaction models employed in pore-pressure interpretation and prediction generally do not consider the effects of changing shale mineralogy with depth. The conversion of smectite to illite reduces the amount of bound water in the shale. This change reduces the equilibrium porosity associated with an effective stress. If dewatering does not accompany the mineralogical transformation, excess pressure develops. This pressure-generating mechanism is independent of the volume change associated with the clay reaction.

Three examples from the United States Gulf Coast show that application of smectitic and illitic compaction models allows improved interpretation of pore-pressure variation with depth. The combination of both porosity and compaction model changing with depth can create a pattern of effective stress and velocity variation that resembles unloading.

INTRODUCTION

Studies of shale porosity/overpressure commonly model shale porosity (Φ) as in equation 1.

$$\Phi = \Phi_0 e^{-B\sigma} \tag{1}$$

where Φ_0 is the surface intercept porosity, B is an empirical constant, and σ is effective stress (Athy, 1930, Rubey and Hubbert, 1959). Hart et al. (1995) used two equations of this form, with different B values, to model shale porosity and pressure in Pleistocene strata of the Gulf of Mexico. Rearrangement of equation 1 allows calculation of effective stress as a function of shale porosity.

One of the features of equation 1 is that increasing effective stress causes shale porosity to approach 0, as shown in Figure 1. The curves displayed in this figure are the two shale porosity-stress relationships from Hart et al. (1995). Note that the two calibrations indicate porosity from 6 to 1% at 50 MPa, which is roughly equivalent to normal pressure at about 4500 m. The deep calibration value is probably unreasonably low for natural materials.

The porosity term in equation 1 includes intergranular porosity and water associated with mineral surfaces (either external or clay interlayer surfaces). The water associated with shale mineral surfaces persists until diagenesis/metamorphism have annealed the minerals into a much coarser grained assemblage. Equation 2,

$$\Phi = \Phi_m + \Phi_0 e^{-B\sigma} \tag{2}$$

allows porosity to decline to a minimum value, Φ_m, based on the grain size/surface area characteristics of the shale. Within this study, Φ_m is referred to as

Lahann, Richard, 2002, Impact of Smectite Diagenesis on Compaction Modeling and Compaction Equilibrium, in A. R. Huffman and G. L. Bowers, eds., Pressure regimes in sedimentary basins and their prediction: AAPG Memoir 76, p. 61–72.

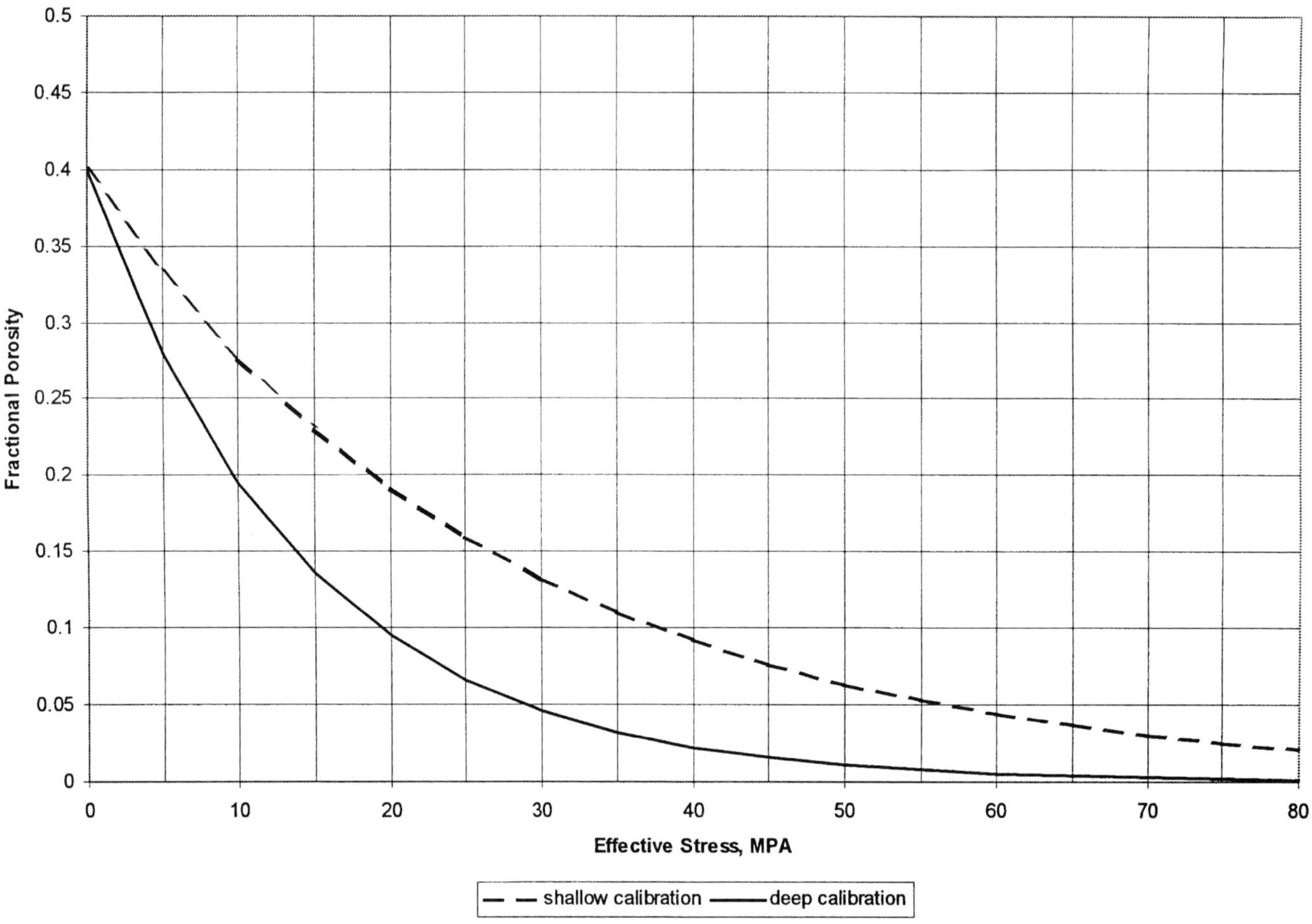

Figure 1. Porosity effective stress plots of shallow and deep shale compaction algorithms published in Hart et al. (1995). Note that both expressions approach zero porosity at large values of effective stress.

"bound" water. The concept of clay-associated (bound) water being distinct from intergranular water was used by Brown and Ransom (1996) in calculating the appropriate porosity for compaction/pressure analysis.

The empirical constants B and Φ_0 are commonly obtained by curve fitting in the shallow, near normally pressured part of a well. For sediments in the Gulf of Mexico the depth of the calibration zone is substantially above the depth at which smectite begins to transform to illite. Thus the Φ_m value in equation 2 should correspond to the surface or bound-water characteristic of a smectitic shale (where calibrated to the shallow part of Gulf of Mexico wells). At greater depth and temperature, following conversion of smectite to illite, the surface or bound water can be expected to be much lower than for a smectitic shale. The value of Φ_m should also be lower as a result.

Several previous works have used the concept of bound water and bound-water release during diagenesis in pressure calculations. Dutta (1986) describes a basin model in which compaction is a function of both stress and temperature. By making the shale equilibrium void ratio (porosity) a function of temperature (as well as stress), Dutta incorporated the conversion of bound water to intergranular water by illite formation in the fluid-pressure calculation. The model described in following sections allows evaluation of the pressure effect of illitization without conducting a basin modeling analysis.

Audet (1995) concluded that conversion of smectite to illite could increase excess pore pressure by as much as 30% by increasing the volume of free water. That analysis maintained the same compaction (stress/porosity) relationship for before and after the bound-water conversion. The pressure contribution could be greater than the 30% value quoted by Audet if the resultant illitic shale is more compactible than its smectitic precursor. This work incorporates probable

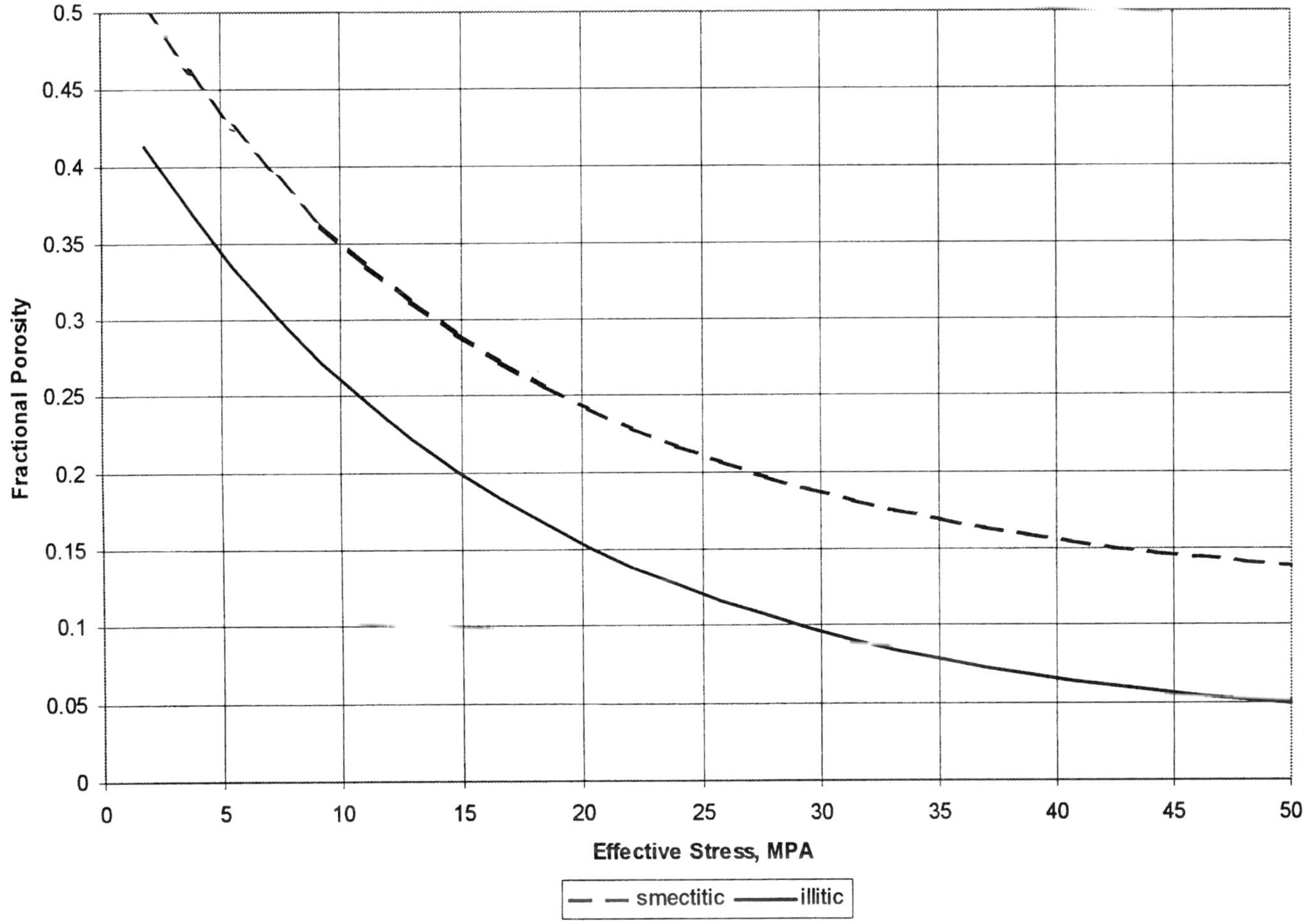

Figure 2. Smectitic and illitic compaction profiles fitted to data from Plumley (1980). Note the substantially lower effective stress associated with the illitic curve.

changes in the compaction relationship as a result of the mineral conversion.

MODEL

Calibration of equation 2 to shallow-water data requires an evaluation of the Φ_m term. This value includes water associated with clay and other mineral surfaces and also water in smectite interlayers. Colten-Bradley (1987) showed that for normal (hydrostatic) pressure conditions, smectite retains two water layers to depths ranging from 1100 to 2200 m. If fluid pressures were greater than hydrostatic pressures, the stability of the two-layer system would extend to greater depths. These depths correspond to temperatures between 50 and 75° C, which are associated with conversion of smectite to illite. Based on these data, I assume that shales above the smectite-illite transition contain smectites with two water layers. A complete analysis would include three-water complexes in the shallow subsurface; this complexity is not dealt with in this chapter.

Hunt et al. (1998) suggest that shales with small internal surface areas stop compacting at about 3% porosity and that shales with mixed-layer water stop compacting at about 10% porosity. The 7% porosity difference cited by Hunt et al. (1998) is close to the value that can be calculated based on preservation of two-water layers in smectite in Gulf of Mexico shales. Data from Hower et al. (1976) suggest that typical, diagenetically immature, Gulf of Mexico shales contain about 50% clay minerals relative to the total volume of solids, and the clays are about 50% smectite. The volume fraction of compacted (and illitized) smectite at 3% intergranular porosity is about 24%. If the illitic clays were expanded by the volume ratio of two-water layer smectite to illite (14/10), the shale volume would increase by 10% because the 24 vol. % of illite would

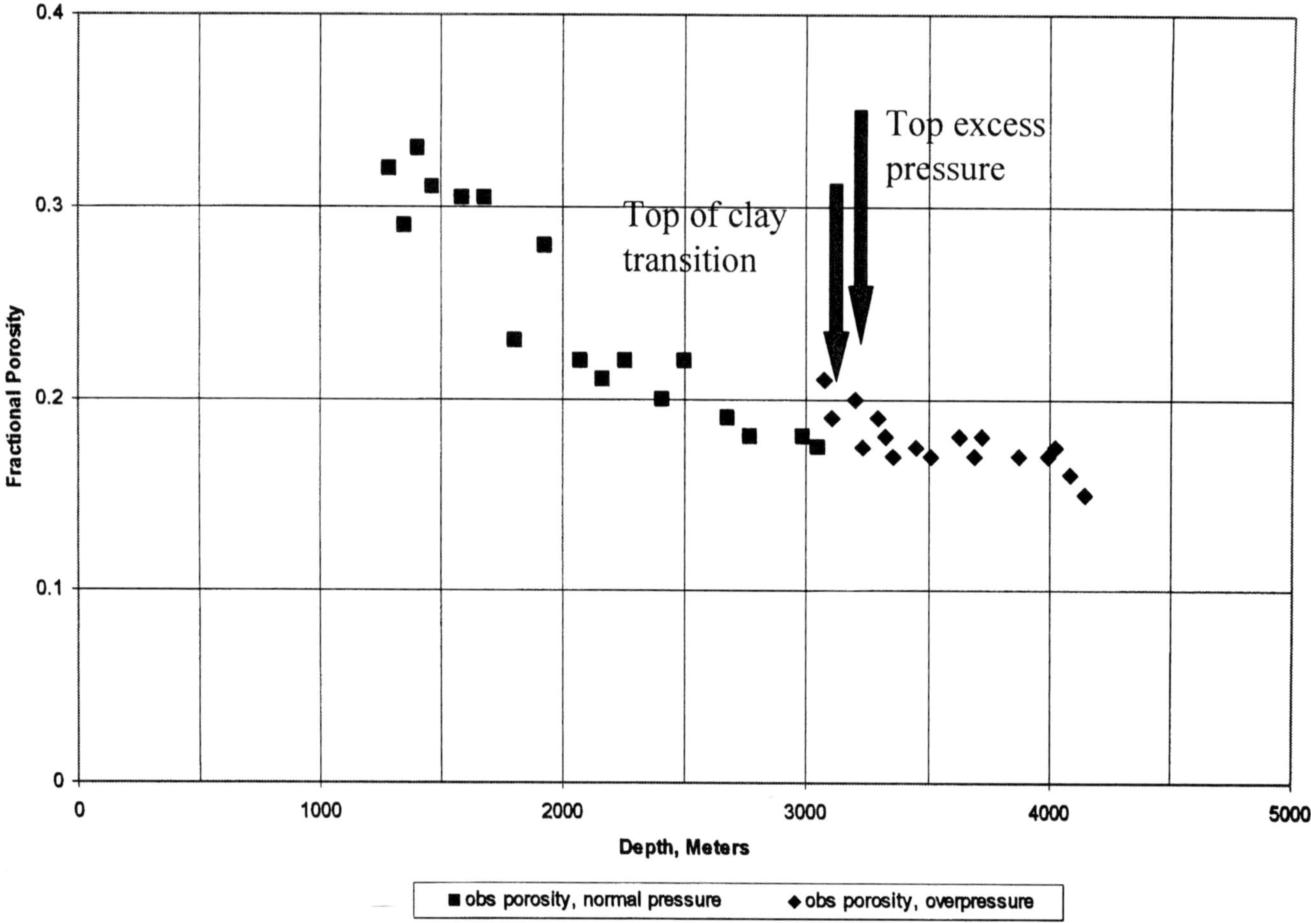

Figure 3. Depth variation of observed porosity for well B described in Plumley (1980). The top of the clay transition and top of excess pressure are also shown. All data replotted from Plumley (1980, figure 5).

now occupy about 34 volume units (24 × (14/10). The porosity fraction of the expanded volume would be 12%, intergranular porosity would be 3%, and interlayer, smectite-associated porosity would be 9%. For this study, the Φ_m value of smectitic shales is fixed at 12%, and the Φ_m value of illitic shales is set at 3%.

The values of B and Φ_0 for smectitic shale are determined empirically by fitting model porosities to measured porosities. For this study, I assume initially that B and Φ_0 are identical for smectitic and illitic shales. This assumption implies that the only differences in the compactional properties of the two shale types is due to the retention of interlayer porosity in the smectitic model.

Figure 2 shows porosity/effective stress plots for smectitic and illitic shales derived from data in Plumley (1980) (discussed in the next section). Note that the effective stress required to produce a given shale porosity is always lower for illitic shale than for smectitic shale. The calculated fluid pressure for a given porosity is lower if the smectitic model is used than if the illitic curve is employed. In terms of Figure 2, the effect of illitization is to move the mineral system horizontally from the smectitic curve toward the illitic curve. The transition from one curve to the next occurs gradationally as the reaction proceeds.

The smectite-illite reaction may be usefully viewed as one of crystal growth and bound-water reduction. At any point along the smectite curve in Figure 2, the porosity is a combination of bound water (the 12% Φ_m value) and intergranular water. The formation of illite eliminates substantial amounts of smectitic interlayer surface, which was hydrated when the clay was a smectite. As a result the amount of bound water (Φ_m) decreases and intergranular water increases. If the smectitic shale was in compactional equilibrium, conversion to illite increases intergranular water. If the intergranular water cannot escape to maintain compactional equilibrium, then excess fluid pressure results.

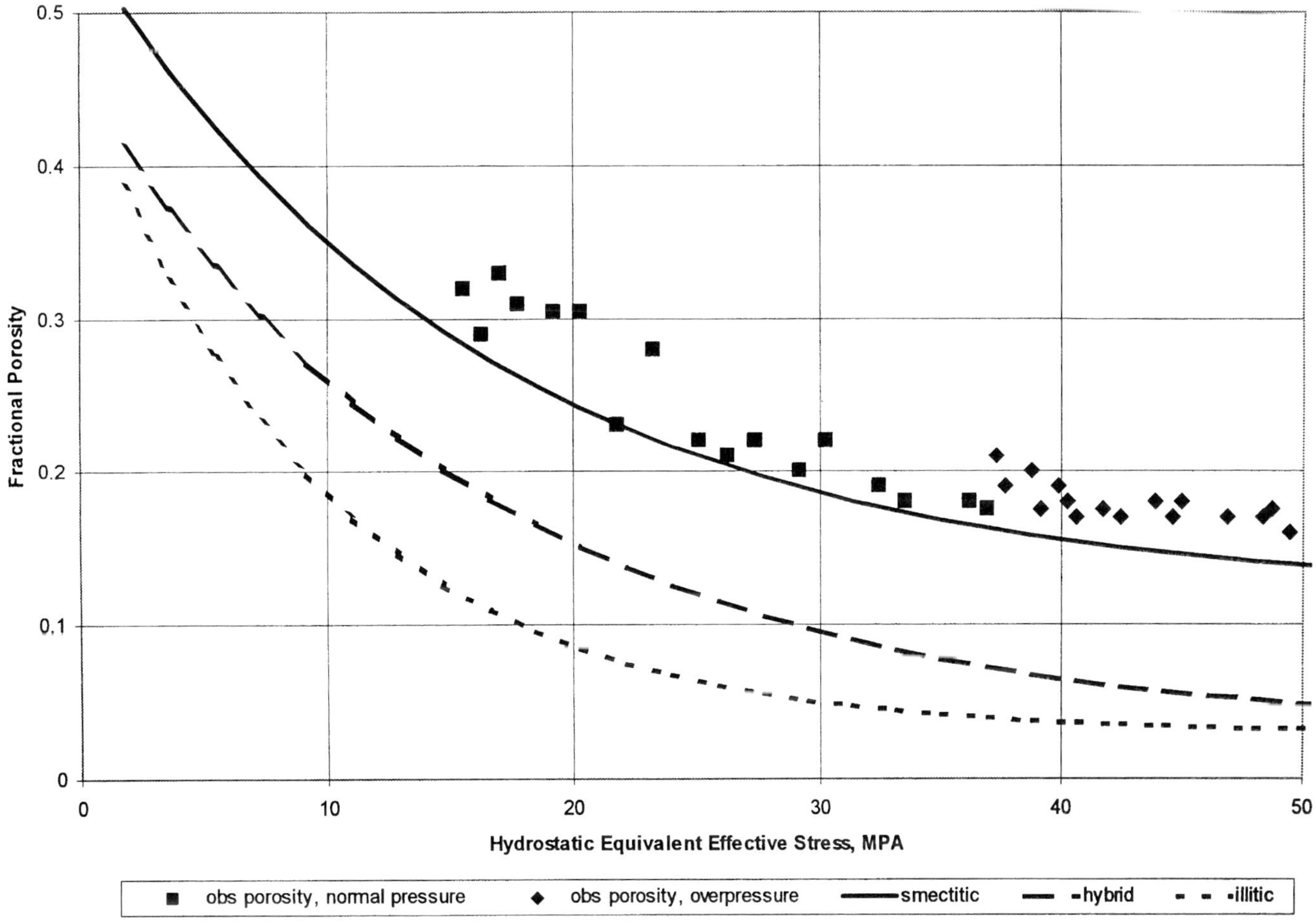

Figure 4. Observed shale porosity (replotted from Plumley, 1980, figure 5), smectitic, hybrid and illitic shale compaction models as function of hydrostatic effective stress, calculated by assuming constant overburden and hydrostatic fluid pressure. Divergence of observed porosity and smectitic model at high porosity is probably due to overestimation of effective stress due to assumption of constant overburden.

The relationship between illitization and effective stress described in Figure 2 is independent of any volume change associated with illitization. If the specific volume of intergranular water is greater than smectitic interlayer water, then the pressure effect of illitization would be greater than shown in Figure 2. Alternatively, a decrease in specific volume would decrease the pressure effect indicated in Figure 2.

Osborne and Swarbrick (1997) argued that the smectite-illite conversion is not a significant pressure source because the maximum likely volume change between interlayer and intergranular water increases the volume of intergranular water by a very small amount. If, for example, both intergranular and interlayer water are 9 vol. %, an expansion of the interlayer water by 10% during illitization increases the porosity from 18 to 18.9%, a change of only 0.9%. Given reasonable values for seal permeability, the pressure increase would be small. If the interlayer water is viewed as bound water, however, illitization increases the volume of intergranular water from 9 to 18%, a far more substantial change. The models in Figure 2 suggest that the difference in fluid pressure can be substantial.

Applications

The pressure/porosity/clay mineralogy concepts developed in previous sections are tested in this section with three data sets taken from the literature. Details of the pressure tests are not known, and so no attempt is made to adjust the pressures measured in the sands for centroid or hydrocarbon column effects. The simplistic assumption is made that the measured pressures reflect shale pressures at that point.

Data Set 1

The data published in Plumley (1980) provide shale porosity data from 1400 to 4200 m, a top of excess fluid

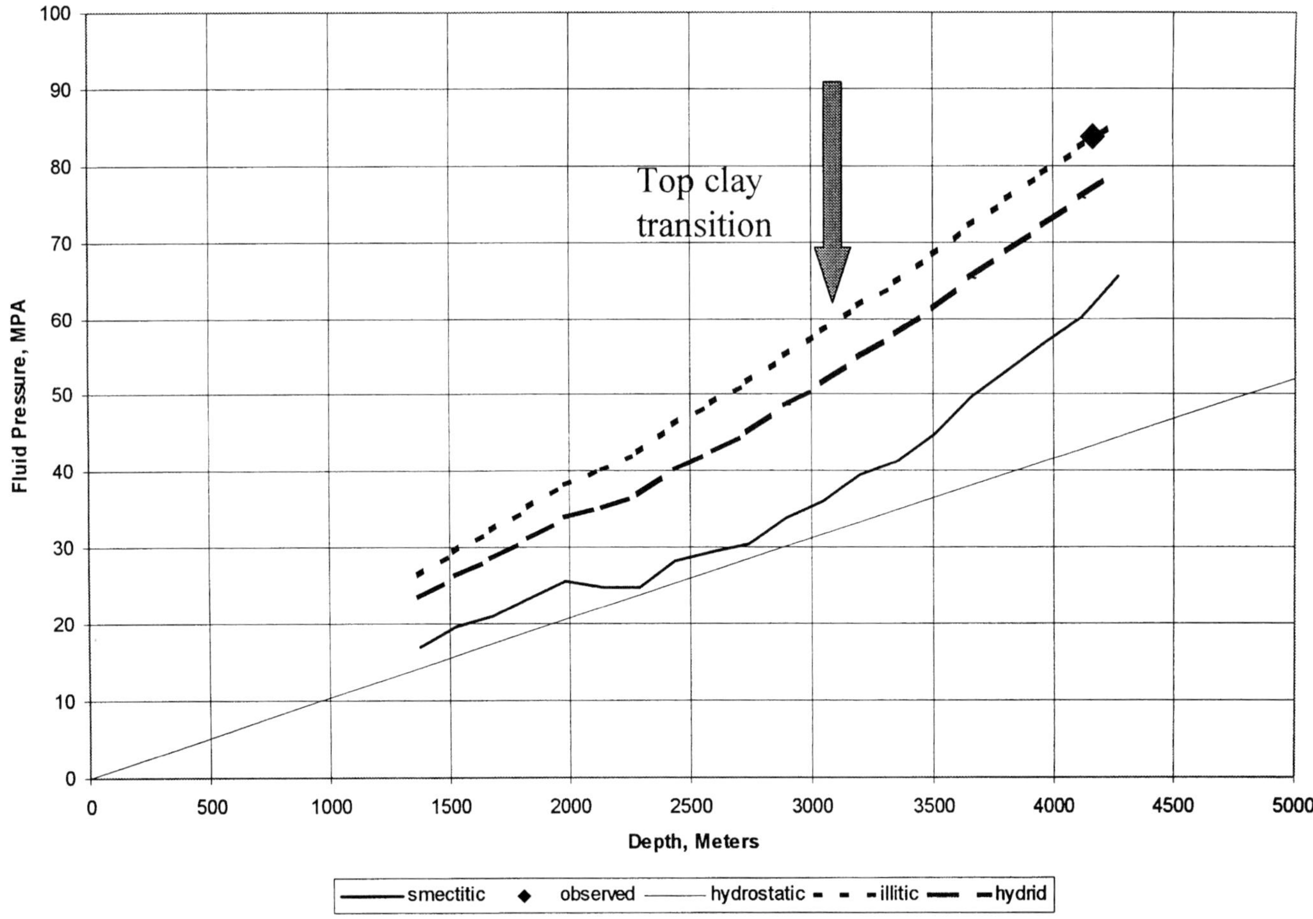

Figure 5. Fluid pressure predicted by smectitic, hybrid, and illitic shale compaction models. Note that the smectitic model predicts minor excess fluid pressure down to about 3000 m. The observed fluid pressure at 4100 m (from Plumley, 1980, figure 5) is greater than predicted by the hybrid model and far in excess of the pressure predicted by the smectitic model.

pressure, a top of clay transition, and a pressure measurement within the excess pressure zone. The porosity/depth data from Plumley are displayed in Figure 3. By assuming a constant overburden gradient (22.6 MPa/km), effective stress values can be calculated corresponding to each of the porosity points. From these data a plot of effective stress vs. porosity can be made, and the values of Φ_0 and B can be solved empirically for a smectitic model. This plot is shown in Figure 4 as the smectitic trend. The mismatch between the smectitic model and data at high porosity is due in part to the assumption of a constant overburden. A better overburden model would reduce the effective stress calculated at shallow depth and result in a better match of model and observation. The porosity effective stress trend generated by changing Φ_0 to 0.34 (9% shift) is identified as the hybrid model; the derivation of the illitic model is discussed subsequently.

Figure 5 displays the calculated pressure profiles, based on the effective stress relationships shown in Figure 4 and the observed porosity profile in Figure 3. From the surface down to about 3050 m, the top of the clay transition, the pressure is best interpreted with the smectitic model (solid line in Figure 5). The smectitic model ($\Phi_m = 0.12$, $\Phi_0 = 0.43$, $B = 0.0624$) agrees reasonably well with the top of excess fluid pressure at 3000 m. The excess pressure predicted by the smectitic model at depths less than 2000 m is probably due to error in estimation of the overburden, as discussed previously.

Below 3000 m (within the clay transition) the model discussed in previous sections predicts that fluid pressure should begin to shift toward the hybrid line. The dashed line in Figure 5 ($\Phi_m = 0.03$, $\Phi_0 = 0.43$, $B = 0.0624$) is termed a hybrid because it has a Φ_m value appropriate for illite and Φ_0 and B values taken from the smectitic model. The pressure measurement at about 4200 m is about 6 MPa greater than predicted by the hybrid model and far in excess of that predicted by the smectitic model.

The pressure measurement at 4200 m is about 1150 m below the top of the clay transition. Based on the

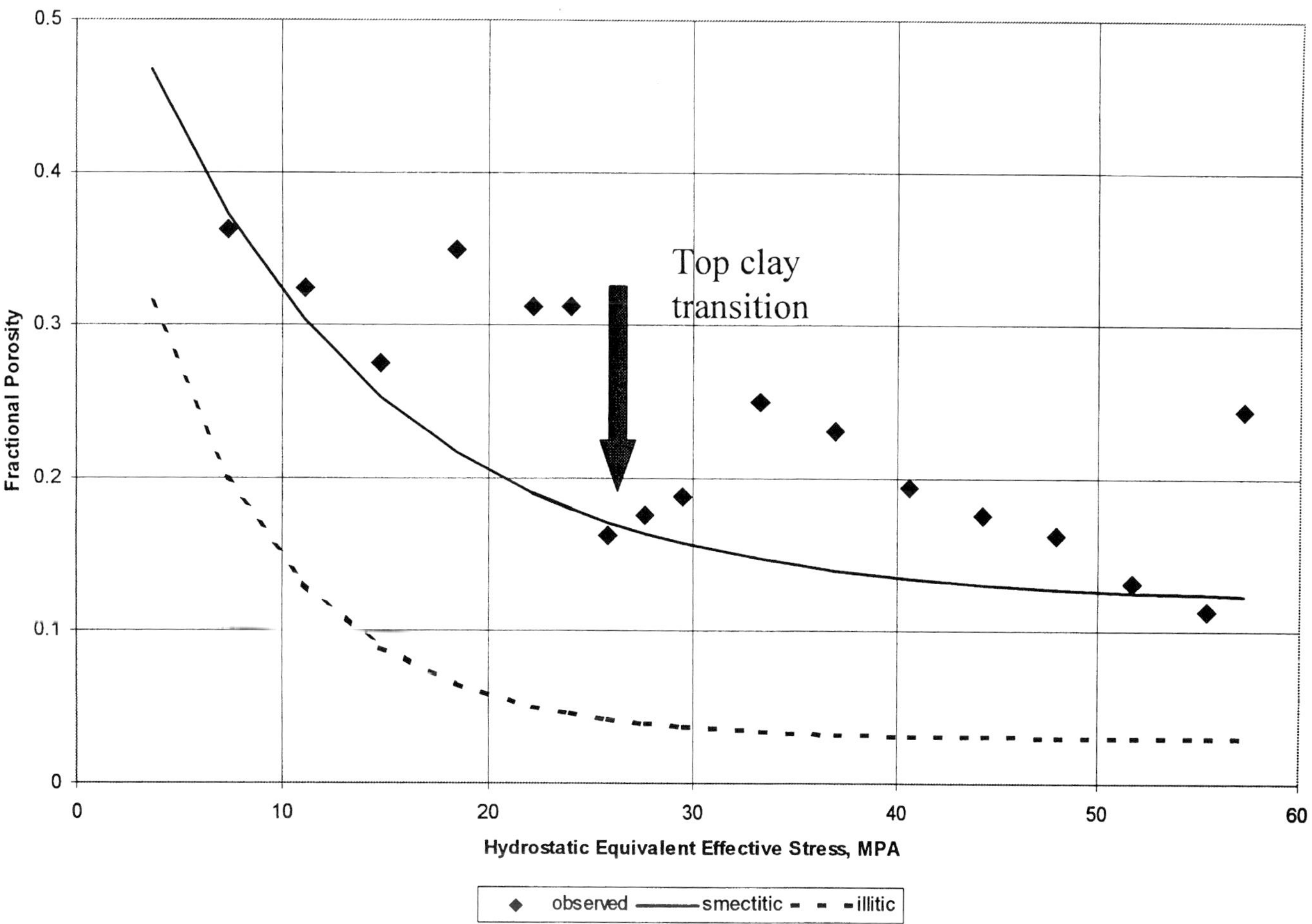

Figure 6. Observed shale porosity calculated from density data in figure 8 of Berg and Habeck (1982), smectitic and illitic shale compaction models as function of hydrostatic effective stress, calculated by assuming constant overburden and hydrostatic fluid pressure. The smectitic model was empirically fitted to match the porosity data at low effective stress.

clay-transition profiles in Hower et al. (1976) and Bruce (1984), 1000 m is a typical thickness for completion of smectite conversion to low expandability illite.

A very good match can be made between an illitic compaction model and the observed pressure if B is increased to 0.1015 (dotted line, illitic model). This model attributes low bound water to the illitic shale, Φ_m = 0.03, and greater compactibility because B is greater for illite than for smectite. Compaction models can be generated that match the observed pressure by reduction of the Φ_0 value to less than 0.03. These models, however, yield unrealistically low porosity values at low effective stress values. These low porosity values also create problems in calculating pressures in systems with high preserved shale porosity at depth.

Data Set 2

Berg and Habeck (1982) provide shale porosity, clay mineralogy, and pressure data from about 600 to 4500 m depth. The data set is from Oligocene sediments in the McAllen ranch area of south Texas. The shale porosity/depth data were used to calibrate (empirically) smectitic and hybrid porosity/effective stress relationships (Figure 6). A good match is possible between a smectitic model (Φ_m = 0.12, Φ_0 = 0.48, and B = 0.087; solid line in Figure 6) and the observed porosities at several points in the shallow part of the section. Hybrid (Φ_m = 0.03, Φ_0 = 0.48, and B = 0.087) and illitic (Φ_m = 0.03, Φ_0 = 0.48, and B = 0.142) effective stress models are also displayed in Figure 6 as dashed and dotted lines, respectively. The illitic model was generated by increasing the B value in the same ratio as was done for the Plumley illitic model.

Figure 7 displays the predicted pressures based on the porosity/effective stress models in Figure 6 and two observed pressures. The smectitic model agrees extremely well with the measured pressure near 2000 m, which is above the top of the transition zone. The illitic pressure agrees within 1 MPa with the observed pressure at 3400 m. Data in Berg and Habeck (1982)

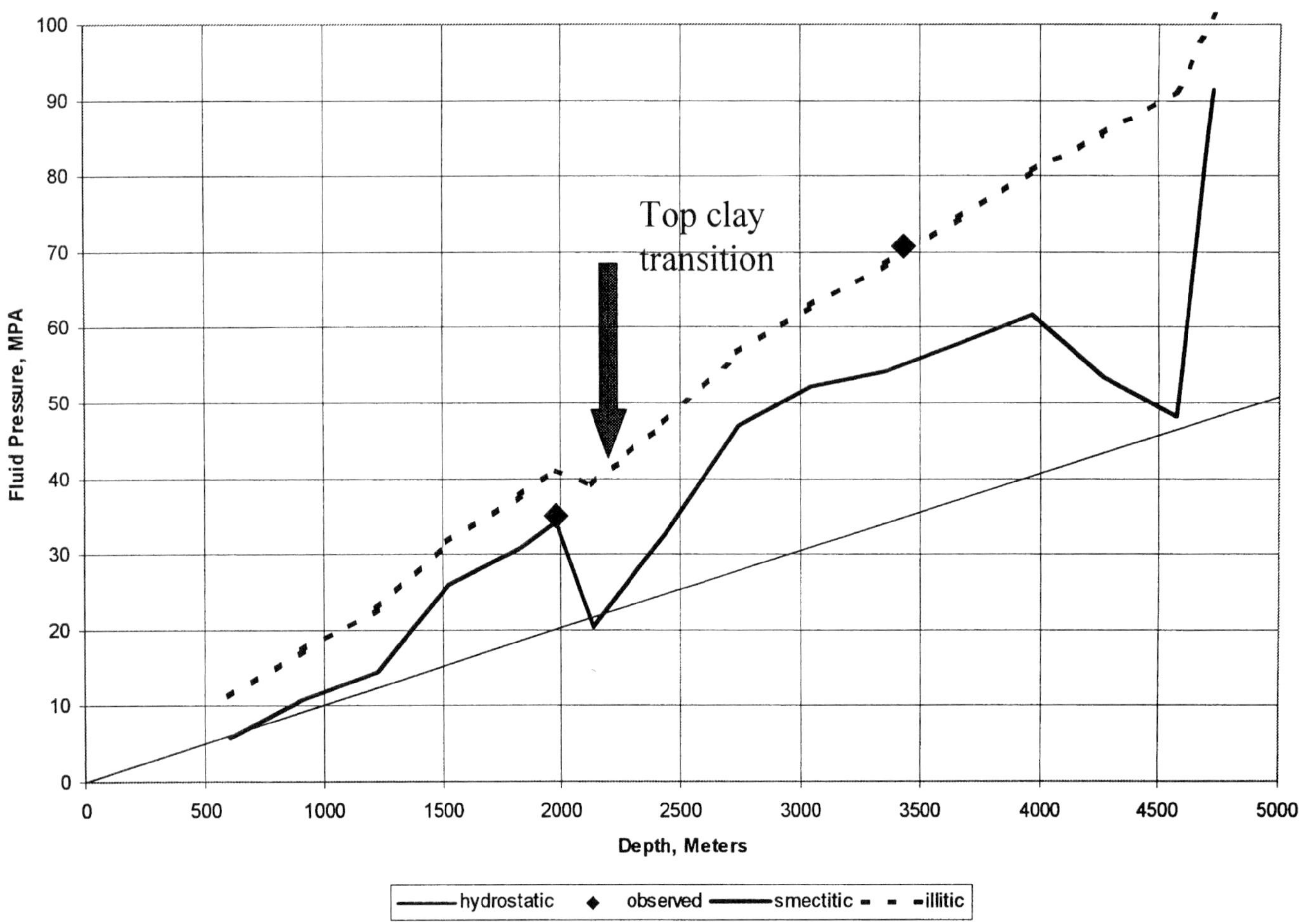

Figure 7. Fluid pressure predicted by smectitic and illitic shale compaction models. Note that the smectitic model agrees extremely well with measured pressure (taken from Berg and Habeck, 1982, figure 8) near 2000 m. The illitic model closely matches the measured pressure at 3400 m.

indicate that the clay transition is nearly complete at this depth.

Data Set 3

Hart et al. (1995) provide shale porosity and eight pressure data points from between 400 and 3600 m from Pleistocene sediments from the Eugene Island Block 330 area. The depth to the top of the clay transition in the Eugene Island 330 A20ST (Pathfinder) well appears to be at about 1500 m (Figure 8), based on data contained in Losh et al. (1999). This depth, as discussed by Losh et al. (1999) is anomalously shallow for the age and temperature of the sediments. Model calculations for the top of the clay transition, using the method described by Gordon and Flemings (1998), place the top of the transition near 2000 m.

The shale porosity and effective stress data from Hart et al. (1995) were used to calibrate smectitic ($\Phi_m = 0.12$, $\Phi_0 = 0.27$, and $B = 0.055$), and illitic porosity/effective stress relationships as in Figure 9. The illitic ($\Phi_m = 0.03$, $\Phi_0 = 0.27$, and $B = 0.088$) model was generated as described previously. The smectitic model provides a close match to the measured porosities in the shallow section of the well. Because the clay composition data (Figure 9) indicates that the most illitic phases present retain about 40% smectite layers, a transition model was generated with Φ_m, Φ_0 and B values ($\Phi_m = 0.075$, $\Phi_0 = 0.27$, and $B = 0.071$) between those used for the smectite and illite models.

Measured pressures and the fluid pressures calculated by the relationships in Figure 9 are displayed in Figure 10. The smectitic model agrees extremely well with the measured pressures at 1300 and 1500 m. The transition model predicts a fluid pressure in excellent agreement with measured pressures at 2100–2200 m. Hart et al. (1995) and Gordon and Flemings (1998) suggest that thermal expansion of pore fluids and clay sourced fluids may account for up to 20% of the excess pressure in Eugene Island wells. The data in Figure 10 account for the pressure increase by changes in the

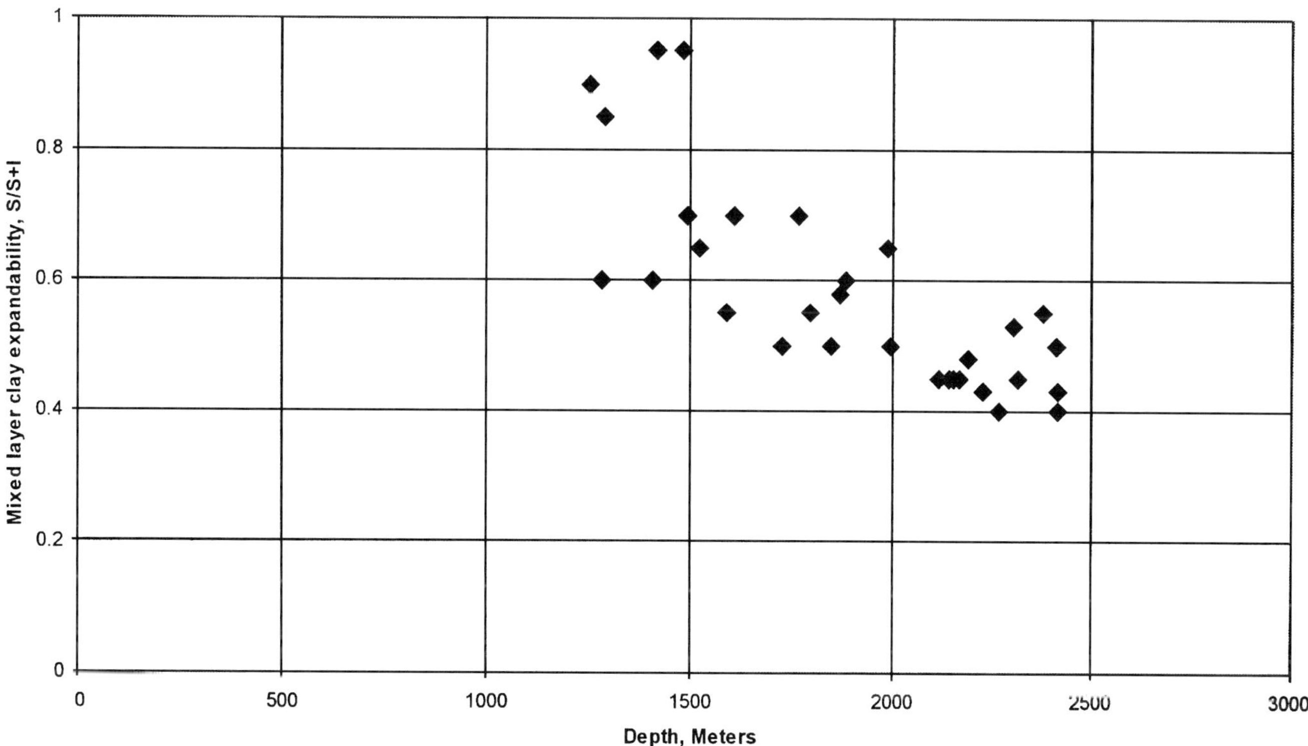

Figure 8. Depth variation of mixed-layer clay expandability, smectite/smectite + illite (S/S + I) for the Pathfinder well. Data from table 1 of Losh et al. (1999). Note that below 1500 m the maximum smectite fraction is .7, substantially less than the maximum value above 1500 m. A top of the clay transition of 1500 m is interpreted for this well.

compaction model for the reacting shales, without appealing to volume expansion during the clay reactions.

Based on these model/pressure correlations, a fluid pressure between 24 and 32 MPa, the values predicted for the smectitic and transition models, is anticipated at 1950 m. The measured pressure at 1950 m is about 23 MPa, well below the anticipated shale pressure and consistent with the smectitic model.

An alternative interpretation for the data in Figure 10 is that the smectitic model matches the observed pressure data down to a depth of 1950 m, followed by a reasonable match between the transition model and measured data at 2100–2200 m. This interpretation requires that the measured clay expandability data (Losh et al., 1999) be in error between 1500 and 1900 m and also requires a dramatic change in shale compaction properties between 1950 and 2100 m.

DISCUSSION AND CONCLUSIONS

A procedure has been developed and demonstrated for using porosity/effective stress relationships in the shallow part of wells to predict fluid-pressure profiles below the clay transition. In the three examples examined in this study, the shallow smectitic part of the well was used to define a smectitic compaction curve with a minimum porosity of 12%. In all three cases, the observed pressure below the top of the clay transition was far greater than could be predicted either from the smectitic model or from a single traditional expression with no minimum porosity (see Plumley, 1980; Hart et al., 1995).

The procedure for generating an illitic model, developed from the Plumley data, provides an excellent estimate of the pressure reported by Berg and Habeck (1982) in the illitic zone of the well. The transitional model generated for the Hart et al. (1995) data provides an excellent pressure match to the deep (2100–2200 m) pressure data in the Pathfinder well. The transitional model is also consistent with the presence of about 50% smectite in the mixed-layer clay at that depth.

In the three cases discussed, the top of the clay transition was known from sample analyses. Ideally, in an exploration mode the top and base of the clay transition would be estimated from thermal modeling (see Gordon and Flemings, 1998) or from offset data. A smectitic compaction profile could then be applied down to the top of the clay transition and an illitic profile from the base of the transition onward. Within the transition zone stepwise application of transitional

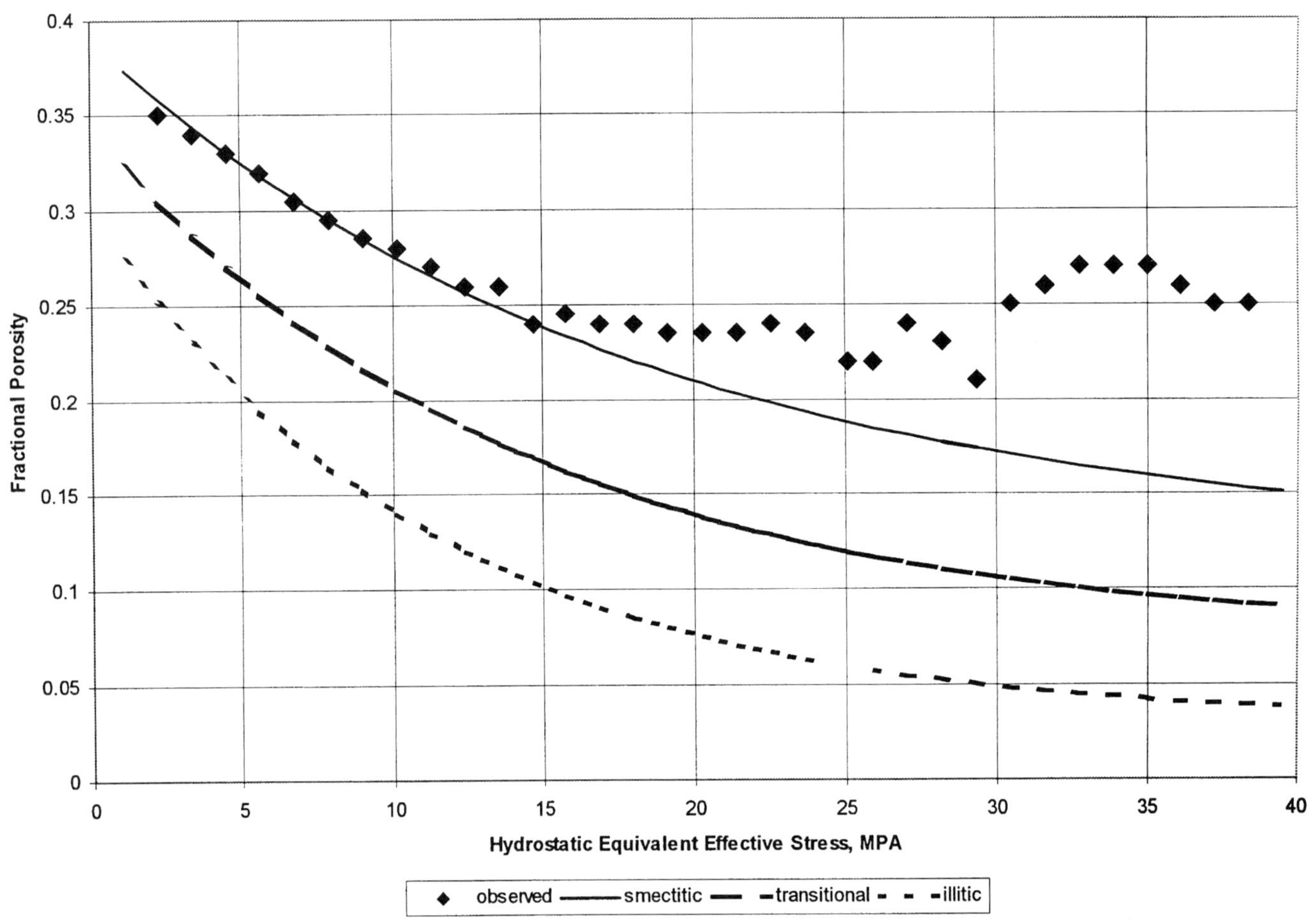

Figure 9. Observed shale porosity (taken from Hart et al., 1995, figure 3), smectitic, transitional and illitic shale compaction models as function of hydrostatic effective stress, calculated by subtracting hydrostatic pressure from overburden data provided by Hart et al. (1995). The smectitic model was empirically fitted to match the low effective stress porosity data.

models can provide reasonably accurate pore-pressure estimates.

The three smectitic models generated in this study had Φ_0 values ranging from 0.27 to 0.48 and B values ranging from 0.055 to 0.087. These ranges may indicate differences in shale/clay properties or may reflect systematic differences in the methods used to calculate porosities. The lowest porosity value is associated with a sonic-log–based porosity transformation (data set 3), whereas the other two models were density-log based. Spatial and temporal variations in clay and shale properties certainly occur in natural geologic materials and contribute uncertainty to the analysis.

Variation in mineralogy and compaction models with depth can produce changes in velocity/effective stress models used to predict pore pressure from seismic data. Figure 11 contains plots of velocity/effective stress for the smectitic, transition, and illitic models in Figure 8. The velocity corresponding to a porosity/effective stress pair was determined with the sonic/porosity relationship of Issler (1992) as detailed in Hart et al. (1995). The progression with depth from a smectitic to a mixed-layer mineralogy requires a shift from the smectitic effective stress velocity relationship toward the transitional relationship.

The depth/porosity relationship for the Pathfinder well (Hart et al., 1995) was combined with a depth/mineralogy relationship to produce the velocity/effective stress relationship shown in Figure 11. In this model effective stress was assumed to be hydrostatic down to 1900 m, which generally agrees with Figure 10. An effective stress model was created, which ranged from smectitic at 1900 m to the transitional model at 2500 m. This model would honor the measured pressure points in Figure 10. The effective stress/velocity trend resembles an unloading curve as discussed in Bowers (1994). In this case the loading curve begins at about 12 MPa because no sonic logs

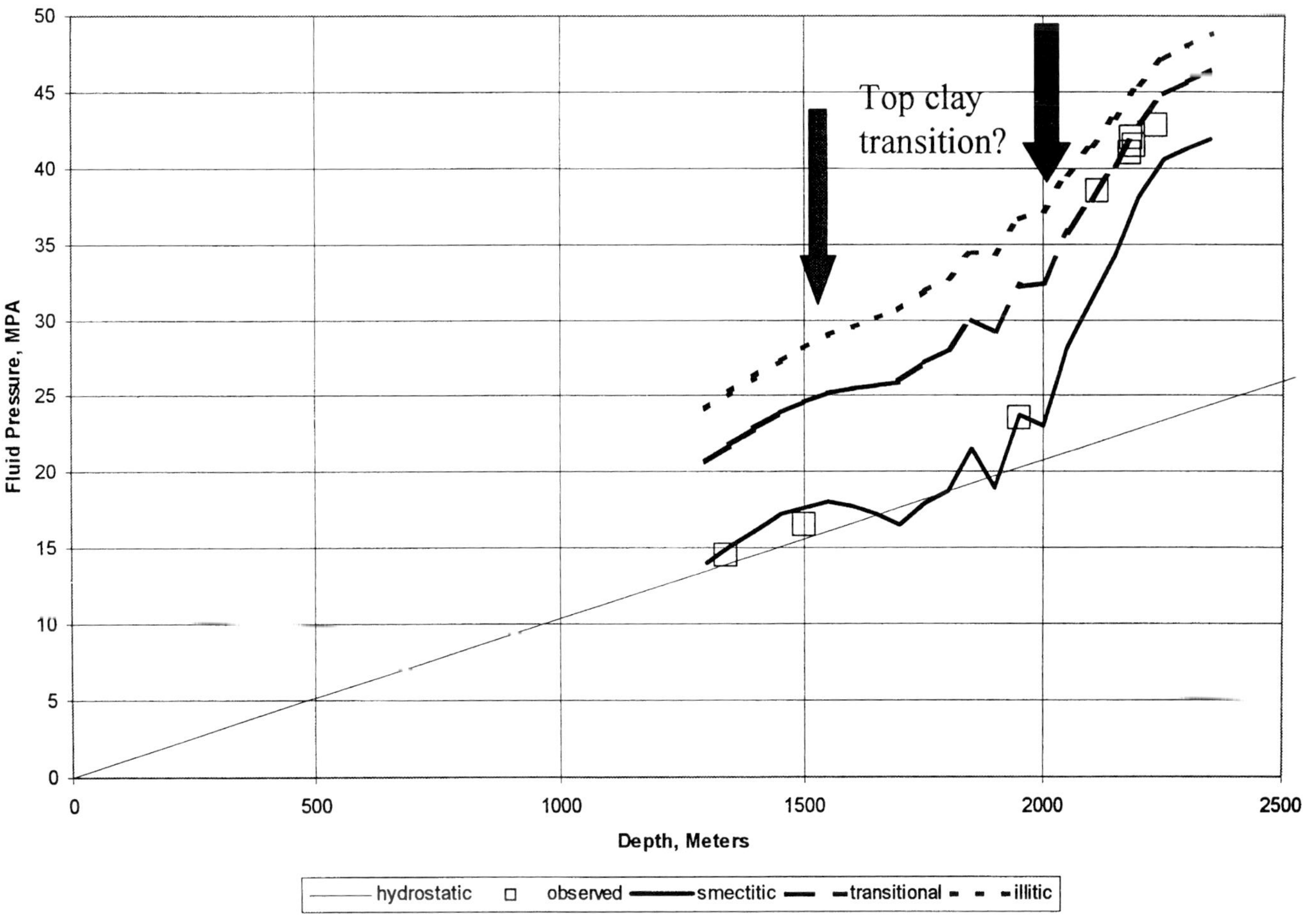

Figure 10. Fluid pressure predicted by smectitic, transitional, and illitic compaction models for the Pathfinder well. Note that the smectitic model agrees well with measured pressures at 1300 and 1500 m and that the transitional model agrees well with measured pressures between 2100 and 2200 m (pressure data from Hart et al., 1995).

are available until almost 2000 m. The greater velocity, for an effective stress value, on the unloading curve reflects the changing mineralogy that requires less effective stress for a given degree of compaction (porosity). The unloading is not due to fluid expansion but load transfer from hydrated solids (smectites) to free water during the mineral transition.

REFERENCES CITED

Athy, L. F., 1930, Density, porosity, and compaction of sedimentary rocks: AAPG Bulletin, v. 56, p. 1–22.

Audet, D. M., 1995, Mathematical modeling of gravitational compaction and clay dehydration in thick sediment layers: Geophysics Journal International, v. 122, p. 283–298.

Berg, R. R., and M. F. Habeck, 1982, Abnormal pressures in the lower Vicksburg, McAllen Ranch field, south Texas: Transactions of the Gulf Coast Association of Geological Societies, v. 32, p. 247–253.

Bowers, G., 1994, Pore pressure estimation from velocity data: accounting for overpressure mechanisms besides undercompaction: International Association of Drilling Contractors/Society of Petroleum Engineers Drilling Conference, no. 27488, p. 515–530.

Brown, K. M., and B. Ransom, 1996, Porosity corrections for smectite-rich sediments: impact on studies of compaction, fluid generation, and tectonic history: Geology, v. 24, p. 843–846.

Bruce, C., 1984, Smectite dehydration—its relation to structural development and hydrocarbon accumulation in northern Gulf of Mexico Basin: AAPG Bulletin, v. 68, p. 673–683.

Colten-Bradley, V. A., 1987, Role of pressure in smectite dehydration—effects on geopressure and smectite-illite transformation: AAPG Bulletin, v. 71, p. 1414–1427.

Dutta, N. C, 1986, Shale compaction, burial diagenesis, and geopressures: a dynamic model, solution and some results, *in* J. Burrus, ed., Thermal modeling in sedimentary basins: Carcans, France, Editions Technip, p. 149–172.

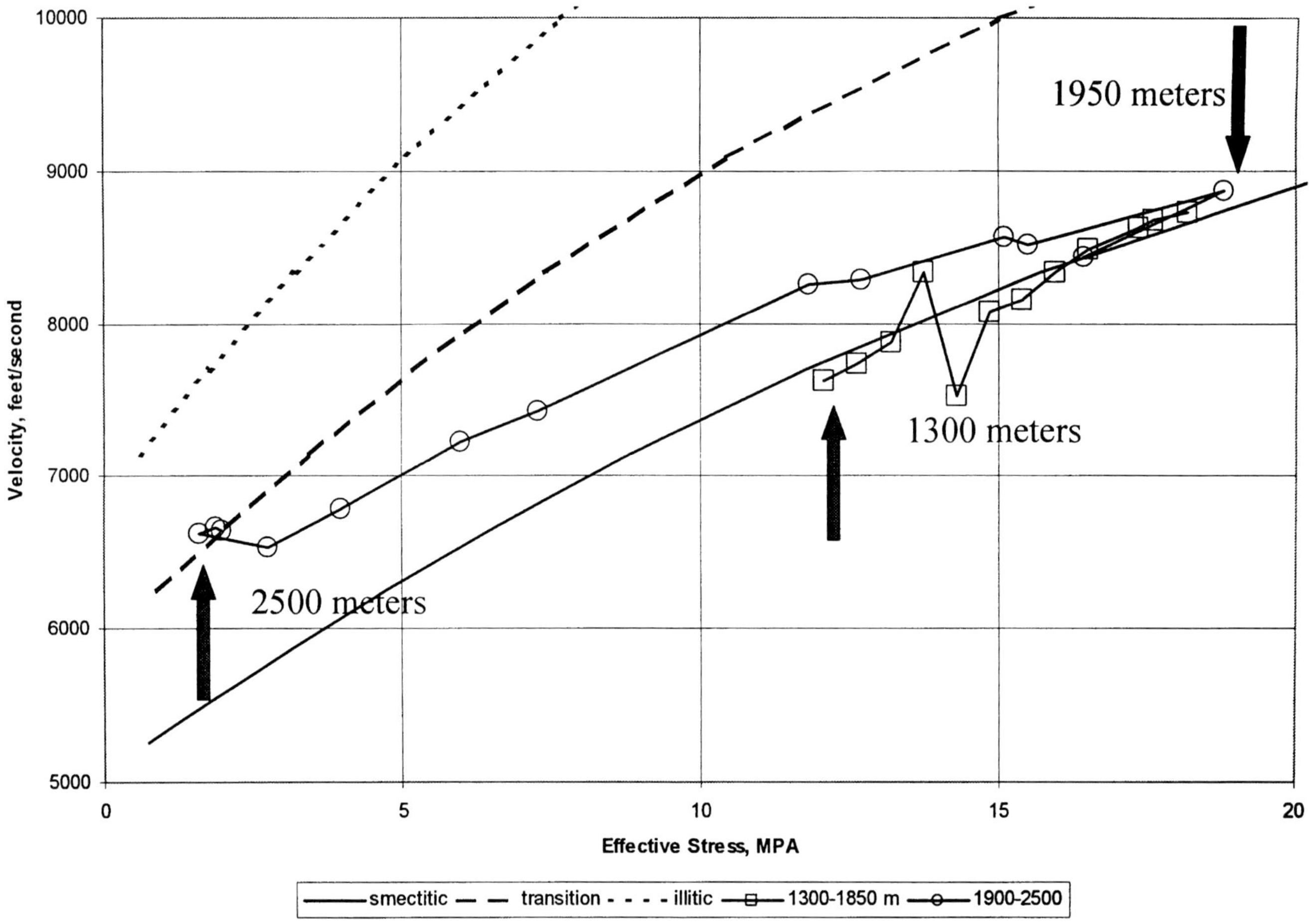

Figure 11. Effective stress/velocity crossplot for smectitic, transition, and illitic models of the Pathfinder well data. Also shown is an interpretation of the effective stress/velocity trend for the Pathfinder well (velocity and stress data taken from Hart et al., 1995), assuming top clay transition occurs at 1900 m. Note that the Pathfinder data resemble an unloading curve as described in Bowers (1994).

Gordon, D. S., and P. B. Flemings, 1998, Generation of overpressure and compaction-driven fluid flow in a Plio-Pleistocene growth-faulted basin, Eugene Island 330, offshore Louisiana: Basin Research, v. 10, p. 177–196.

Hart, B. S., P. B. Flemings, and A. Deshpande, 1995, Porosity and pressure: role of compaction disequilibrium in the development of geopressures in a Gulf Coast Pleistocene basin: Geology, v. 23, p. 45–48.

Hower, J., E. V. Eslinger, M. E. Hower, and E. A. Perry, 1976, Mechanism of burial metamorphism, 1: mineralogical and chemical evidence: Geological Society of America Bulletin, v. 87, p. 725–737.

Hunt, J. M., J. K. Whelan, L. B. Eglinton, and L. M. Cathles III, 1998, Relation of shale porosities, gas generation, and compaction to deep overpressures in the U.S. Gulf Coast, *in* B. E. Law, G. F. Ulmishek, and V. I. Slavin, eds., Abnormal pressures in hydrocarbon environments: AAPG Memoir 70, p. 87–104.

Issler, D. R., 1992, A new approach to shale compaction and stratigraphic restoration, Beaufort-MacKenzie basin and MacKenzie Corridor, northern Canada: AAPG Bulletin, v. 76, p. 289–300.

Losh, S., L. Eglinton, M. Schoell, and J. Wood, 1999, Vertical and lateral fluid flow related to a large growth fault, South Eugene Island Block 330 field, offshore Louisiana: AAPG Bulletin, v. 83, p. 244–276.

Osborne, M. J., and R. E. Swarbrick, 1997, Mechanisms for generating overpressure in sedimentary basins: a reevaluation: AAPG Bulletin, v. 81, p. 1023–1041.

Plumley, W. J., 1980, Abnormally high fluid pressures: survey of some basic principles: AAPG Bulletin, v. 64, p. 414–422.

Rubey, W. W., and M. K. Hubbert, 1959, Overthrust belt in geosynclinal area of western Wyoming in light of fluid-pressure hypothesis, 2: role of fluid pressure in mechanics of overthrust faulting: Geological Society of America Bulletin, v. 70, p. 167–205.

7

Effect of Gas on Poroelastic Response to Burial or Erosion

K. W. Katahara
BP America Inc., Houston, Texas

J. D. Corrigan
Williams Energy Services, Tulsa, Oklahoma

ABSTRACT

Where hydraulically isolated rock is either buried or erosionally unloaded, its pore pressure depends on differences in compressibility and thermal expansivity between the pore fluid and the solid skeleton. Pore pressure in gas-saturated rock changes relatively little compared to water-saturated rock because gas has a much higher compressibility than water. For the same reason, escape of pore fluid through surrounding seals affects pressure less in gas-filled than in water-filled rocks. This effect is accentuated where the rock is well consolidated and has a stiff skeleton. One implication is that erosional unloading may contribute to overpressuring in tight-gas sandstones, where gas saturation is high, the rock is stiff, and the surrounding rocks have low permeability.

INTRODUCTION

Barker (1972) suggested that thermal expansion of pore water during burial could cause overpressure. This suggestion stimulated a series of articles on the thermoelastic response of rock to burial or erosion (for instance, see Dutta [1987] for a brief review and references; Neuzil and Pollock, 1982; Shi and Wang, 1986; Luo and Vasseur, 1992, 1993; Miller and Luk, 1993). These articles primarily consider only water as the pore fluid. A notable exception was Barker (1987), who recognized that reservoirs charged early with biogenic gas would behave differently under burial than would water-saturated reservoirs. In this chapter, we use laboratory data on rock compressibility to evaluate how gas affects the development of abnormal pressures in well-consolidated rocks during burial or erosion. We conclude that the high compressibility of gas has strong effects on how pore pressure responds to changes in overburden and that these effects should be considered in models of pore-pressure evolution in tight-gas sands.

The basic idea can be illustrated by the following thought experiment. Consider a well-consolidated sandstone that is initially at 3000 m depth, at hydrostatic pore pressure (~31 MPa or 4500 psi), and encased in essentially impermeable shales. If 1500 m of overburden is eroded away, we might expect the pore volume of the sandstone to increase slightly, say by 1% more than its initial value. Ignore thermal effects for the moment, and assume that the pore space is filled with water. Because the compressibility of water is about 0.36 GPa^{-1} (2×10^{-6} psi^{-1}), a 1% increase in water volume implies a decrease in pore pressure of about 28 MPa (4000 psi). So the pore pressure at 5000 ft (1524 m) is only about 3 MPa (500 psi). If the pores are filled with gas of compressibility 10^{-4} psi^{-1}, however, the pore pressure decreases by about 1 MPa, to 30 MPa (about 4400 psi). Thus the erosion causes subnormal pressure if water is

Katahara, K. W., and J. D. Corrigan, 2002, Effect of Gas on Poroelastic Response to Burial or Erosion, in A. R. Huffman and G. L. Bowers, eds., Pressure regimes in sedimentary basins and their prediction: AAPG Memoir 76, p. 73–78.

the pore fluid and overpressure if gas is the pore fluid. Furthermore, because the compressibility of gas is so high, leakage of a small volume of pore fluid through the surrounding shales has much less effect on pressure in the gas-filled sand as compared to the water-filled sand.

ROCK COMPRESSIBILITY AND SKEMPTON'S COEFFICIENT

Assume that the rocks are undrained, that is, that changes in pore pressure occur at a rate much faster than can be relaxed by flow to or from surrounding rocks. According to Biot's theory of poroelasticity (Biot, 1941; Rice and Cleary, 1976), a change in confining stress, ΔS, in an undrained rock is accompanied by a change in pore pressure, ΔP, according to

$$\Delta P = B\ \Delta S \tag{1}$$

The parameter B is the undrained pore-pressure buildup coefficient, called "Skempton's coefficient" in soil mechanics. The value of B depends on the properties of the rock skeleton and pore fluid, and it can range from nearly 0 to about 1. In the earth, a typical overburden gradient is 3.5 kPa/m (1 psi/ft), and a typical hydrostatic pressure gradient is 1.6 kPa/m (0.45 psi/ft). A B value of 0.45 then implies that a change in overburden stress causes the pore pressure to change along a near-hydrostatic gradient, even if the rock is encapsulated in a perfect seal. If B is much less than 0.45, then the pore pressure in an isolated rock is nearly independent of overburden. If B is approximately equal to 1, then changes in pore pressure are the same as changes in overburden stress. These scenarios are shown schematically in Figure 1.

Ignore thermal effects for the moment. The isothermal coefficient B_T depends on pore-fluid properties and degree of consolidation. A B_T value equal to 1 is a good approximation in unconsolidated water-saturated sediments, but B_T can be significantly lower for well-lithified rock. Berge (1998) lists values from 0.55 to 0.99 for water-saturated sandstones. Furthermore B_T must approach 1 with vanishing porosity. Figure 2 shows B_T values estimated from laboratory measurements of sandstone static compressibility, as described in the Appendix.

The value of B_T decreases with increasing compressibility of the pore fluid and is generally between 0.5 and 1 when water is the pore fluid. The value of B_T, however, is much smaller when gas is present, as shown by the open squares in Figure 2. Oil saturation implies values intermediate between water and gas.

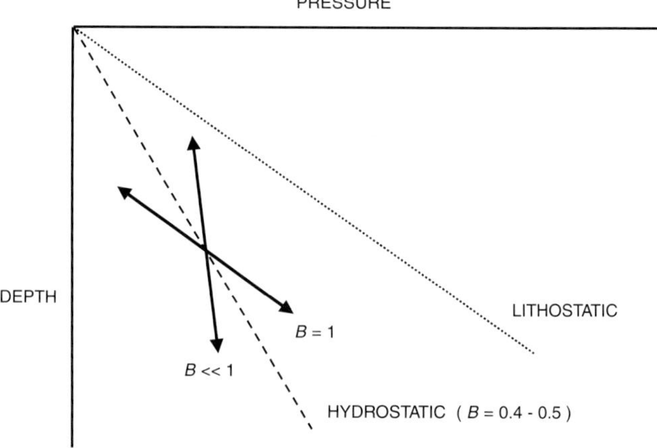

Figure 1. Schematic diagram of pressure variations with burial or erosion in undrained rocks. Where only poroelastic processes are considered, pressure is governed by Skempton's B coefficient. A low B value implies that pore pressure varies little with burial or erosion. A B value equal to 1 implies that the pore pressure varies along the lithostatic gradient. Where B is equal to the ratio of the hydrostatic gradient to the lithostatic gradient, typically 0.4–0.5, the pore pressure varies along a hydrostatic gradient.

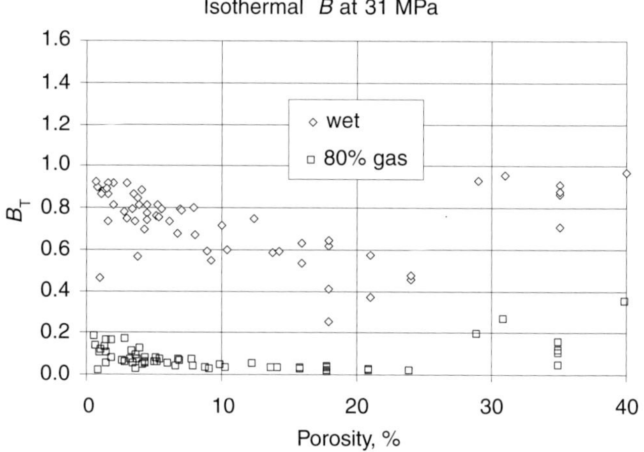

Figure 2. Skempton's isothermal coefficient B at a pore pressure of 31 MPa (4500 psi) and a temperature of 93°C (200°F) plotted against porosity. Where thermal effects are ignored, B is small for gas-bearing sands with low porosity. Because B falls below the hydrostatic range, 0.4–0.5, for tight gas sands, erosional unloading causes overpressure. For wet sands, B is generally above the hydrostatic range, which implies that unloading causes subnormal pressure. These values have been computed from laboratory data as described in the Appendix. Most of the data are for tight-gas sands. A few points have been added for unconsolidated sands to illustrate that B approaches 1 at high porosity. The B value also approaches 1 in the limit of vanishing porosity.

When the pore pressure is low, the gas compressibility is very high, so only a small fraction of the pore space needs to be occupied by gas in order for B_T to be nearly the same as for a completely gas-saturated rock. For much the same reason, seismic compressional velocities respond similarly to low gas saturation as to high gas saturation. Unlike the dynamic seismic case, however, the static pressure response described in this chapter is the same whether the gas exists as a gas cap over water or is distributed throughout the reservoir.

Now consider thermal effects. An increase in overburden stress is generally accompanied by an increase in temperature. Thermal expansion of the pore fluid, which is greater than thermal expansion of the granular skeleton, augments any pore-pressure increment due to confining stress alone. Thus B is greater along a geotherm than along an isotherm. The Appendix defines a modified coefficient, B_G, for a geothermal temperature gradient. The value of B_G has been computed for two pore-pressure (P) and temperature (T) states: first for P = 31 MPa (4500 psi) and T = 93°C (200°F), and second for P = 16 MPa (2250 psi) and T = 57°C (135°F). The overburden gradient is assumed to be 3.5 kPa/m (1 psi/ft), and the temperature gradient is assumed to be 24°C/km (1.3°F/100 ft). Figures 3 and 4 show B_G values computed for these two states.

Figure 3 shows that, at 31 MPa (4500 psi) and 93°C (200°F), B_G is roughly 0.2 for tight gas sands. This value is well below the hydrostatic range (0.4–0.5). In other words, temperature effects only partially cancel out the effects of high gas compressibility for these gas-bearing sandstones. The undrained thermoelastic response to erosion is that pore pressure decreases at about half the hydrostatic gradient. At the lower pressure and temperature shown in Figure 4, B_G is even smaller, about 0.1 for tight gas sands, well below the hydrostatic range. Undrained gas sands with low pore pressure have steeper pressure gradients than gas sands with high pore pressure.

Finally, Figure 5 shows some modeled pore-pressure trajectories for a tight-gas sand during erosion. Each trajectory starts at hydrostatic pressure at a different depth. The sand is assumed to have 12% porosity and a static drained bulk modulus of 12 GPa, independent of depth. As depth decreases with increasing erosion, pore pressure decreases along each trajectory. The rate of pressure decrease is less than hydrostatic, so excess pressure increases. The trajectories become less steep with increasing pressure because the gas compressibility decreases.

DISCUSSION AND CONCLUSIONS

Poroelastic response in gas-bearing sands is very different from the response of water-saturated sands. Although erosional unloading tends to induce subnormal pressure in isolated water sands, it tends to induce overpressure in gas sands. We have only considered the undrained case in this chapter. Obviously there must be some fluid leakage through surrounding formations. A small fluid loss (or gain) causes a large pressure drop (or increase) in water-saturated rock and may largely negate poroelastic effects. But a small

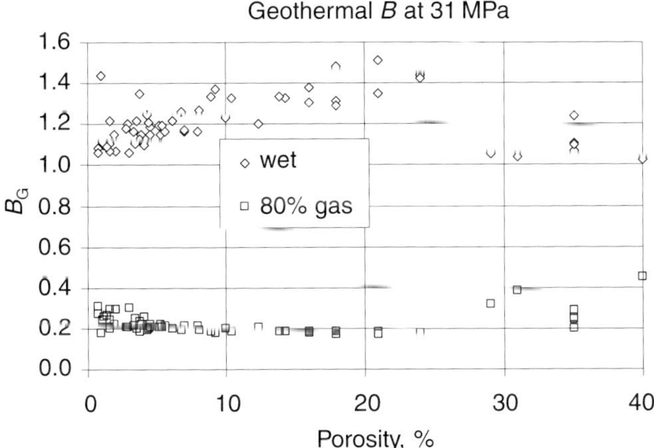

Figure 3. Skempton's coefficient B along a geotherm, at a pore pressure of 31 MPa (4500 psi) and a temperature of 93°C (200°F). In comparison to Figure 2, B values for gas sands are higher but still well below the hydrostatic range of 0.4–0.5. At this pressure, thermal effects reduce abnormal pressure in gas sands due to erosional unloading.

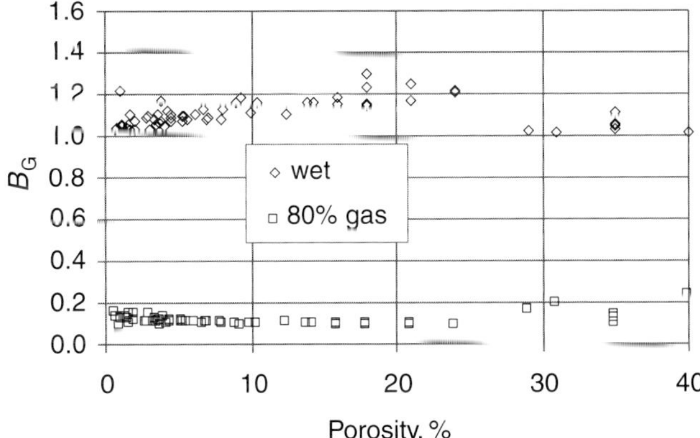

Figure 4. Skempton's coefficient, B, along a geotherm at a pore pressure of 16 MPa (2250 psi) and at a temperature of 57°C (135°F). At this pressure the gas compressibility is high enough that thermal effects are minor in comparison. For tight gas sands B is about 0.1, which is significantly below the hydrostatic range (0.4–0.5).

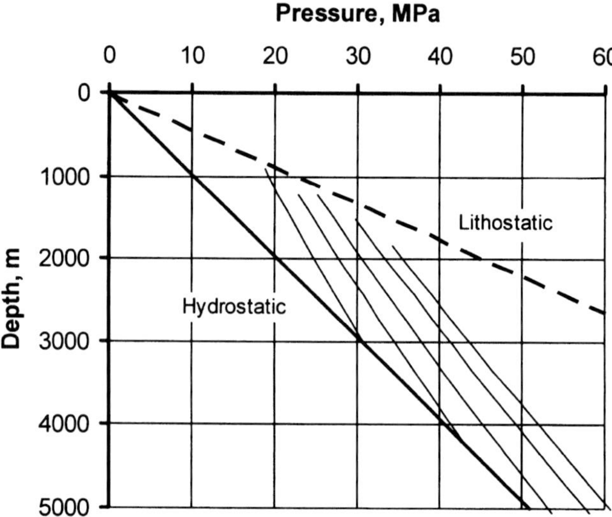

Figure 5. Pore pressure during undrained erosion for a model tight-gas sand with 12% porosity and a static drained bulk modulus of 12 GPa. Several cases are shown for different starting depths. Also shown for reference are hydrostatic (dashed) and lithostatic (heavy solid) lines. In all cases the excess pore pressure increases with erosion. The increase is steepest where the starting pore pressure is lowest because the gas compressibility is highest. The pore pressure could conceivably increase to the point where hydraulic fractures occur. Temperature was assumed to linearly increase at 0.013°F per foot from a surface temperature of 70°F.

fluid-volume loss causes relatively little pressure change in a gas reservoir because of the much greater compressibility of the gas.

We conclude that poroelastic effects in gas sands deserve further consideration. For instance, overpressure is common in tight-gas sandstone reservoirs throughout basins in the Rocky Mountain region (Spencer, 1985; 1989). Active charging of these reservoirs by ongoing generation of gas from adjacent coal or carbonaceous shale formations has been invoked to explain the development of overpressure in such reservoirs (e.g., Law and Dickinson, 1985; Spencer, 1987; Surdam et al., 1994). Rocky Mountain area basins, however, have experienced net erosion during the post-Laramide (post-Eocene) (Burgess et al., 1997). Estimates of overburden removed for some of these basins are on the order of several thousand feet (e.g., Nuncio and Johnson, 1984; Nuncio, 1990; Naeser, 1992). Active gas generation from source facies would have essentially shut down as a result of decreasing formation temperature upon commencement of erosion. Consequently, development and maintenance of overpressure via active, or recently active, gas charging seems unlikely for tight-gas sands in these basins. Poroelastic effects may contribute to, or largely be responsible for, overpressuring developed in tight-gas sands in these and other basins subjected to recent erosional unloading. This effect would be most accentuated in gas-bearing, well-consolidated sandstone bodies with low B values encased by very low permeability shale that have been subjected to rapid decrease in overburden stress. Poroelastic underpressuring effects are less likely to be significant during burial because effective stress tends to increase to the point where inelastic compaction occurs.

We have ignored several effects that should be considered in a more complete treatment. These include pressure- and temperature-dependent solubility of gas in brine, the difference between overburden stress and mean confining stress, and possible differences in B value at laboratory vs. geological strain rates. A realistic treatment requires careful consideration of other mechanisms such as gas generation and disequilibrium compaction and of how rates of erosion or burial balance off against fluid flow through sealing formations.

ACKNOWLEDGMENTS

We thank Vastar and ARCO for permission to publish, Alton Brown, Chuan Yin, Steve Crews, and Bill Kilsdonk for valuable comments, and Pat Berge for a timely preprint.

APPENDIX

The parameter B in equation 1 is defined (Rice and Cleary, 1976) to be

$$B_T = \frac{1}{1 + \phi\left(\frac{\beta_f - \beta_\phi}{\beta - \beta_s}\right)} \tag{2}$$

where β_f = isothermal pore-fluid compressibility; β_s = isothermal solid grain compressibility; β_ϕ = isothermal pore-space compressibility at constant differential stress; β = isothermal bulk rock drained compressibility. The parameter B_T is known as Skempton's coefficient, or as the undrained isothermal pore-pressure buildup coefficient. In the special case where the rock is monomineralic and the only mineral is elastically isotropic, $\beta_\phi = \beta_s$, so that

$$B_T = \frac{1}{1 + \phi\left(\frac{\beta_f - \beta_s}{\beta - \beta_s}\right)} \tag{3}$$

For well-lithified rocks of primary interest in this chapter, $\beta_s < \beta < \beta_f$. Where gas is present, $\beta << \beta_f$, which implies $B_T << 1$, except when porosities approach 0.

Recent studies indicate that β_ϕ may be significantly higher than β_s (e.g., Berge et al., 1993; Berge, 1998). Thus B_T values computed using $\beta_\phi = \beta_s$ may be too low. Although this effect may be significant for wet rocks, our computations indicate that the effects of $\beta_\phi > \beta_s$ are negligible for gas-bearing rocks.

Because gas compressibility varies strongly with temperature and pressure, B_T for gas-bearing rocks also depends on pressure and temperature. It will be higher for lower temperatures and higher pressures.

Where temperature is not constant but varies along a geotherm, a modified geothermal B coefficient can be defined as (e.g., see Miller, 1995, equation 4)

$$B_G = B_T\left(1 + \phi\frac{\alpha_f - \alpha}{\beta - \beta_s}\frac{\Delta T}{\Delta S}\right) \tag{4}$$

where α_f is the thermal expansivity of the fluid, α is the thermal expansivity of the drained rock, and ΔT is the temperature increment corresponding to an increment of confining stress, ΔS, during burial or erosion.

Measured β values were from Jizba (1991), Tutuncu and Sharma (1992), Tutuncu et al., (1993), G. G. Ramos (1995, unpublished data), and T. E. Scott (1995, unpublished data). Fluid properties were computed as in Batzle and Wang (1992), assuming a gas of gravity 0.6 and water salinity of 50,000 ppm NaCl. The expression $\alpha = 3 \times 10^{-5}{}^\circ C^{-1}$ was assumed for sandstones (Miller, 1995).

The previous computations have also neglected the solubility of gas in water or oil and the dependence of that solubility on pressure and temperature. Solubility of gas in water increases with increasing pressure and decreasing temperature.

REFERENCES CITED

Barker, C., 1972, Aquathermal pressuring: role of temperature in development of abnormal pressure zones: AAPG Bulletin, v. 56, p. 2068–2071.

Barker, C., 1987, Development of abnormal and subnormal pressures in reservoirs containing bacterially generated gas: AAPG Bulletin, v. 71, p. 1404–1413.

Batzle, M. L., and Z. Wang, 1992, Seismic properties of pore fluids: Geophysics, v. 57, p. 1396–1408.

Berge, P. A., 1998, Pore compressibility in rocks *in* J.-F. Thimus, Y. Abousleiman, A. H.-D. Cheng, O. Coussy, and E. Detournay, eds., Poromechanics: Rotteram, Balkema, p. 351–356.

Berge, P. A., H. F. Wang, B. P. Bonner, 1993, Pore pressure buildup coefficient in synthetic and natural sandstones: International Journal of Rock Mechanics and Mining Sciences, v. 30, p. 1135–1141.

Biot, M. A., 1941, General theory of three-dimensional consolidation: Journal of Applied Physics, v. 12, p. 155–164.

Burgess, P. M., M. Gurnis, and L. Moresi, 1997, Formation of sequences in the cratonic interior of North America by interaction between mantle, eustatic, and stratigraphic processes: Geological Society of America Bulletin, v. 108, p. 1515–1535.

Dutta, N. C., 1987, Editor's introduction to chapter 2, *in* N. C. Dutta, ed., Geopressure: Tulsa, Society of Exploration Geophysicists, p. 19–83.

Jizba, D. L., 1991, Mechanical and acoustical properties of sandstones and shales: Ph.D. dissertation, Stanford University, Stanford, California.

Law, B. E., and W. W. Dickinson, 1985, Conceptual model for origin of abnormally pressured gas accumulations in low-permeability reservoirs: AAPG Bulletin, v. 69, p. 1295–1304.

Luo, X., and G. Vasseur, 1992, Contributions of compaction and aquathermal pressuring to geopressure and the influence of environmental conditions: AAPG Bulletin, v. 76, p. 1550–1559.

Luo, X., and G. Vasseur, 1993, Contributions of compaction and aquathermal pressuring to geopressure and the influence of environmental conditions: reply: AAPG Bulletin, v. 77, p. 2011–2014.

Miller, T. W., 1995, New insights on natural hydraulic fractures induced by abnormally high pore pressures: AAPG Bulletin, v. 79, p. 1005–1018.

Miller, T. W., and C. H. Luk, 1993, Contributions of compaction and aquathermal pressuring to geopressure and the influence of environmental conditions: discussion: AAPG Bulletin, v. 77, p. 2006–2010.

Naeser, N. D., 1992, Miocene cooling in the southwestern Powder River basin, Wyoming: preliminary evidence from apatite fission-track analysis: U.S. Geological Survey Bulletin, B 1917-O, p. O1–O17.

Neuzil, C. E., and D. W. Pollock, 1982, Erosional unloading and fluid pressures in hydraulically "tight" rock: Journal of Geology, v. 91, p. 179–193.

Nuncio, V. F., 1990, Burial, thermal, and petroleum generation history of the Upper Cretaceous Steele Member of the Cody Shale (Shannon Sandstone Bed horizon), Powder River basin, Wyoming: U.S. Geological Survey Bulletin, B 1917-A, p. A1–A17.

Nuncio, V. F., and R. C. Johnson, 1984, Thermal maturation and burial history of the Upper Cretaceous Mesaverde Group, including the Multiwell Experiment (MWX), Piceance Creek basin, Colorado, *in* C. W. Spencer and C. W. Keighin, eds., Geologic studies in support of the U.S. Department of Energy Multiwell Experiment, Garfield County, Colorado: U.S. Geological Survey Open-File Report 84-757, p. 102–109.

Rice, J. R., and M. P. Cleary, 1976, Some basic stress diffusion solutions for fluid-saturated elastic porous media with compressible constituents: Reviews of Geophysics and Space Physics, v. 14, p. 227–241.

Shi, Y., and C.-Y. Wang, 1986, Pore pressure generation in sedimentary basins: overloading versus aquathermal: Journal of Geophysical Research, v. 91, p. 2153–2162.

Spencer, C. W., 1985, Geologic aspects of tight gas reservoirs in the Rocky Mountain region: Journal of Petroleum Technology, v. 37, p. 1308–1314.

Spencer, C. W., 1987, Hydrocarbon generation as a mechanism for overpressuring in Rocky Mountain region: AAPG Bulletin, v. 71, p. 368–388.

Spencer, C. W., 1989, Review of characteristics of low-permeability gas reservoirs in western United States: AAPG Bulletin, v. 73, p. 613–629.

Surdam, R. C., Z. S. Jiao, and R. S. Martinsen, 1994, The regional pressure regime in Cretaceous sandstones and shales in the Powder River basin, *in* P. Ortoleva and Z. Al-Shaieb, eds., Pressure compartments and seals: AAPG Memoir 61, p. 213–233.

Tutuncu, A. N., and M. M. Sharma, 1992, Relating static and ultrasonic laboratory measurements to acoustic log measurements in tight gas sands: 67th Annual Technical Conference and Exhibition of the Society of Petroleum Engineers, Society of Petroleum Engineers paper 24689, p. 299–311.

Tutuncu, A. N., A. L. Podio, and M. M. Sharma, 1993, Strain amplitude and stress dependence of static moduli in sandstones and limestones, *in* P. P. Nelson and S. E. Laubach, eds., Rock mechanics: Rotterdam, Balkema, p. 489–496.

8

Relationships between Pore Pressure and Stress in Different Tectonic Settings

Najwa Yassir
Rijswijk, Netherlands
formerly with CSIRO Petroleum,
Melbourne, Australia

M. Anthony Addis
Shell, SIEP,
Rijswijk, Netherlands;
formerly with CSIRO Petroleum,
Melbourne, Australia

ABSTRACT

This chapter discusses the effect of different overpressure mechanisms and tectonic settings on the pore-pressure–stress relationship. We demonstrate that the two are intrinsically linked and that a change of one affects the other. Vertical and horizontal stress increases result in overpressuring in normally consolidated low-permeability sediments; overpressuring due to tectonic stresses can be far higher than that generated by rapid sedimentary loading. Pore-pressure increase, however, causes a change in the stresses and fracture gradient if deformation is constrained in any direction. We present a theoretical model, based on field observations, which suggests that overpressuring by a fluid source tends to render the stresses more isotropic. The variability of porosity and pore-pressure–stress relationships for different overpressure mechanisms and tectonic settings means that methods that consider risk and uncertainty in pressure-fracture gradient prediction need to be developed for geologically complex areas.

INTRODUCTION

The knowledge of pore-pressure and fracture gradient (minimum principal stress) is crucial in all aspects of the upstream petroleum industry. In exploration, a major consideration in bidding for a permit in an overpressured area is the so-called window between the pressure and fracture gradients. If the pressure is high enough, it can approach the minimum stress and result in seal breach by fracture reopening or possibly by natural hydraulic fracturing. In drilling, it is important to define the window between the two to design casing points and a safe mud weight that will prevent blowouts (during underbalanced drilling) and mud losses (during overbalanced drilling). An understanding of the relationship between pore pressure and stress is also important during production. Depletion of overpressured reservoirs results in dramatic stress changes that have an impact on reservoir productivity, subsidence, seismicity, and, in some cases, well integrity.

Overpressuring is known to occur by several different mechanisms related to burial, tectonism, hydrocarbon generation, mineral transformation, and temperature

Yassir, Najwa, and M. Anthony Addis, 2002, Relationships between Pore Pressure and Stress in Different Tectonic Settings, in A. R. Huffman and G. L. Bowers, eds., Pressure regimes in sedimentary basins and their prediction: AAPG Memoir 76, p. 79–88.

increase, among other sources. The geological details of most of the individual mechanisms are exhaustively described in the literature and are not discussed in detail here (see reviews by Fertl, 1976; Mouchet and Mitchell, 1989; Yassir, 1989; Osborne and Swarbrick, 1997). The main topic of this chapter is the coupled relationship between the pore-fluid pressure (P_P) and fracture gradient (FG), which varies according to overpressure mechanism. The impact of this relationship on sediment porosity is also discussed.

Overpressure mechanisms are divided into two categories in this chapter: overpressuring caused by stress (including sedimentary loading and tectonic loading) and overpressuring caused by a fluid source or fluid expansion (including hydrocarbon generation, osmosis, smectite dehydration, and thermal pressuring).

STRESSES CAUSING OVERPRESSURE

Two types of overpressure are caused by increases in stresses: undercompaction that results from rapid vertical loading (commonly sedimentary), and tectonic loading, leading to undrained shear. Undercompaction is by far the best understood of the overpressure mechanisms and is predominantly used to explain and quantify overpressures. Tectonic loading, however, is poorly understood and has been underrated in the literature, mainly because it is a difficult mechanism to quantify. Emphasis in this chapter is therefore placed on how overpressure generated by tectonic loading is different from undercompaction and on its possible manifestations in exploration.

Rapid Vertical Loading

Overpressuring is commonly associated with Tertiary basins where rapid deposition and subsidence occur, such as the Mississippi, Orinoco, and Niger delta regions (type areas for the development of pore-pressure prediction methods). In these regions, the fluid pressure (P_P) increases in response to an increase in total vertical stress (σ_V) in low-permeability sediments. According to Terzaghi's equation, an increase in σ_V is taken up partly by the rock matrix and partly by the pore fluid (Terzaghi and Peck, 1948),

$$\sigma_V = \sigma_V' + P_P \tag{1}$$

assuming a Biot constant of 1, where σ_V' is the effective vertical stress.

In the extreme case, where vertical loading occurs without any fluid escape, the load is taken up totally by the pore fluid, and the pressure response is expressed as

$$C = \Delta P_P \ / \ \Delta\sigma_V \tag{2}$$

C is a constant for uniaxial strain conditions, which is related to sediment saturation and system compressibility (see Lambe and Whitman, 1979). C commonly has values of 1 for saturated clays (Lambe and Whitman, 1979) and greater than 0.95 for shales (Yassir, 1989).

This mechanism of overpressuring is most common in young sediments where anomalously high porosities for depth of burial are preserved. Older, normally consolidated, sediments can become overpressured in a similar manner, however. Rapid vertical loading by tectonic subsidence or overthrusting is mechanistically akin to undercompaction in that the addition of vertical load is reflected by a pore-pressure increase. Although the original porosity may have been normal, the sudden addition of load renders the sediment technically undercompacted for its depth of burial. If the mechanism also involves changes in the lateral stresses (and therefore the shear stress in the rock), however, this too will have an important effect on the pore pressure, as discussed in the following section.

Tectonic Loading

Tectonic loading, long recognized as a potential cause of overpressuring (e.g., Higgins and Saunders, 1974; Unruh et al., 1992), is far less considered in the literature as an overpressure mechanism than sedimentary loading, even in tectonically active basins. Yet global occurrences of overpressuring show, with a few notable exceptions, a strong relationship between overpressure and present-day compressional tectonics (Figure 1). Some examples include Trinidad (Higgins and Saunders, 1974), Papua New Guinea (Hennig et al., 2002), California (Unruh et al., 1992) and the Gulf or Alaska (Hottman et al., 1979). In the Gulf of Alaska, the onset of overpressuring coincides with a stress reversal from normal faulting (or strike slip) to compressional faulting (Hottman et al., 1979). The pressures in the San Andreas fault region have a distribution that is strongly associated with the geometry of the fault and is thought by some workers to be related to fault activity (e.g., Unruh et al., 1992). The superlithostatic ($P_P > \sigma_V$) overpressures observed in tectonically active regions (e.g., Bigelow, 1994) and the eruption of deep-seated mud volcanoes in earthquake regions (Figure 1) are further indications of the strong relationship between tectonic stress and overpressure.

Figure 1. Schematic map of global overpressure occurrences (shading), areas of Cenozoic folding (lines), and deep-seated mud volcanoes (triangles). Modified from Mouchet and Mitchell (1989), Yassir (1989), and Higgins and Saunders (1974).

Recently, overpressure models have begun to consider tectonic stresses (Van Balen and Cloetingh, 1995). To date, however, the models treat the pressure response to a horizontal load in the same manner described in equations 1 and 2, that is, the pressure increase is caused by a unidirectional increase in stress, in this case horizontal rather than vertical. The important difference between the two mechanisms, however, is the shear stress—the difference between maximum and minimum total stresses ($\sigma_{MAX} - \sigma_{MIN}$). In the case of sedimentary loading in a passive basin, the sediments are laterally constrained by the basin, so an increase in vertical stress (σ_V) is countered by a partial increase in horizontal stress (σ_H). Under such conditions, the sediments do not experience critical shear stresses (shear failure) with increasing vertical stress (Figure 2a). With lateral tectonic loading, however, the sediments experience very high shear stresses because the horizontal stress increases without significant constraint in the vertical stress (overburden) direction. The shear stresses are capable of generating pore pressures greater than those generated by undercompaction (Figure 2b). To our knowledge, the relationship between shear stress and pore pressure is not considered in overpressure models.

Pressure Response to Shear Stress and the Concept of *A* Values

In a compressional basin ($\sigma_1 = \sigma_H$), the sediments undergo changes in the shear stress that, if undrained, also result in a pore-pressure change described by

$$A = (\Delta P_P - \Delta\sigma_3)/(\Delta\sigma_1 - \Delta\sigma_3) \quad (3)$$

Or, if we assume that the overburden weight remains constant during shearing, the shear stress can increase (to failure) through an increase in σ_H, and the relationship simplifies to

$$A = \Delta P_P \; / \; \Delta\sigma_H \quad (4)$$

where A is Skempton's parameter describing the pore-pressure response to shear stress (see Lambe and Whitman, 1979). The parameter A is not a constant, and, unlike the parameter C for most rocks, it commonly exceeds 1, that is, the pore-pressure increment can exceed the increment of applied load ($\Delta\sigma$). Where a low-permeability sediment is subjected to shearing, the sediment structure or matrix deforms without significant fluid escape. This transfers some of the load to the fluid and results in overpressuring (Yassir, 1990). The higher the porosity, the greater the potential for the rock matrix to rearrange itself under shear and to generate larger pore-pressure responses.

Figure 3 shows typical experimentally derived values of A for different sediment types: claystones (Yassir, 1990), a smectitic shale (Wu, 1987), a cemented shale (Ohtsuki et al., 1981), high porosity chalks (Leddra and Jones, 1989), and sand (Bishop et al., 1965). In these tests, the sediment was consolidated isotropically ($\sigma_1' = \sigma_3'$) or anisotropically ($\sigma_3'/\sigma_1' = K$) to a certain mean effective stress before being subjected to shear stresses in an undrained state (no fluid escape) until failure is reached. Each point in the figure plots the ratio of the total pore-pressure change divided by the total applied shear stress versus the initial mean stress.

All the sediment types show a positive A value, which means that the application of shear stress to a

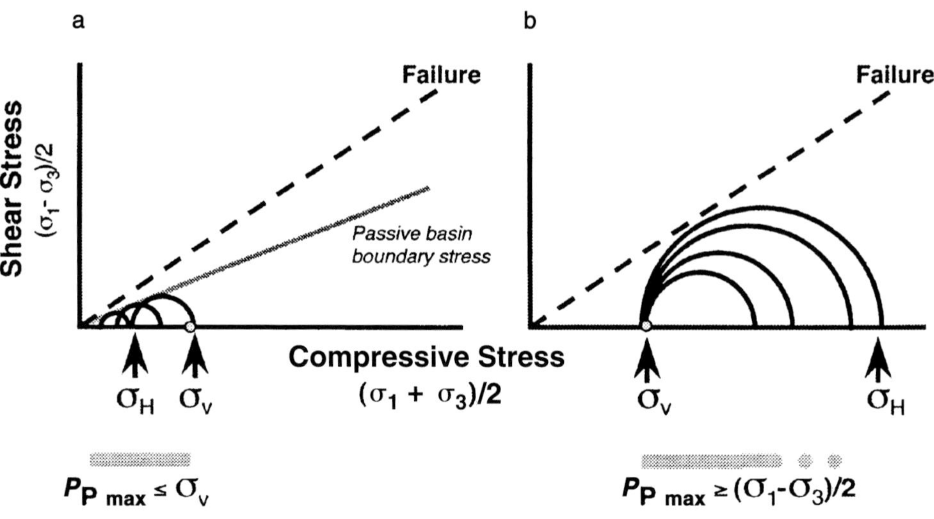

Figure 2. Schematic illustration of the different shear stresses ($\sigma_{MAX} - \sigma_{MIN}$) acting in a passive basin and in a compressional basin. (a) Sedimentary loading; (b) tectonic loading. In the passive basin, the increase in both vertical and horizontal stress means that no critical shear stresses are reached and the pore pressures are controlled by the maximum vertical stress. In the compressional basin, the increase in horizontal stress and the relatively constant vertical stress means that the sediment can reach critical shear stresses. The pore-pressure increase in this case is related to these shear stresses and can therefore be very high.

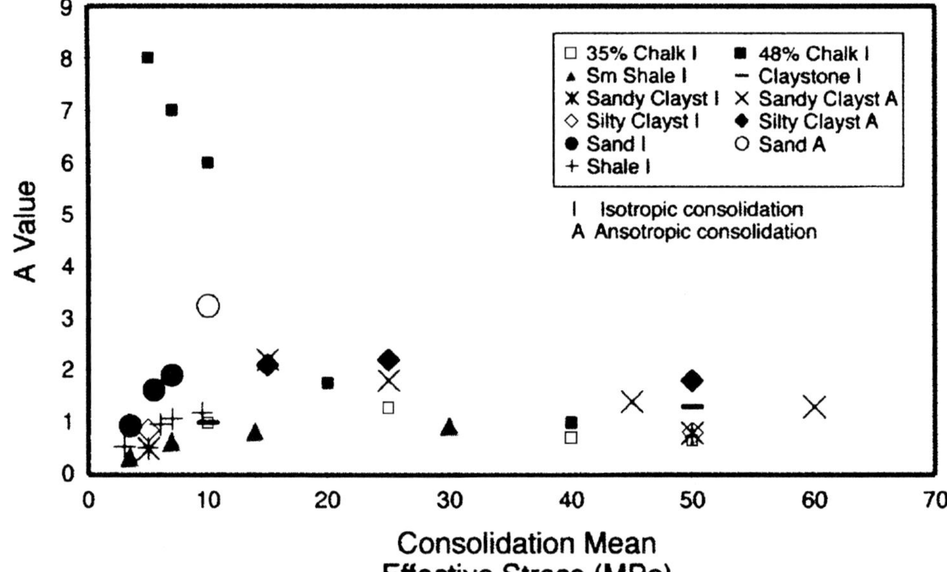

Figure 3. Variation in A values with mean effective stress for a variety of sediment types: claystones (Yassir, 1990), an overconsolidated smectitic shale (Wu, 1987), a cemented shale (Ohtsuki et al., 1981), high-porosity chalks (Leddra and Jones, 1989), and sand (Bishop et al., 1965).

range of sediments in an undrained state results in overpressuring. Many of the results in Figure 3 give A values far greater than 1, that is, an increment of pore-pressure increase greater than the increment of applied shear stress. In the case of the chalks at low mean effective stresses (high porosities), extremely high A values are recorded (up to 8), reverting back to around 1 at the higher consolidation stresses. Figure 3 clearly shows that overpressuring is achieved not only in high-porosity sediments but also in low-porosity sediments consolidated to high mean effective stresses. A comparison between isotropic and anisotropic loading in the sand and the claystones indicates that the stress path during consolidation also has an effect on the A value, the anisotropically loaded samples giving the higher A values (>1).

An important implication of the data in Figure 3 is that large overpressures can be generated by shearing, even in low-porosity normally consolidated materials. Bearing in mind that the stresses in a compressional sedimentary basin are far higher than those in a passive basin, the shear-induced pore-pressure magnitudes relative to the overburden stress are higher.

To illustrate this, take a normally consolidated sediment buried to a depth of 1500 m in a passive basin; the vertical stress is approximately 33.9 MPa and the hydrostatic pressure is 15.3 MPa. If undercompaction starts at this depth the overpressure increases monotonically with the increase in the vertical total stress ($\Delta P_P/\Delta\sigma_V = C = 1$; equation 2); it never reaches the total vertical (overburden) stress. The application of a horizontal stress, which takes the basin to inversion, however, results in an increase in shear stress and accompanying pore pressure at a constant depth. If we take a lower bound fault friction angle of 15° and a Poisson's ratio of 0.35 for the shale, the expected horizontal stress increase is $1.15\sigma_V'$ (Addis et al., 1996). Using an A value of 1, the overpressures at 1500 m depth are 36.9 MPa, that is, exceeding the overburden (vertical) stress at that depth. For the undercompaction case to reach the same overpressure, undrained burial to 2460 m is required.

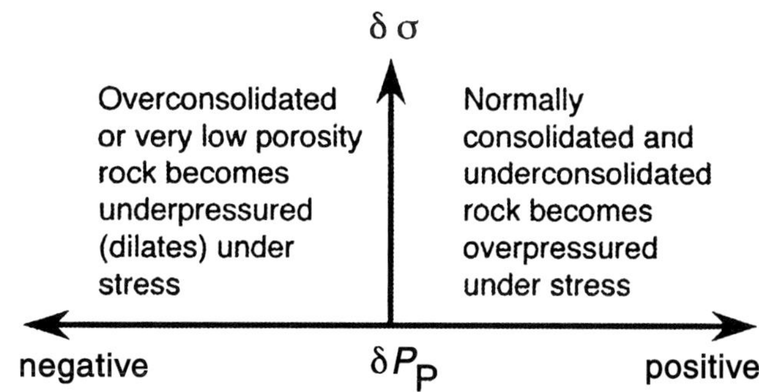

Figure 4. Schematic illustration of the relationship between stress and pore pressure during undrained loading. If the sediment is normally or underconsolidated, the pore-pressure response is positive; if it is overconsolidated, the response is negative, leading to dilation.

Liquefaction: A Manifestation of Overpressure by Shear Stress

A values far exceeding 1 are observed in Figure 3. This behavior is related to the phenomenon of pore collapse (Addis, 1987), where the matrix of a high-porosity, commonly cemented, sediment collapses under stress. Where this occurs in an undrained state, or under rapid cyclic loading, such as in an earthquake, all the stress carried by the structure or matrix is transferred to the pore fluid, generating overpressures and causing liquefaction (see Lambe and Whitman, 1979). At the surface, liquefaction is observed in loose sands and sensitive clays (see Lambe and Whitman, 1979), and the excess pressure is quickly dissipated. At depth, however, where permeabilities decrease significantly, the generated overpressure could be preserved. Evidence of liquefaction at depth includes chalks in the North Sea (Addis, 1987; Leddra and Jones, 1989). It is also the mechanism by which many deep-seated mud volcanoes erupt (Yassir, 1989). Deep-seated mud volcanoes are almost always associated with areas of Cenozoic folding (Higgins and Saunders, 1974). They mostly appear along compressional faults and fold axes, for example, Trinidad (Higgins and Saunders, 1974), New Zealand (Ridd, 1970), and Azerbaijan (Jakubov et al., 1971). Their eruption is also commonly coincident with earthquakes (Ridd, 1970; Jakubov et al., 1971).

A Note on Loading of Overconsolidated Sediments

The previously mentioned relationships between loading and pore-pressure increase apply only to normally consolidated or underconsolidated sediments. Sediments that have experienced higher stresses/depths than present have abnormally low porosities for depth of burial and are termed overconsolidated. These sediments cannot be overpressured by rapid loading or undrained shear unless the imposed stress change exceeds the historic stresses. In fact, undrained shear of overconsolidated sediment results in dilation and therefore a reduction in pore pressure (Figure 4) (see Lambe and Whitman, 1979).

The rock mechanics literature has, until recently, largely concentrated on strong, low-porosity rocks, which display this dilatant behavior. This has, unfortunately, colored our view of the importance of shearing as an overpressure mechanism in sedimentary basins.

The previous section dealt with overpressure mechanisms where a stress increase results in overpressuring. The next section deals with the opposite phenomenon of overpressuring by a fluid source causing an increase in fracture gradient.

OVERPRESSURE CAUSING STRESS CHANGES

Fluid source overpressure mechanisms are related to fluid generation at depth, either by hydrocarbon generation, smectite dehydration, or any such internal pressuring mechanism. These overpressure mechanisms also show a pressure-stress relationship, but unlike the sedimentary and tectonic loading cases, the pressure is not generated by stress, but instead causes a change in the total stress regime (Yassir and Bell, 1994, 1996; Engelder and Fischer, 1994). According to Terzaghi's effective stress equation (equation 1), a change in pore pressure causes a change in the effective stress, without affecting the total stress, if strains are allowed in all directions. Under field boundary conditions where the rock is constrained in a direction, however, the pore-pressure changes can also affect the total stresses. The simplest example is the elastic zero-lateral strain ($\varepsilon_H = 0$) boundary assumption commonly applied to passive basins. The assumption is not entirely accurate for passive basins, but it is used to illustrate the effect of purely vertical compaction. In this case, the increase in pore pressure results in a partial increase in horizontal stress, as defined by the relationship in equation 5

$$\sigma_h = \left(\frac{\nu}{1-\nu}\right)\sigma_v + P_P\left(\frac{1-2\nu}{1-\nu}\right) \quad (5)$$

where ν = Poisson's ratio.

The horizontal stress (commonly the fracture gradient in sedimentary basins) increases in this case because the rock cannot strain (or so-called bulge) in the horizontal direction to accommodate the pore-pressure increase and the accompanying volume increase in the rock unit. Vertically, the strain boundary is free, and, in the absence of structural effects, there is no corresponding increase in total vertical stress.

Qualitatively there is a great deal of evidence in the literature to suggest that overpressured rock in sedimentary basins is associated with a minimum horizontal stress increase (Breckels and van Eekelen, 1981; Yassir and Bell, 1994; Addis et al., 1996; Engelder and Fischer, 1994). Evidence also exists that underpressured rock is associated with abnormally low horizontal stress (Breckels and van Eekelen, 1981). Field measurements of reservoir pressures and fracture gradients during depletion further illustrate the decrease in fracture gradient with decreasing pore pressure (Teufel et al., 1991; Addis, 1997).

The passive-basin equation (equation 5) assumes that the horizontal stresses are isotropic. Yet it is known that in most sedimentary basins, there is significant anisotropy in the horizontal stresses ($\sigma_H > \sigma_h$), as indicated by the occurrence of faulting. Here we consider the effect of overpressuring by a fluid source at depth on normal faulting (stress system causing fault: $\sigma_v > \sigma_H > \sigma_h$) and thrust faulting (stress system causing fault: $\sigma_H > \sigma_h > \sigma_V$). In both cases, it is assumed that the rock block is bounded by an active fault, with an additional tectonic component of the horizontal stress resulting from fault friction, which has to be overcome before the fault is mobilized. The same pressure in the fault plane and intact rock, a plane strain boundary normal to the fault and a constant vertical stress are also assumed. Using elasticity theory and the Mohr-Coulomb failure criterion for slippage on the fault, the relationship between horizontal stresses and pore pressure during mobilization of normal and thrust faulting is (Addis et al., 1996)

$$\text{Normal faulting:} \quad \frac{\sigma_H}{\sigma_h} = \frac{2\nu}{1-\sin\Phi}\left(1-\frac{P_P}{\sigma_h}\right)+\frac{P_P}{\sigma_h} \quad (6)$$

$$\text{Thrust faulting:} \quad \frac{\sigma_H}{\sigma_h} = \frac{1}{\nu(K_p+1)}\left\{K_p-\frac{P_P}{\sigma_h}[K_p(1-\nu)-\nu]\right\} \quad (7)$$

where Φ = internal friction angle of the fault; K_p = Mohr-Coulomb passive coefficient = $(1 + \sin\Phi)/(1 - \sin\Phi)$; (Fault cohesion is assumed to be zero in both equations).

The horizontal stress–pore-pressure relationship is illustrated for both normal and thrust faulting in Figure 5a and b, respectively. In both cases, the ratio of horizontal stresses becomes more isotropic with increasing pore pressure until $P_P = \sigma_V$, at which point all stresses become isotropic. In the normal faulting case, the stresses increase with pore-pressure increase, but, depending on the fault friction angle and on the orientation of σ_H, the maximum horizontal stress can be parallel or perpendicular with the fault (Figure 5a). In the thrust-faulting case, the stresses decrease with pore-pressure increase, to converge with the vertical stress (Figure 5b).

This illustrates, using a simple mechanical model, the potential integration of pore-pressure and fracture gradients in faulted regimes.

LIMITS TO PORE-PRESSURE INCREASE AND FLUID MIGRATION

The minimum stress is generally accepted as the upper limit to the pore pressure in a rock (Grauls, 1997). The minimum stress is the minimum horizontal stress (σ_h) in normal and strike-slip faulting regimes and can be σ_V or σ_h in a thrust-fault regime (Addis et al., 1996). As illustrated previously for normal-faulting and strike-slip faulting regimes, however, where assuming rock elasticity, the pore pressure in a layer can increase to the value of the vertical stress, causing a corresponding increase in σ_h in that layer. This suggests that it is potentially difficult for the pore pressure to reach the fracture pressure. The theory is qualitatively supported by field observations of high fracture gradients in association with high pore-pressure gradients in tectonically passive basins (Breckels and van Eekelen, 1981; Yassir and Bell, 1994). Yet, it is known from leak-off and hydraulic fracture tests that rock can be hydraulically fractured at pressures far lower than σ_V, and from field studies, that fluid pressures can frac-

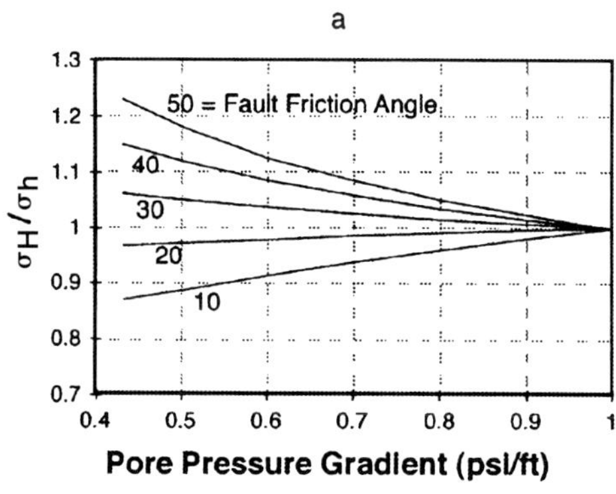

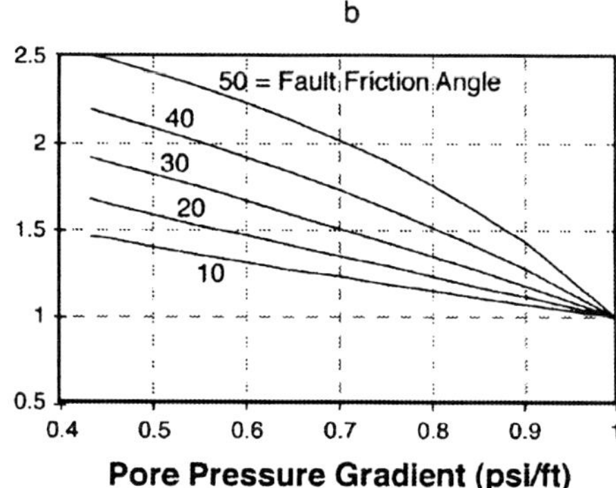

Figure 5. The relationship between horizontal stress ratio (σ_H/σ_h) and pressure gradient for (a) normal faulting and (b) thrust faulting for different values of fault friction angle (Φ). Here, σ_H is assumed to be the stress parallel with the fault plane; under certain friction angles, it can become smaller than the horizontal stress normal to the fault (σ_h) (Addis et al., 1996). σ_V gradient = 1 psi/ft; Poisson's ratio $\nu = 0.3$.

ture, or open fractures, in the overburden, leading to massive fluid migration.

In this discussion, it is important to distinguish between P_P-FG relationships (1) on a regional vs. a localized scale and (2) in different sediment layers. In a leak-off test a local rapid increase in drilling mud pressure in the wellbore does not affect the far-field minimum stress; a fracture opens where the minimum stress (fracture pressure) is reached, as long as the tensile strength of the rock is exceeded (Enever et al., 1996). If the pore pressure increases in a sedimentary layer, by hydrocarbon generation, for example, this results in a corresponding increase in the fracture gradient in that layer. The fracture gradient of the normally pressured overlying rocks can remain unaffected. This means that hydraulic fracturing of the overburden occurs if the overpressure in the layer exceeds the minimum stress and tensile strength of the overburden. This is evoked as an effective method of hydrocarbon leakage in petroleum basins (Grauls, 1997). To summarize, although pore pressure is coupled with the fracture gradient within one layer, it is uncoupled with the fracture gradient of the overburden. It is the minimum stress in the overburden that controls the maximum pore pressure in a layer.

Furthermore, on a regional scale, the horizontal stress is controlled not only by the pore pressure, but also by a tectonic (or structural) component. The minimum stress in the overburden can be tectonically reduced, causing fracturing, seal breach, and massive migration of overpressured fluids (J. Cosgrove, 1996, personal communication).

A note should be made here of references in the literature to pressures exceeding the overburden stress. For example, in the Himalayan foothills in Pakistan, pressures reach 7000 psi at a depth of 1646 m (Bigelow, 1994), amounting to a gradient of 1.3 psi/ft—potentially 1.3 times the expected overburden gradient. These superlithostatic pressures are unstable and can only be localized (i.e., they are impossible on a regional scale). They can be accounted for in a thrust-fault stress regime, where $\sigma_V = \sigma_3$, the minimum stress, and then, only if the tensile strength of the rock exceeds the pore pressure and the vertical stress. In these cases, however, the vertical stress can be greater than the weight of the overburden due to tectonic flexure.

OVERPRESSURE DETECTION IN DIFFERENT TECTONIC SETTINGS

Standard pore-pressure prediction methods, predominantly developed and calibrated in the soft underconsolidated Tertiary sediments of the Gulf of Mexico, are based on finding a porosity anomaly in the overpressured sediment, manifested in deviation of velocity, resistivity, and so on, from a normal trend with depth. The previous discussion illustrated that high pore pressure can be generated in completely different stress environments and that the overpressure can influence the stress regime. Each overpressure mechanism therefore has a unique geomechanical signature, which is manifested in its porosity.

Porosity–Effective Stress Relationships for Different Overpressure Mechanisms

In the absence of shear strain, porosity reduction only occurs with an increase in effective stress (Terzaghi and Peck, 1948). If the effective stress remains constant during loading, so does the porosity. This is the principle used in pore-pressure estimation. Further assumed in most pressure prediction methodologies is that overpressure preserves a porosity that is uniquely related to the maximum effective stress experienced by the sediment, that is, that unloading (by uplift for example) has little effect on porosity. Knowledge of the normal compaction trend for the sediment is therefore deemed sufficient for the estimation of the pore pressure from porosity. Figure 6 demonstrates that these assumptions do not apply to all overpressure mechanisms.

Rapid Vertical Loading

The relationship between porosity and effective stress is defined by the normal compaction curve for the sediment (Figure 6). If no compaction occurs during burial, the sediment remains at the same point on its compaction curve (Figure 6), that is, the pore pressure and porosity become uncommonly high for depth of burial.

Fluid Source

The sediment is buried to depth with normal compaction (i.e., normal pressure), then is injected with a fluid to a certain overpressure. Here, we can have the same pore pressure at the same depth as the previous case, but the conditions are different. The load on the sediment is constant (no additional burial), but the pore pressure increases internally. This decreases the effective stress, which is manifested in unloading of the sediment. Because compaction is an inelastic, irreversible process, porosity rebound is below the compaction trend (Figure 6). Bowers (1994) addressed this unloading effect on pore-pressure prediction in basins where a fluid source is the overpressure mechanism and noted the less pronounced porosity anomaly associated with this mechanism. Yassir and Bell (1996) made the same observation for the Scotian Shelf.

Undrained Tectonic Shearing

The sediment is normally consolidated to depth, and then it is sheared by tectonic forces, resulting in overpressure. The shearing occurs without drainage (low permeability), so the sediment experiences little or no porosity change with reduction of effective stress (Figure 6) (Yassir and Bell, 1996). The porosity response to a change in effective stress is not expected in this case because it involves shear strains (Terzaghi and Peck, 1948). This mechanism can therefore achieve overpressuring without a significant porosity anomaly.

Even if the three overpressure mechanisms result in the same pressure at the same depth, the porosity is different for the three cases. In the rapid loading case, the porosity can be used to estimate the effective stress (and therefore pore pressure) using the normal compaction trend for the sediment. In the case of a fluid source, a porosity anomaly is registered but it gives an underestimate of pore pressure with standard prediction techniques (Bowers, 1994; Yassir and Bell, 1996). In the tectonic shearing case, however, the porosity can potentially remain constant so that the overpressure can go undetected (Yassir and Bell, 1996).

Pore-Pressure Prediction in Sheared Sediments

The previous discussion suggests that the lack of porosity anomaly in sediments sheared in a low-permeability environment renders prediction difficult. So far, however, the different overpressure mechanisms have been discussed independently of one another, whereas more than one mechanism can act in one area. Overpressuring by undrained shear is particularly effective in sediments with high porosity (Figure 3), which renders abruptly deposited undercompacted sediments in a compressional basin ideal candidates for this mechanism (Yassir, 1989). Furthermore, compression is locally associated with overthrusting (e.g., Barbados Ridge complex, Westbrook and Smith, 1983), which involves the addition of overburden load in a manner not unlike rapid sedimentary loading. This should be possible to detect by geophysical anomalies. In other areas (e.g., Trinidad) (Yassir, 1989), the compression is accompanied by significant hydrocarbon generation, which also contributes to an overpressure signature.

If the shearing mechanism is acting alone, detection is difficult. Even so, there are cases in which it is possible: shearing can occur under drained or partially drained conditions, resulting in fluid migration and potential overpressuring of adjacent sediments. This is illustrated in Figure 6, which shows the relationship between pore pressure and shear stress for the two limiting cases (drained/undrained) and the expected corresponding porosity–effective stress relationship. Figure 6 shows that the porosity can remain constant, or it can be further reduced by shear-related compaction. In other words, overpressuring can also be associated with abnormally low porosity. Evidence for this was presented by Fertl and Chilingarian (1989) in the Pripyatskiy Deep, Byelorussia. They found that the Burejskiy shales, which overlie overpressured reservoirs, are highly compacted with an associated increase in resistivity. Theoretically, therefore, tectonic overpressures can, in some lithologies, be detected by a departure from the compaction curve that is opposite to that expected from the other overpressure mechanisms.

Relationships between Pressure and Fracture Gradients for Prediction

The fracture gradient in a sedimentary basin is commonly estimated from the pore-pressure and overburden gradients. This is done by making an assumption

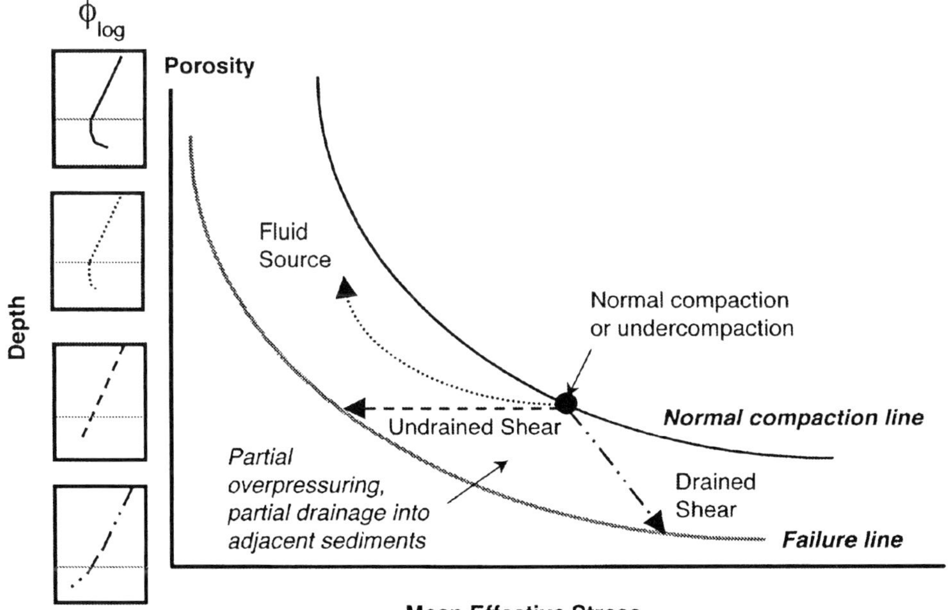

Figure 6. The relationship between porosity and effective stress for different pore-pressure mechanisms. Undercompaction is the only mechanism that results in a correct identification of overpressure magnitude. A fluid source mechanism shows a log anomaly, but if undercompaction is assumed, pressure magnitude is underestimated (Bowers, 1994; Yassir and Bell, 1996). Undrained tectonic shear can result in overpressuring without an impact on shale log properties (Hennig et al., 2002). If tectonic shearing is occurring with partial drainage, this results in an abnormally low porosity anomaly in association with overpressure (Fertl and Chilingarian, 1989; Yassir, 1998).

on the stress regime (commonly a variation on the passive-basin assumption, equation 5) or by obtaining an empirical relationship between measured P_P and σ_h (e.g., Breckels and van Eekelen, 1981; Gaarenstrom et al., 1993; and see review in Mouchet and Mitchell, 1989). Either approach is valid where confirmed by repeatable measurements in a particular basin. The previous discussion, however, demonstrates that pore-pressure–stress relationships are highly dependent on the stress regime and the field boundary conditions. Addis et al. (1996) use field data from Breckels and van Eekelen (1981) to compare the effect of different stress conditions on the in-situ P_P/σ_h ratio. They find that the commonly used passive-basin assumption can give unrealistic stress values in some cases. Yassir and Bell (1996) further demonstrate that, for the same basin with identical passive-basin conditions (equation 5), the same pressure increase by sedimentary loading or a fluid source mechanism results in different P_P/σ_h ratios. In that context, methods of estimation of one parameter from the other should consider the geology, stress system, stress history, and mechanisms of overpressure.

CONCLUDING REMARKS

This chapter illustrates that the pore-pressure and fracture gradient in a sediment are intrinsically coupled so that an increase in one results in an increase in the other. Special emphasis was placed on undrained shearing as a largely ignored yet potentially important overpressure mechanism not only in tectonically active basins, but in basins experiencing high shear stresses caused by inversion, diapirism, or overthrusting. This chapter also illustrates that sediment porosity and the P_P-FG relationship vary according to tectonic setting and overpressure mechanism. In an era of exploration in less accessible and more complex geology, therefore, new methods of quantifying the risk and uncertainty in P_P-FG prediction need to be developed.

REFERENCES CITED

Addis, M. A., 1987, Mechanisms of sedimentary compaction responsible for oilfield subsidence: Ph.D. thesis, University of London, London, United Kingdom, 561 p.

Addis, M. A., 1997, The stress-depletion response of reservoirs: Annual Society of Petroleum Engineers Conference and Exhibition, Society of Petroleum Engineers Paper 38720, p. 55–65.

Addis, M. A., N. Last, and N. Yassir, 1996, Estimation of horizontal stresses at depth in faulted regions and their relationship to pore pressure variations: Society of Petroleum Engineers Formation Evaluation, v. 48, no. 3, p. 11–18.

Bigelow, E. L., 1994, Global occurrences of abnormal pressures, *in* W. H. Fertl, R. E. Chapman, and R. F. Hotz, eds., Studies in abnormal pressures: Amsterdam, Elsevier, p. 1–16.

Bishop, A., D. Webb, and A. Skinner, 1965. Triaxial tests on soil at elevated cell pressures: Proceedings of the Sixth International Conference on Soil Mechanics and Civil Engineering, p. 170–174.

Bowers, G. L., 1994, Pore pressure estimation from velocity data: accounting for overpressure mechanisms besides undercompaction: International Association of Drilling Contractors/Society of Petroleum Engineers Drilling Conference, Society of Petroleum Engineers Paper 27488, p. 515–530.

Breckels, I., and H. van Eekelen, 1981, Relationship between horizontal stress and depth in sedimentary basins: Annual Society of Petroleum Engineers Conference and Exhibition, Society of Petroleum Engineers Paper 10336, p. 1–19.

Enever, J. R., N. Yassir, D. R. Willoughby, and M. A. Addis, 1996, Recent experiences with extended leak-off tests for in-situ stress measurement in Australia: Australian Petroleum Production and Exploration Association Journal, p. 528–535.

Engelder, T., and M. P. Fischer, 1994, Influence of poroelastic behaviour on the magnitude of minimum horizontal stress, S_h, in overpressured parts of sedimentary basins: Geology, v. 22, p. 949–952.

Fertl, W. H., 1976, Abnormal formation pressures: Amsterdam, Elsevier, Developments in Petroleum Science 2, 382 p.

Fertl, W. H., and G. V. Chilingarian, 1989, Prediction of tectonically-caused overpressuring by using resistivity and density measurements of associated shales: Journal of Petroleum Science and Engineering, v. 3, p. 203–208.

Gaarenstrom, L., R. A. J. Tromp, M. C. de Jong, and A. M. Brandenburg, 1993, Overpressures in the central North Sea: implication for trap integrity and drilling safety, *in* J. R. Parker, ed., Petroleum geology of northwest Europe: Petroleum Geology 86, Proceedings of the 4th Conference of the Geological Society, v. 2, p. 1305–1313.

Grauls, D., 1997, Minimum principal stress as a control of overpressures in sedimentary basins (abs.), *in* J. P. Hendry, P. F. Carey, J. Parnell, A. H. Ruffell, and R. H. Worden, eds., Geofluids II '97: Second International Conference on Fluid Evolution, Migration and Interaction in Sedimentary Basins in Orogenic Belts, Belfast, p. 219–222.

Hennig, A., M. A. Addis, N. Yassir, and A. H. Warrington, 2002, Pore-pressure estimation in an active thrust region and its impact on exploration and drilling, *in* A. R. Huffman and G. L. Bowers, eds., Pressure regimes in sedimentary basins and their prediction: AAPG Memoir 76, p. 89–105.

Higgins, G., and J. Saunders, 1974, Mud volcanoes—their nature and origin, contributions to the geology and palaeobiology of the Caribbean and adjacent areas: Naturfoschende Gesellschaft Basel, v. 84, p. 101–152.

Hottman, C. E., J. H. Smith, and W. R. Purcell, 1979, Relationship among earth stresses, pore pressure and drilling problems offshore Gulf of Alaska: Journal of Petroleum Technology, v. 31, p. 1477–1484.

Jakubov, A., A. Ali-Zade, and M. Zeinalov, 1971, Mud volcanoes of the Azerbaijan, SSR: Baku, Publishing House of the Academy of Sciences of the Azerbaijan SSR, 256 p.

Lambe, T. W., and R. V. Whitman, 1979, Soil mechanics, SI version: New York, John Wiley and Sons, 553 p.

Leddra, M., and M. E. Jones, 1989, Steady state flow during undrained loading of chalk: Proceedings of the International Chalk Symposium, p. 245–252.

Mouchet, J. P., and A. Mitchell, 1989, Abnormal pressures while drilling: Elf Aquitaine Manuels Techniques 2, 255 p.

Ohtsuki, H., K. Nishi, T. Okamoto, and S. Tanaka, 1981, Time dependent characteristics of strength and deformation of a mudstone: Proceedings of the Symposium on Weak Rock, v. 1, p. 119–124.

Osborne, M. J., and R. E. Swarbrick, 1997, Mechanisms for generating overpressure in sedimentary basins: a reevaluation: AAPG Bulletin, v. 81, p. 1023–1041.

Ridd, M., 1970, Mud volcanism in New Zealand: AAPG Bulletin, v. 54, p. 601–616.

Terzaghi, K., and R. B. Peck, 1948, Soil mechanics in engineering practice: New York, John Wiley, 566 p.

Teufel, L. W., D. W. Rhett, and H. E. Farrell, 1991, Effect of reservoir depletion and pore pressure drawdown on in-situ stress and deformation in the Ekofisk field, North Sea, *in* J.-C. Roegiers, ed., Proceedings of the 32nd United States Symposium on Rock Mechanics: Rotterdam, Balkema, p. 63–72.

Unruh, J., M. Davisson, K. Criss, and E. Moores, 1992, Implications of perennial saline springs for abnormally high fluid pressures and active thrusting in western California: Geology, v. 20, p. 431–434.

Van Balen, R., and S. Cloetingh, 1995, Neural network analyses of stress-induced overpressures in the Pannonian Basin: Geophysical Journal International, v. 121, p. 532–544.

Westbrook, G. K., and M. J. Smith, 1983, Long decollements and mud volcanoes: evidence from the Barbados Ridge complex for the role of high pore fluid pressure in the development of an accretionary complex: Geology, v. 11, p. 279–283.

Wu, B., 1987, Investigations into the mechanical behaviour of soft rocks: Ph.D. thesis, University of London, London, United Kingdom, 485 p.

Yassir, N., 1989, Mud volcanoes and the behaviour of overpressured clays and silts: Ph.D. thesis, University of London, London, United Kingdom, 249 p.

Yassir, N., 1990, Undrained shear characteristics of clay at high total stresses, *in* V. Maury and D. Fourmaintraux, eds., Rock at great depth: Rotterdam, Balkema, v. 2, p. 907–913.

Yassir, N., 1998, Overpressuring in compressional regimes—causes and detection, *in* A. Mitchell and D. Grauls, eds., Overpressures in petroleum exploration: Bulletin Centre Recherche Elf Exploration and Production, Memoir 22, p. 13–18.

Yassir, N., and J. S. Bell, 1994, Relationships between pore pressure, stresses and present-day geodynamics in the Scotian Shelf, offshore eastern Canada: AAPG Bulletin, v. 78, p. 1863–1880.

Yassir, N., and J. S. Bell, 1996, Abnormally high fluid pressures and associated porosities and stress regimes in sedimentary basins: Society of Petroleum Engineers Formation Evaluation, v. 48, no. 3, p. 5–10.

9

Pore-Pressure Estimation in an Active Thrust Region and Its Impact on Exploration and Drilling

Allison Hennig
CSIRO Petroleum,
Perth, Australia

Najwa Yassir
CSIRO Petroleum,
Melbourne, Australia

M. Anthony Addis
Shell, SIEP,
Rijswijk, Netherlands;
formerly, CSIRO Petroleum,
Melbourne, Australia

Andrew Warrington
BP Developments Australia Ltd.,
Melbourne, Australia

ABSTRACT

A study of overpressuring is presented for the fold belt and foreland basin of Papua New Guinea (PNG), where pore pressures are known to be highly variable and compartmentalized. The project was initiated to identify a methodology for predicting pore pressures in PNG, as all the standard approaches to pore-pressure prediction had failed to provide adequate estimates. Ten wells were selected for this study, including normally pressured and highly overpressured wells. Central to this study were the Hides field wells. Pore-pressure data are presented from formation pressure tests (repeat formation tests and drillstem tests) and from kicks calculated from mud weights and shut-in drill-pipe pressures. By designing an interactive database, the pressure data were analyzed with respect to topographic variations, the corresponding geology, drilling, and electrical logs and by using drilling events.

The pressure regimes in the overburden and reservoir sections were generally found to be unrelated. The reservoir pressures were analyzed using different methods both to evaluate the source of the pressures and to quantify the pressures.

The problems encountered with pressure prediction in the shales in this part of PNG are common to argillaceous sedimentary rocks in uplifted regions. These include the absence of a recognizable normal compaction trend from log data, the presence of uplift, the potential influence of fracturing and shearing on overpressure development and on the compaction trend, and a water table several hundred

Hennig, Allison, Najwa Yassir, M. Anthony Addis, and Andrew Warrington, 2002, Pore-Pressure Estimation in an Active Thrust Region and Its Impact on Exploration and Drilling, in A. R. Huffman and G. L. Bowers, eds., Pressure regimes in sedimentary basins and their prediction: AAPG Memoir 76, p. 89–105.

meters below the drill floor. Poor-quality log data, due to the presence of wellbore damage, exacerbates the difficulties of overpressure detection and quantification in the shale formations. Data from the Hides wells are used to illustrate the effect of wellbore instability on pore-pressure prediction from electrical logs.

No obvious correlation was observed between the pore pressures and the shale sonic transit times, resistivity, or other electrical and drilling logs because the sedimentary rocks where kicks have been observed during drilling do not display significant log anomalies. Standard methodologies of pressure detection based predominantly on a porosity-related anomaly cannot therefore be applied effectively in this region.

Comparing well data was complicated by differences of well elevations, formation depths, and thicknesses. The use of the appropriate datum for comparing pressure data is crucial in identifying the cause of the overpressures. A geomechanical approach was used to assess the sensitivity of electrical and drilling logs to the effective stress in the shales in an attempt to circumvent the problems raised by elevation differences and highly variable formation thicknesses. The advantage of using effective stress is that it corrects for the effect of topography on the pore-pressure data. Preliminary results show that the logs have a weak relationship with effective stress.

The important lesson from this study is that conventional pore-pressure detection techniques in shales cannot be used with confidence in tectonically active regions. The development of an interactive database that captures significant events and conditions in offset wells has proved invaluable in understanding the complexity of the pore-pressure regime in the Hides region and its impact on drilling. The approaches adopted here move some way toward more effective pressure-detection methodology in complex geological areas.

INTRODUCTION

One of the major challenges for exploration and drilling in the Highlands of Papua New Guinea (PNG) is to understand the variability and distribution of the fluid pressures in the area. Both the overburden section (Ieru formation) and the reservoir (Toro formation) can be highly overpressured, but the fluid pressures are variable and compartmentalized. The standard methodologies for overpressure detection and prediction have not worked well in the area because they rely on the identification of anomalous porosities, principally associated with overpressuring by rapid burial and undercompaction. In an area like PNG, tectonic activity and related faulting and fracturing are likely to have an overriding effect on pressure distributions, and the sediments might not be expected to display significant porosity anomalies. In such environments, standard approaches to overpressure detection and quantification may be inappropriate. Poor overpressure detection leads to problems in well design (mud weight and casing point selection) that, in turn, results in large uncertainties and a greater risk of experiencing kicks. Furthermore, it impacts significantly on assessments of seal breach risk and fluid migration pathways.

The scope of this study is to comprehensively review data and drilling experiences for 10 wells in the general vicinity of the Hides field (Figure 1). Most of the wells are located in the Papuan fold belt (Hides-1, Hides-2, Hides-3, Angore-1A, SE Mananda-1X, SE Mananda-2X, Paua-1X, and Kutubu-1X) but two foreland basin wells (Elevala-1 and Ketu-1) were also included for comparison (Figure 1). To this end, a well database was created that includes pressure data (kicks/flows, repeat formation tests [RFTs], drillstem tests [DSTs]), fracture data (leak-off tests, formation integrity tests), lithology, stratigraphy, biostratigraphy, and all available electrical logs and mud logs, as well as drilling events. The database was designed to allow visual comparison among all the different parameters. The aim of the database was to capture the experiences from offset wells, as well as to identify correlations (if any) between overpressuring, geology, logs, and/or significant drilling events.

The chapter discusses variations in pressure distribution in the shaly Ieru formation and the Toro/Imburu reservoir sections and the difficulties experienced in detecting and quantifying the overpressures. The correlations between pressures and logs for pore-pressure prediction purposes are reviewed, and possible controls on the observed overpressures are discussed.

GEOLOGY OF THE AREA

The Papuan fold belt is a northwest-southeast–trending mountain range that rises to about 3500 m and faces southwest toward a lowland foreland basin sitting approximately at sea level. The fold belt was created by the Pliocene–Holocene oblique collision of the

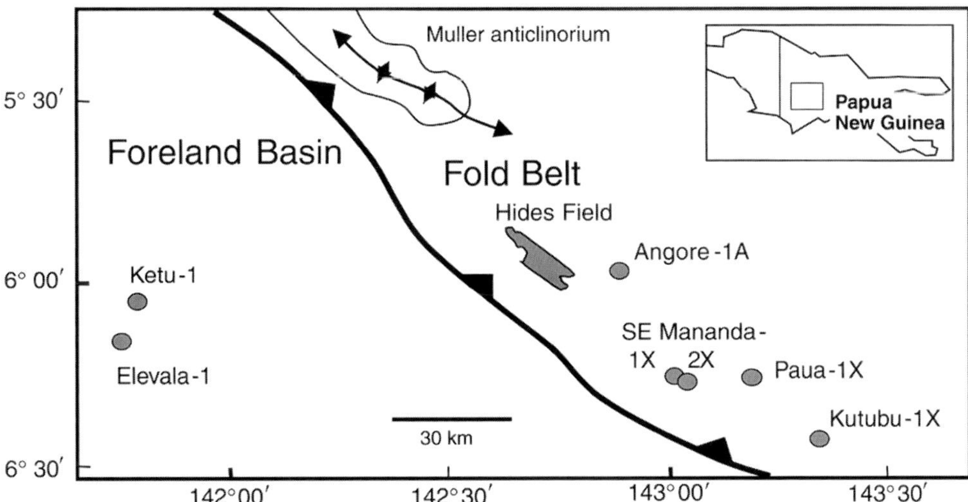

Figure 1. Map of the study area showing general well locations.

Australian and Pacific plates (Hill, 1991; Eisenberg, 1993). The collision and accompanying deformation caused the inversion of the Papuan sedimentary basin; a Mesozoic rift/passive margin marine sequence and a Cenozoic foreland basin (Hill et al., 1993; Phelps and Denison, 1993; Eisenberg et al., 1994). The major commercial hydrocarbon discoveries of PNG are associated with this deformation belt and are located in a series of thrust-related northwest- to southeast-trending en echelon anticlines formed during this period. Figure 2 is a simplified, representative cross section through the fold belt, showing the direction of thrust faulting in the study area.

The major units in the region include the Toro formation, which is the main hydrocarbon target, the Ieru formation, and Darai limestone formation. This sequence is overlain by Quaternary clastics in some localities.

Toro Formation

The Toro formation comprises a series of stacked, areally extensive, gas-rich sand bodies deposited in a wave-dominated shelf environment. This formation has been informally separated into three members. The upper and lower members consist predominantly of fine- to coarse-grained sandstone, interbedded with mudstone and siltstone. The middle member is predominantly mudstone and siltstone, with minor interbedded sandstone. The Toro is known at outcrop in several locations and elevations throughout the fold belt and foreland basin (Hill, 1991; Eisenberg, 1993). In the study area it crops out in the Muller anticlinorium (Figure 1). The Toro formation is overpressured in some areas and normally pressured in others, which implies that the reservoir has been compartmentalized by the thrust faulting. The Toro formation is underlain by the shaly Imburu formation, which is not discussed in this chapter.

Ieru Formation

The Cretaceous Ieru formation conformably overlies the Toro reservoir and acts as a regional seal. It consists of interbeds of mudstone, siltstone, and sandstone. The lower section is divided into three members, which in decreasing age are the Alene, the Juha, and the Bawia. These are predominantly comprised of mudstone lithologies with interbedded siltstone and occasional minor sandstones. The upper section of the Ieru is divided into the Giero, Ubea, and the Haito members, again with decreasing age.

The Giero member is further subdivided into the Giero A, B, and C units. The upper section is sandier but is generally dominated by mudstone with interbedded sandstone and siltstone. Only the Giero C unit is consistently sandstone. The mudstones and siltstones are generally massive or blocky and soft to firm in texture. They range from moderately to very calcareous and are generally glauconitic. The sandstones are moderately to well sorted, very fine to fine-grained quartz. Divisions in the Ieru are based on palynological assemblages and lithology changes.

The Ieru formation was deposited during the Turonian and Berriasian prior to the opening of the Coral Sea and the collision of the Australian and Pacific plates. During the collision the Ieru formation was uplifted and exposed to erosion. It is known to outcrop in places within the fold belt and unconformably underlies the Darai limestone (Phelps and Denison, 1993).

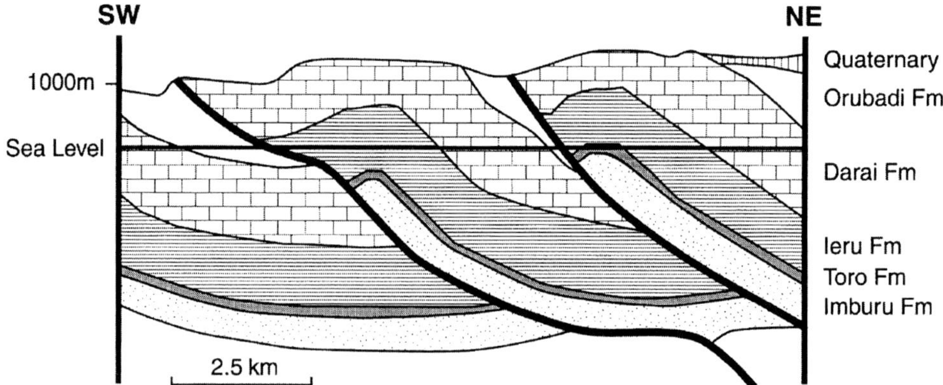

Figure 2. Schematic structural cross section through the Papuan fold belt (modified after British Petroleum).

The Ieru formation is commonly associated with severe drilling problems, including gas shows, kicks, wellbore instability, and mud losses.

Darai Formation

The Darai limestone is a thick (up to 2500 m), shallow-marine sequence deposited during the middle to late Miocene period; it unconformably overlies the Ieru formation (Daniels, 1993). Deposition commenced prior to the collision of the Australian and Pacific plates and continued during the subsequent inversion of the Papuan basin that led to the development of the fold and thrust belt (Phelps and Denison, 1993). At outcrop, this limestone unit is deeply karstified, which has, until recently, prevented acquisition of high-quality seismic data in the fold belt (Hill et al., 1996). Thus outcrop and well information have formed the basis for structural models and have provided the only means of understanding and predicting subsurface pressure distributions.

PRESSURE REGIMES

The pressure data measured in the 10 wells selected for this study (Figure 1; Table 1) illustrate the significant pressure variability across the area, both within individual formations and between overburden and reservoir. Pressure data are presented in MPa (1 MPa = 145 psi), and pressure gradients (equivalent mud weights [EMWs]) are presented in terms of their specific gravity (s.g.). Note that 1 s.g. unit = 9.81 kPa/m = 0.433 psi/ft.

Summary of Well Pressures

In the Hides field and Angore-1X, the Ieru formation is more overpressured than the reservoir. Kicks have been recorded in the Giero B and C units in these wells, but not in the lower part of the formation (Juha/Alene members) in which the pressures are unknown. The uncertainty regarding the continuation of overpressuring in the lower Ieru section in these wells renders casing design problematic.

The pressure system in the southeast Mananda wells is very different: the Ieru formation is sandier and seems to be only slightly overpressured, with moderate overpressuring in the reservoir. The same applies to Paua-1X, but reservoir pressures reach a value of 2.19 s.g.—the highest recorded pressure gradient in this study (Table 1). Further to the southeast (Figure 1), Kutubu-1X shows severe overpressuring in the Ieru formation again, but the pressures appear to be normal down to the top of the Juha/Alene members, where kicks occur at pressures exceeding 2 s.g. The high overpressures continue into the Toro formation in this well (Table 1).

Considerable variability in the pressure profiles of the wells in the fold belt is apparent. By contrast, the two foreland wells have thin, sand-rich, normally to mildly overpressured Ieru formations and a normally pressured reservoir (Table 1).

In the following section, we examine the pressures in the overburden and the reservoir separately. As is illustrated in the following section, comparisons between wells in this type of terrain are difficult, as the selection of the appropriate datum is not obvious. For example, choosing a kelly bushing (or drill floor) level datum is inappropriate for comparing pressure data if the overpressures have an artesian origin (connection to an elevated outcrop). Depending on the datum used, pressure-depth plots of the composite well data can lead to very different interpretations. The identification of a correct datum depth at which to plot pressure data is therefore crucial for an understanding of the source of the overpressures. Pressure analyses are further complicated by steeply dipping beds, three-

Table 1. Elevation and Formation Thickness for Wells in the Study Area and Their Equivalent Formation Pressure*

Well Name	Latitude (S)	Longitude (E)	Kelly Bushing Elevation (m AMSL)	Depth of Water Table (m RKB)	Depth to Ieru Formation (m RKB)	Depth to Toro Formation (m RKB)	Max. Pressure Gradient in Ieru Formation (s.g.)	Max. Pressure Gradient in Toro/Imburu Formation (s.g.)
Fold Belt								
Angore-1A	−5° 58′ 3.55″	142° 53′ 3.66″	1610	200	2547	3994	1.73	1.26
Hides-1	−5° 55′ 51.76″	142° 42′ 49.97″	2698	962	1094	2999	1.64	1.31
Hides-2	−5° 56′ 56.18″	142° 43′ 52.26″	2454	665	1254	2725	1.68	1.4
Hides-3	−5° 56′ 59.51″	142° 44′ 40.76″	2264	632	1232	3019	1.85	1.31
Kutubu-1X	−6° 27′ 55.8″	143° 20′ 57.6″	1260	525	847	2062	2.03	1.99
Paua-1X	−6° 14′ 54.06″	143° 10′ 25.14″	1594	520	454	2759	1.42	2.19
SE Mananda-1X	−6° 16′ 19.46″	143° 2′ 30.63″	1541	350	1045	2024	1.13	1.34
SE Mananda-2X	−6° 15′ 56.87″	143° 0′ 59.71″	1708	617	1064	2179	1.10	1.26
Foreland Basin								
Elevala-1	−6° 9′ 4.92″	141° 45′ 57.42″	64	20	2329	3102	1.34	1.02
Ketu-1	−6° 2′ 15.62″	141° 47′ 20.34″	101	7	2451	3334	0.99	1.00

*AMSL = above mean sea level; RKB = relative to kelly bushing; 1 s.g. = 0.43 psi/ft.

dimensional structural closures, reservoir outcrop at several elevations, and pressure compartmentalization.

A Note on Datum Selection

In such mountainous terrain, difficulties arise in directly comparing pressures between two wells. There is variability in topographic elevation, depth to various formations and their thickness, dip, structure, and even in the top of the saturated zone (water table—with the additional complexity of seasonally changing water-table elevations). Close attention therefore needs to be paid to the datum used in pressure analysis. This can be a formation top, sea level, ground level, or the water table, depending on the type of analysis required and the prevailing geological conditions.

Reservoir Pressures

To illustrate the importance of datum selection, Figure 3 is a plot of RFT/DST pressures in the Toro and Imburu formations plotted relative to kelly bushing (Figure 3a) and to sea level (Figure 3b). Ieru formation pressures in Elevala-1 and Paua-1X were measured by RFTs, and these are also noted for comparison.

Figure 3a shows clearly that the Kutubu-1X and Paua-1X reservoir sections had to be drilled with high mud weights (>2 s.g.). Note that the Ieru formation pressures in Paua-1X are subhydrostatic (<1 s.g.), which can be explained by the low water table, recorded at 520 m below kelly bushing, (i.e., within the Ieru formation, Table 1). The Ketu-1 and Elevala-1 wells are hydrostatically pressured in the Toro, whereas the remaining well data indicate only moderately overpressured reservoirs, with pressure gradients less than 1.5 s.g.

Where the data are plotted relative to sea level (Figure 3b), the interpretation is different. The wells in the foreland basin Ketu-1 and Elevala-1 still lie on the freshwater hydrostatic gradient because their kelly bushing is close to sea level. This may indicate that these wells are connected to a Toro outcrop lying close to sea level. In the fold belt the Toro formation outcrops at much higher elevations. The outcrop at the Muller anticlinorium (Figure 1) occurs at approximately 2350 m above mean sea level (AMSL) (Grainge, 1993), Figure 3b. Southeast Mananda-1X and -2X pressures seem to be hydrostatic and are possibly in pressure communication with the outcrop, as suggested by the hydrostatic water gradient between them (Figure 3b). The pressures measured in the Hides and Kutubu-1X wells are similar in magnitude and plot at the same depth relative to sea level (Figure 3b). Because of the difference in well elevations, however, the Toro formation pressures in the Hides wells appeared only mildly overpressured during drilling, whereas in Kutubu-1X, they are associated with very high mud weights (1.8 s.g.) (Figure 3a). Figure 3b also clearly shows that the Toro pressures in the foreland wells (Ketu-1 and Elevala-1) appear unrelated to the pressures in the fold belt.

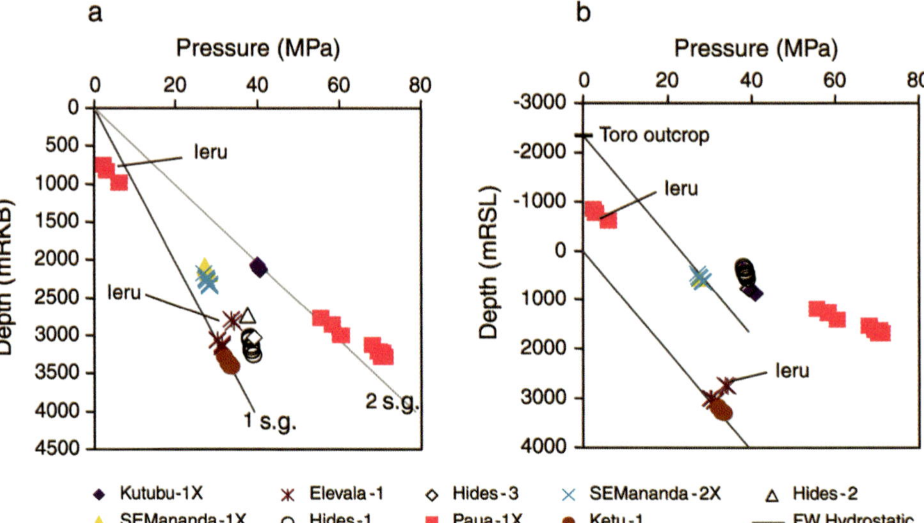

Figure 3. Reservoir RFT/DST pressures in all wells plotted relative to (a) kelly bushing (RKB) and (b) sea level (RSL). Where Ieru formation pressures are plotted, these are labeled (1 MPa = 145 psi).

Pore Pressures in the Overburden

Pressure estimation in argillaceous formations is notoriously difficult because of the low permeability of the rock. Accurate pressure measurements are always sparse and associated with high-permeability stringers. The Ieru formation is no doubt overpressured in many wells, but the degree and extent of overpressuring is unknown except where a kick is taken when drilling a sandy layer and, uncommonly, where RFT/DST measurements are taken (Elevala-1, Paua-1X). Many of the recorded influxes are gas, which indicates that the Ieru is not completely water saturated. This renders the application of pressure detection technology more difficult because (1) it is dependent on the assumption of a saturated, undercompacted shale sequence, and (2) some electrical logs used for detection may need to be corrected for the presence of gas.

Pressure determination is further complicated by the occurrence of stress-induced wellbore instability (Twynam et al., 1994; Addis et al., 1998), which can be mistaken for overpressuring.

All kicks and flows recorded in the wells included in this study are listed in Table 2. The pressures recorded during the kicks in the Ieru formation are plotted relative to the kelly bushing in Figure 4a and relative to the top of the Ieru in Figure 4b to illustrate the lack of any relationship between the observed pressures and the datum.

The pattern of overpressuring in the shales is inconsistent even for wells within the same field. The three Hides wells all experienced kicks while being drilled in the Ieru. In these wells, the depth to the top of the reservoir and the thickness of the overburden varies by only 200 m, with Hides-1 being the deepest and Hides-3 the shallowest (Table 1). The kicks range in value between 1.64 s.g and 1.90 s.g. The kicks taken in the Giero C sandstone in Hides-1 and Hides-3 were 1.64 and 1.85 s.g., respectively, although the two wells are located in similar locations, on the crest of the Hides anticline.

In contrast, the foreland basin well, Elevala-1, which has a similar depth to the top of the reservoir (Table 1), was drilled with much lower mud weights and experienced a kick of 1.34 s.g. in the Ieru.

Overpressures in the overburden have been estimated and discussed here based on kicks only. Mud weights have not been used because they are not reliable indications of the pore pressures where wellbore instability occurs.

PORE-PRESSURE PREDICTION IN PNG FROM LOGS

The pressures measured in stringers are normally taken as representative of the pressures in the juxtaposed argillaceous sediments. In the absence of such data, the pressures in argillaceous sedimentary rocks need to be understood and evaluated by other means. This is commonly achieved through the use of drilling information and log data. Where seismic data are available, interval velocities can be correlated to the drilling and log data to estimate the likely pore pressures. No seismic data were available for this study.

The use of the log-based data to achieve an understanding of the overpressures in the shale and mudstone lithologies of the PNG fold belt is now discussed. The structural complexity and the topographical variations of the region again make interpretation of pore pressures in the shales difficult.

Table 2. Kick/Flow Pressures and Parameters for Wells in the Study Area*

Well Name	Kick Depth (m RKB)	Kick EMW (s.g.)	Kick Pressure (MPa)	Formation	Influx Fluid
Angore-1A	3511	1.73	59.47	Ieru/Giero C	Gas
	4107	1.26	50.67	Toro	?Gas
	4226	1.33	55.03	Imburu	?Gas
Elevala-1	2426	1.34	31.83	Imburu	Gas
Hides-1	2305	1.64	37.01	Ieru/Giero C	Gas
	3002	1.31	38.50	Toro	Gas
Hides-2	1906	1.68	31.35	Ieru/Giero B	Gas
Hides-3	2038	1.90	37.91	Ieru/Giero B	Water
	2258	1.85	40.90	Ieru/Giero C	Water
Hides-3 sidetrack	3000	1.40	41.06	Ieru/Alene	Gas
Kutubu-1X	1857	1.88	34.18	Ieru/Juha	Water
	2029	2.03	40.33	Ieru/Alene	?
	2065	1.99	40.23	Toro	?
Paua-1X	2767	1.68>>1.99	45.51>>53.91	Toro	Water
	2888	2.08	58.58	Imburu	?
	3138.7	2.19	66.83	Imburu	?

*RKB = relative to kelly bushing; EMW = equivalent mud weight; ? = unknown.

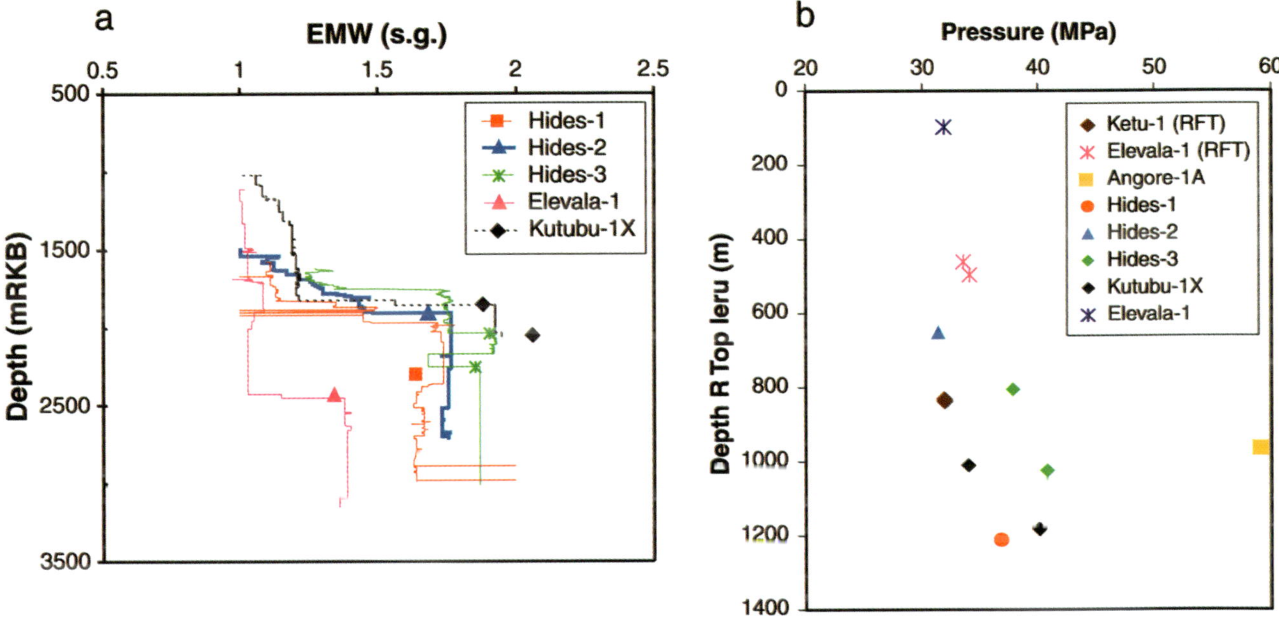

Figure 4. Ieru formation pressures from kicks and some RFT data, plotted (a) with mud weight as EMW (equivalent mud weight) vs. depth relative to kelly bushing; (b) as pressure vs. depth relative to the top of the Ieru.

Electrical Log Anomalies

Pore-pressure prediction from logs traditionally relies on the identification of a normal compaction trend in a lithologically uniform and thick shale sequence. A normal compaction curve based on local data, or a generic global curve, can be used for the analysis (Mouchet and Mitchell, 1989). The latter normally requires local calibration to well data. These methods of pressure detection identify the onset of overpressuring with the occurrence of an anomalously high porosity (undercompaction) for the depth. Overpressuring appears as a reversal or deviation from the normal compaction trend line. The extent of the deviation from the normal trend is used to quantify the overpressure. This methodology works well in young basins where the depositional environment is stable and continuous. The method commonly relies on the presence of a uniform shale lithology—where only one compaction curve needs to be identified. Where sediment sources have varied over time and led to differing shale lithologies and mineral assemblages, the compaction behavior (Aplin and Yang, 1995) and the resultant log response with depth varies. Accurate pore-pressure prediction then relies on identifying numerous compaction trends—one for each shale or mudstone lithology.

In the PNG fold belt the validity of using a compaction trend-line approach to pore-pressure prediction in shales is questionable, as many of the criteria underlying this pore-pressure prediction methodology are violated:

1. Discontinuous shale and mudstone deposition, with variable lithologies
2. A thick (up to 1000 m) limestone overlies the shale, which would have significantly enhanced compaction
3. The rocks have been heavily deformed and uplifted, which leads to pervasive fracturing, and affects not only formation porosity but fluid pressures as well (Yassir and Addis, 2002)

As mentioned previously, the Cretaceous Ieru formation, although thick in some areas, comprises interbedded shale and thick sandstone units and is known to have undergone major tectonic uplift and deformation in the late Tertiary. Under the circumstances, the overpressuring observed in the Ieru formation may not be the result of undercompaction. Any study of pore-pressure estimation in shales based on wire-line logs is further complicated by the occurrence of wellbore instability (breakouts), particularly in tectonic regions. In the Ieru, wellbore instability is known to have occurred during drilling. Nevertheless, the ability to estimate the pore pressures in the Ieru shales from log data was addressed along with an assessment of the potential errors introduced into the analysis due to the presence of wellbore instability and wellbore damage.

Figure 5 shows drilling and pressure data for Hides-1 and the corresponding wire-line log curves that are commonly used to determine normal compaction gradients for pore-pressure evaluation: resistivity and transit time, and a gamma-ray log for shale identification. The well location does not contain a thick section of pure shales from which to define a normal compaction trend, a problem common to all of the wells in this study. The transit time increases with depth into the Ieru formation then abruptly decreases with depth at around 2000 m relative to kelly bushing. The initial increase in transit time (slowness) could be interpreted as the onset of overpressuring; however, it corresponds to wellbore instability problems encountered in the upper Ieru sections (Figure 5). Note also that the transit-time response decreases with depth toward the Giero C unit, which is known to be highly overpressured in this well. This is contrary to what would be expected. A similar observation was made for the resistivity logs (Figure 5).

Generic Compaction Curves for Pore-Pressure Prediction

The absence of a normal compaction trend line in log-based data has led to the use of generic trend lines, or compaction curves, which are considered to be globally applicable. This approach to pore-pressure prediction has been widely adopted (Eaton, 1972; Mouchet and Mitchell, 1989; Holbrook, 1995).

Generic compaction curves define the change of some porosity-related property; resistivity, sonic transit time, and so on, with effective stress. The variation of the parameter from the generic curves indicates lower effective stress acting on the rock, and, by implication, higher pore pressures. Because the detection and quantification of pore pressure is based on the presence of a porosity anomaly, this method is most successful in detecting overpressures caused by undercompaction (Bowers, 1995; Yassir and Addis, 2002). The generic compaction curves are lithology specific, and large variations in the back-calculated pore pressures can occur, depending upon the mineralogical assemblage (Skempton, 1970; Baldwin and Butler, 1985). As a result, calibration of these global techniques to local well experience is commonly necessary. The use of generic compaction curves for pore-pressure detection and evaluation has the same underlying assumptions as the compaction trend-line–based approach that uses local data. Consequently, the use of generic compaction curves to calculate pore pressures is theoretically inconsistent with the geological development of a tectonic region, where uplift and lateral shear are dominant geological processes.

These potential shortcomings were illustrated where the pore pressures in the Hides wells were evaluated using the Eaton method, the most commonly used of the generic relationships. Sonic-log compaction trend lines were superimposed on the data from Hides-1, Hides-3, and Elevala-1. The Eaton curves were calibrated to kick data observed in the Hides-3 (Figure 6a) and Elevala-1 (Figure 6b) wells. Shale sonic values were used that are within 1–2 m from the kick location, with the assumption that the shale layers have the same pore pressure as calculated for the kick.

Three wells are shown in Figure 6: Hides-1, Hides-3, and Elevala-1. The Hides-3 data are used to calibrate the Eaton curves, in preference to Hides-1, as the for-

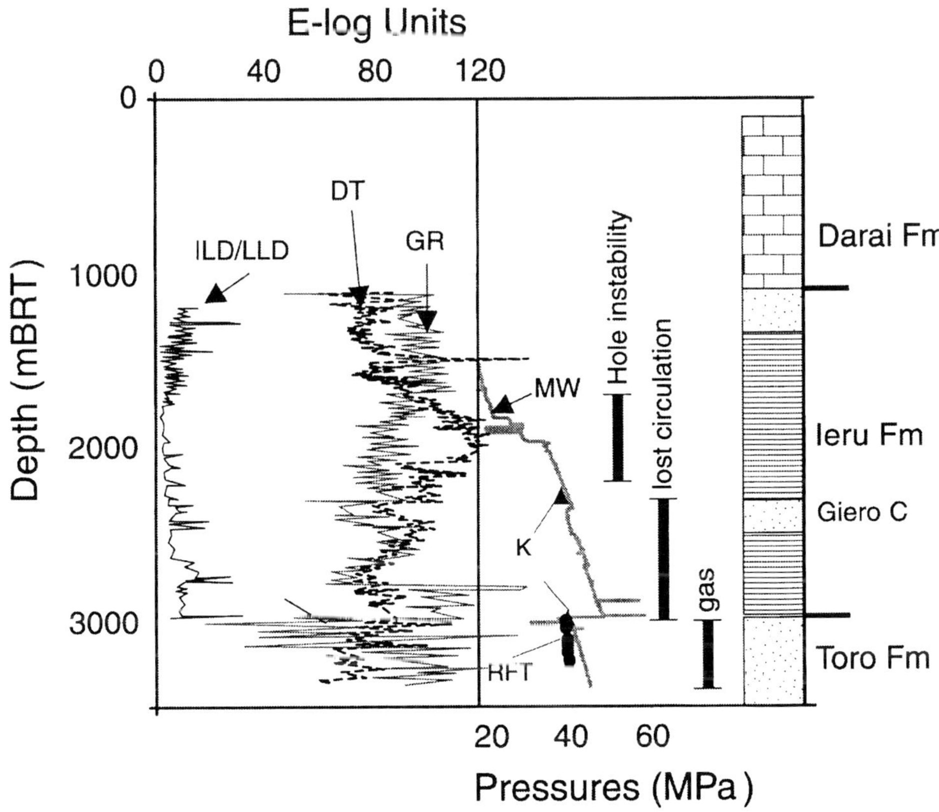

Figure 5. Hides-1 generalized stratigraphy showing pressure measurements and e-logs. Delta transit time (DT) in μs/ft; induction resistivity deep (ILD)/laterolog deep (LLD) in ohms; gamma ray (GR) in API units. K denotes kick (an influx of formation fluid).

mer contains fewer artifacts arising from wellbore instability. The data from these three wells allow a comparison between pore-pressure prediction in the uplifted wells of Hides and the Elevala-1 well, which has undergone less uplift. The data also allow a comparison between the Hides-1 and Hides-3 wells, which are the most geologically similar on the Hides structure. Using these comparisons the likely success of the compaction-curve approach for pore pressure estimation in this area was assessed.

The location of the normal compaction trend line varies quite considerably depending upon the choice of the pore pressure taken for the calibration. Using the Elevala-1 kick to construct the normal compaction curve results in large errors in pore-pressure predictions in the uplifted Hides wells. More seriously, constructing the compaction trend line based on the kick in Hides-3 leads to large errors when predicting the pore pressures in the adjacent Hides-1 well. Errors are in the order of 0.2–0.3 s.g., which are unacceptable for well design and for drilling purposes. The shortcomings of this approach are not expected to be limited to Eaton's curves but to the use of any generic compaction curve that relies on a porosity-related anomaly as a measure of the overpressure.

The use of standard log-based methods for pore-pressure prediction in the Hides area by BP has been unsuccessful and was the basis for initiating the project. The approach described here is consistent with BP's experience. Consequently, a broader approach to identifying pore pressures and hole problems while drilling was adopted by generating a database of events.

Pore-Pressure Prediction in Uplifted Shale Formations

Pore-pressure prediction in areas that have experienced uplift requires corrections to standard pore-pressure methodologies and techniques. Using the standard trend line or a generic compaction-curve approach, the correction entails displacing the compaction curves to different depths. This is equivalent to unloading with no relaxation—zero porosity change—which is a conservative approach, particularly where the uplift has been accompanied by major tectonic shearing, faulting, and natural fracturing. We argue that, having gone through uplift and shearing, the sediments no longer resemble the preuplift and preshear sediments in terms of their compaction state, mechanical properties, and porosity profile.

Pore-pressure evaluations that incorporate the effects of unloading have been discussed by Bowers (1995) and Ward et al. (1995). Unloading curves were developed that account for increases in porosity with

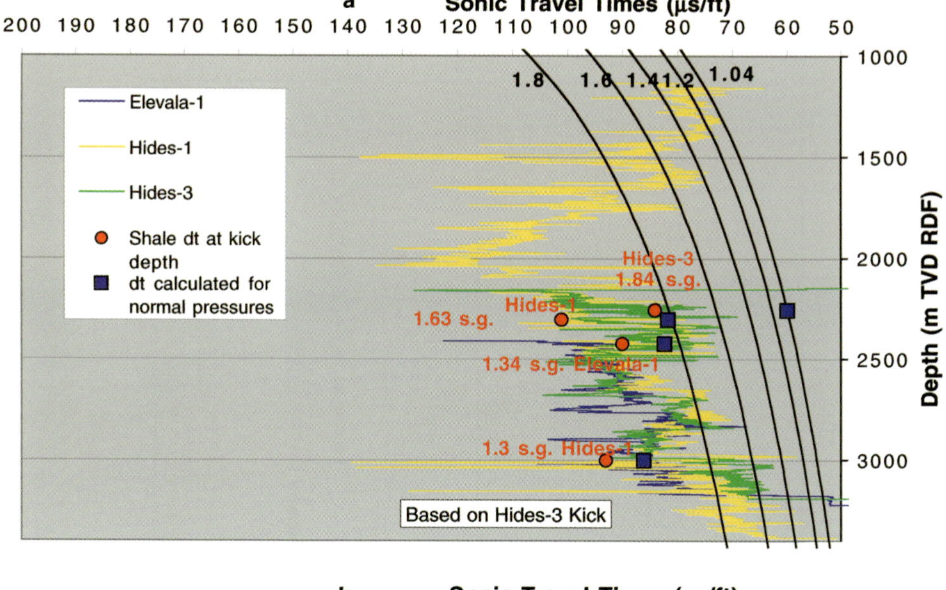

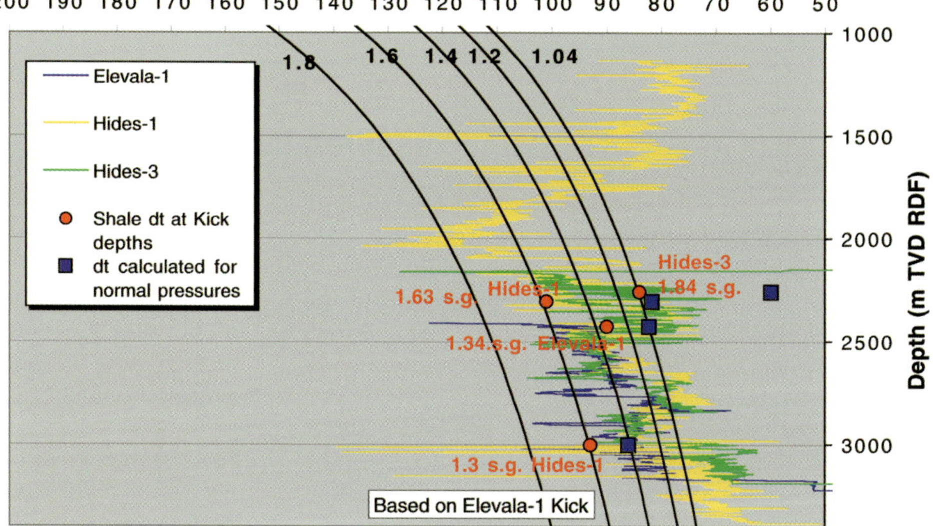

Figure 6. Comparison of pressure prediction using Eaton's trend lines calibrated with (a) Hides-3 and (b) Elevala-1 data with observed kicks.

reduction of effective stress. Both studies require that regional unloading responses of the logs are defined for the overpressured sections. Bowers (1995) and Ward et al. (1995) consider mechanical unloading–effective stress decreases resulting from overpressure generation at depth. The physical unloading experienced in PNG requires a different approach, especially as the tectonic forces driving the uplift physically change rocks and their mechanical behavior.

To summarize, the detection of overpressures using compaction trend lines based on a monotonically decreasing porosity with stress (or depth) appears inappropriate in the Papuan fold belt. Depth-corrected compaction curves, which aim to consider the effects of physical uplift, inherently assume that the magnitudes of overpressure can be quantified using the undercompaction-based curves. This ignores any geological processes that have accompanied uplift. Bowers (1995) shows the shortcomings of this approach for less complicated unloading scenarios and geological environments.

GEOMECHANICAL APPROACH TO PRESSURE EVALUATION

The discussion of pore-pressure detection has centered on the difficulties experienced when working in a mountainous area that has been uplifted, where the wells have different elevations, where formation thicknesses vary, and where the cause of overpressuring is unknown. The use of undercompaction-based methodologies for quantifying the magnitudes of the overpressures has been hampered by the lack of any normal compaction trend line and is deemed inappropriate for this complex geological region. To start to address some of this complexity and specifically the problem of appropriate datum selection, a geomechanical approach was used to assess whether any improvement in the analysis could be achieved.

Rock properties such as porosity are highly dependent on overburden thickness and stress, and this has to be corrected for when comparing wells in this region. Using the vertical effective stress (overburden total stress minus pore pressure) normalizes the results with respect to the overburden. This enables the most direct comparison of data between wells of different elevations. Mud weights and kick pressures for the Ieru formation were used to calculate the vertical effective stress in the Hides wells, Elevala-1, and Kutubu-1X, using an overburden gradient of 1 psi/ft as an estimate. Wells with the highest pore pressures have the lowest effective stress, and those with the lowest pressures have the highest effective stress. Because of highly variable Ieru thickness, the datum depth used in Figure 7 is the center of the Ieru formation, which allows direct comparison of the log and drilling data between wells. Sand units were removed from the transit-time curves, leaving only the mudstones and siltstones for evaluation.

The estimated effective stress values presented in Figure 7 show that, as expected, the normally to slightly overpressured Elevala-1 has the highest effective stress, the overpressured Hides wells have the lowest, and Kutubu-1X, which is also normally to mildly overpressured, lies between. In contrast to this, the sonic response for the Hides and Elevala-1 wells are similar throughout the Ieru formation regardless of the significant variation in effective stress. The transit time of Kutubu-1X is lower (faster) and almost constant with depth. It would be expected that if the sonic logs were accurately representing the pressures found in the Ieru formation, the relative positions of the transit-time curves would correspond better with the effective stress data. Instead, this is not seen because the sonic response of three of the fold-belt wells is almost overlain by the sonic response of the foreland wells, despite the significant difference in the calculated effective stress.

Figure 8 is a plot of effective stress based on pressure measurements in shale vs. sonic transit time for all of the wells in this study. Despite the previous findings, Figure 8 does show a weak relationship of decreasing effective stress with increasing transit time (higher porosity). This suggests that the sonic log may have some sensitivity to the overpressure; however, the relationship cannot as yet be used in a predictive sense because of the significant scatter of the sonic response. The same finding was confirmed for the density logs and the neutron logs.

The failure of the sonic logs to reflect the measured pore pressures suggests that overpressuring is not necessarily associated with a porosity-related anomaly in this area. Consequently, the factor of undercompaction in overpressure generation is considered to be minor. This is consistent with overpressures produced

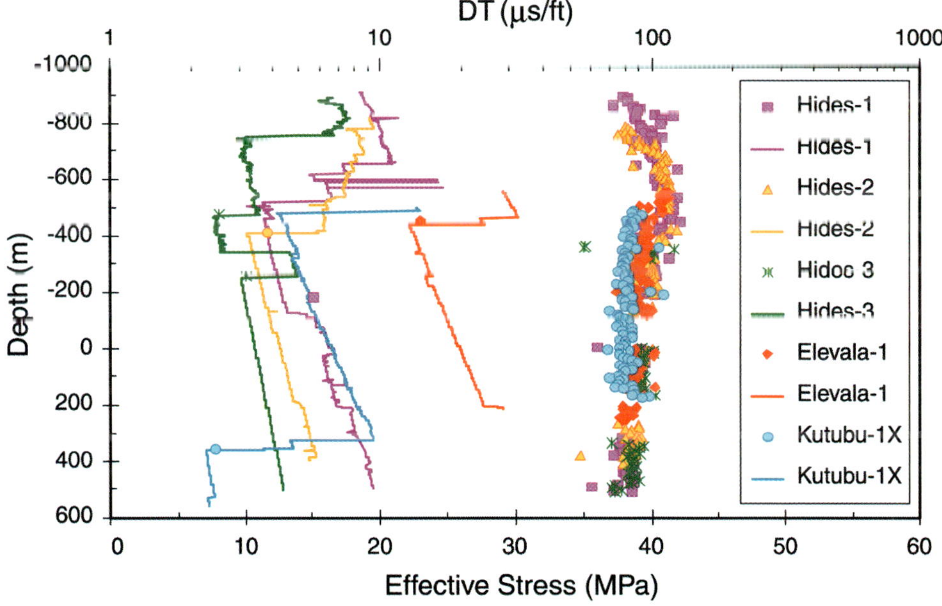

Figure 7. Effective stresses (curves) from kicks and EMW plotted with delta transit time (DT, shown as symbols) against normalized depth (top Bawia shale, Ieru formation) for five wells from the study area.

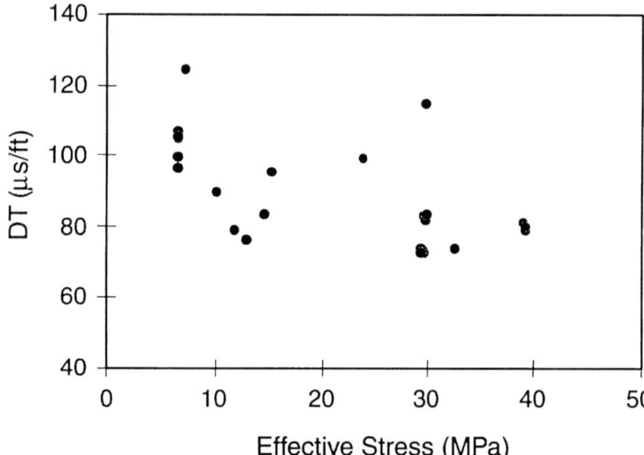

Figure 8. Effective stress values from shales in the Ieru for all wells from the study area plotted against delta transit time (DT).

by tectonic shearing, which commonly go undetected (Yassir and Addis, 2002). Although standard pore-pressure analysis and prediction techniques do not work well in the study area, rigorous geomechanical correlation between wire-line logs and effective stresses may be of value in improving pressure prediction.

EFFECT OF WELLBORE INSTABILITY ON PRESSURE DETECTION FROM ELECTRICAL LOGS

A further complication in pore-pressure detection in shale sections is the occurrence of wellbore instability (McLean and Addis, 1990). Instability occurs in the majority of shale formations in hydrocarbon exploration wells and particularly in areas of high stress, such as PNG. Wellbore instability is recognized in most compressive areas throughout the world, for example, in PNG (Twynam et al., 1994; Addis et al., 1998), offshore Bali (Ramos et al., 1998), in the Canadian Rockies (Woodland, 1990), and in Colombia (Addis et al., 1993; Last and McLean, 1995; Last et al., 1996).

Wellbore instability commonly manifests itself as breakouts: elliptically shaped, overgauge holes created by anisotropic compressive stresses, which are accompanied by the development of cavings. Wellbore instability occurs during both overbalanced and underbalanced drilling and has the same signature in both cases: shale sloughing, cavings, mud-weight increases, and gas increases. Consequently, none of these can be used as true pore-pressure indicators in shale formations. More significantly for pore-pressure detection, electrical logs, even where corrected for hole size, are highly sensitive to borehole condition (Addis et al., 1998). This has been recognized in the Colombian foothills, where stress-related hole problems were initially attributed to overpressures. Electric logs are corrected for borehole size but not for the change in the rock fabric in the unstable and damaged wellbore wall (Addis et al., 1990). Wellbore damage has been recognized in both in-gauge holes and out-of-gauge holes. There is a wealth of literature describing the electric-log response due to wellbore damage and mud-shale interaction (Hornby and Chang, 1985; Wu et al., 1993; Winkler, 1997; Boonen et al., 1998; Hsu et al., 1998).

Two questions are addressed in this chapter. (1) How significant is the effect of wellbore damage on pore-pressure detection and quantification? (2) Does wellbore damage influence the pore-pressure evaluation in PNG?

To address the first question, time-lapse logging data from five separate logging runs of a long-spaced sonic tool were used to determine the potential error in pore-pressure estimation resulting from borehole damage (Figure 9). The data come from an offshore well where wellbore stresses and mud-shale interaction have affected the wellbore (Blakeman, 1982). Changes in the interval transit-time response of up to 23 μs/ft (0.3 m) were recorded over a period of 35 days. If these altered sonic data were used for pore-pressure evaluation a significant error would be introduced. This is illustrated using Eaton's relationships to quantify the resultant apparent pore pressures. The alteration of the sonic values of 5–20 μs/ft (0.3 m), due to wellbore damage, results in a normally pressured formation having an apparent pore pressure of 1.19 to 1.54 s.g. Electric logs are not commonly rerun repeatedly over the same interval with significant time intervals between each run, which makes removing such artifacts from pore-pressure analysis difficult. As stated previously, the mud weights offer little help in distinguishing between pore pressures and instability-related log anomalies.

The occurrence of wellbore instability in the Papuan fold belt is well known, and the geological stresses acting in the area, as well as the mud types used in the earlier wells, are know to have contributed to the instability (Twynam et al., 1994). The effect of instability on the log response for Hides-1 is shown in Figure 10a, which is a plot of the sonic log against the caliper data in the 8.5 in. section. Hides-1 is known to have experienced wellbore instability problems, confirmed by the elliptical-shaped hole on the four-arm caliper data. Figure 10a shows a scatter of data with an increase in the minimum sonic traveltime with larger borehole diameters. Figure 10b shows the corresponding sonic-log values for Hides-3, which experienced few hole prob-

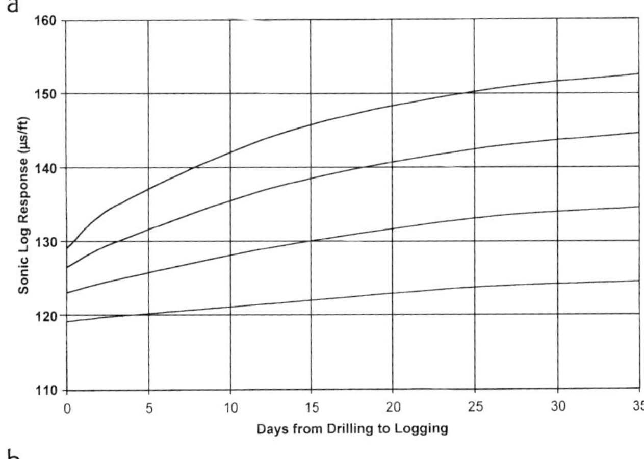

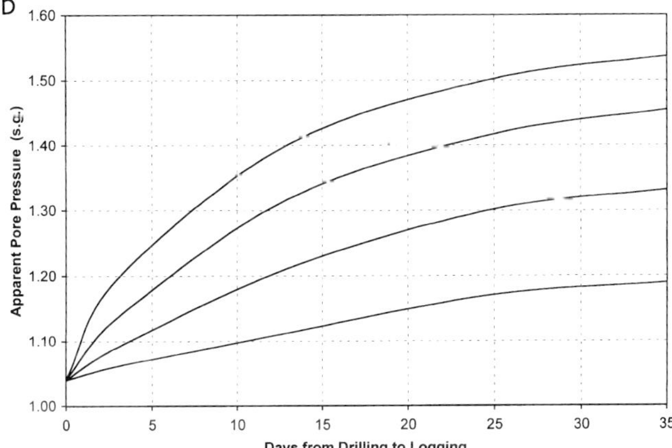

Figure 9. Effect of wellbore alteration with time for four sections of a well on the (a) sonic response of a long-spaced sonic tool (from Blakeman, 1982) and (b) the calculated apparent pore pressures.

lems. A similar response is also observed in the resistivity log for the same hole sections for these two wells.

Figure 11 shows the difference in the distributions of sonic values from Hides-1 to Hides-3. The two wells are considered to be very similar geologically and are geographically the closest wells in the Hides area. The shift in the sonic transit-time distribution to higher values in Hides-1 is considered to be a result of the wellbore damage and instability. This shift in sonic values of approximately 20 µs/ft (0.3 m) from 91 to 112 µs/ft (0.3 m) gives rise to apparent pore pressures. These shifts are consistent with altered sonic traveltimes reported by Hornby and Chang (1985). Using the same approach as previously, for a true pore-pressure gradient in the Ieru shales of 1.8 s.g., the shift in the sonic-log response of 20 µs/ft (0.3 m) would give rise to an apparent pore pressure of approximately 2.05 s.g.

The deviation of transit times from a normal compaction trend, which is used to identify overpressuring, can therefore sometimes be purely an artifact of wellbore instability. The occurrence of wellbore instability and the effect of wellbore damage on the sonic-log response give rise to large errors in pore-pressure prediction. In-gauge boreholes can contain incipient breakouts (Brereton and Evans, 1987; Hornby and Chang, 1985; Addis et al., 1990), which also result in a porosity increase of the borehole wall. These effects would be particularly seen in areas of high tectonic stresses.

For the reliable use of porosity-based log data (e.g., sonic, resistivity, density) the data quality of the log data can only be assured by eliminating the possibility of wellbore instability or wellbore damage. This is also essential where mud-weight data are used as an indicator of pore pressure, particularly in a tectonically active region such as PNG.

POSSIBLE CAUSES OF OVERPRESSURING IN THE PAPUAN FOLD BELT

Several overpressure mechanisms are briefly considered in this chapter as possible causes of overpressuring in PNG. We also refer readers to a detailed review in Kota et al. (1996). Note that the driving mechanisms behind the pressures in the Ieru may be different from those defining the Toro pressures. The following sections constitute preliminary views based on the limited data set used in this project.

Reservoir connection to a recharge area at high topographic elevation is the most obvious potential cause of overpressure in the reservoir in a mountainous terrain. This possibly explains the SE Mananda reservoir pressures (Figure 3b). Reservoir connection to a recharge area at high topographic elevation is also the generally accepted interpretation of the reservoir pressures in the Hides-Angore structure (Grainge, 1993; Eisenberg, 1993; Kotaka, 1996), although the gas-water contact (GWC) has not yet been penetrated. Maximum reservoir water heads in Kutubu-1X and Paua-1X, however, are 3000 m and more than 5500 m, respectively. In the absence of Toro outcrops at these elevations, the pressures would be controlled by other factors.

Compaction disequilibrium associated with rapid sedimentary loading is a common cause of overpressuring in young sedimentary basins. Unlikely, however, is that the high overpressures observed in some of the wells in this study are related to this mechanism, because the geological conditions necessary for this form of overpressuring are absent. Ideally, compaction disequilibrium occurs in thick, abruptly deposited, preferably Tertiary, shale sequences. The Cretaceous Ieru section can be thick in places but contains dominant

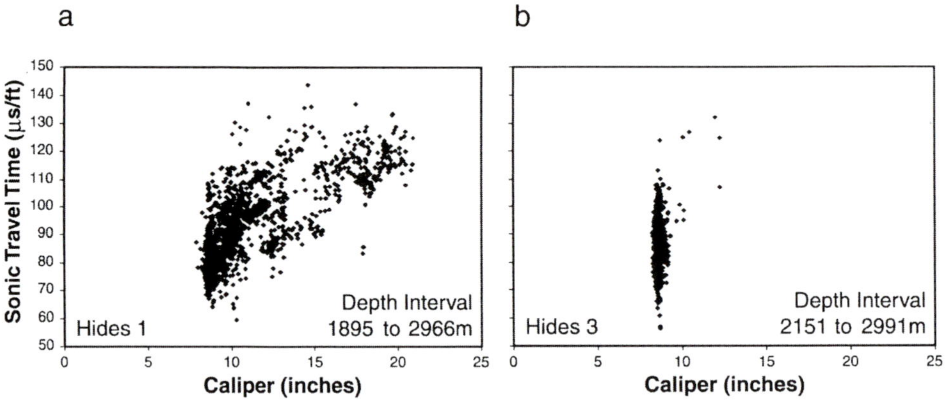

Figure 10. The relationship between the sonic transit time (slowness) and the average hole size encountered in (a) Hides-1 and (b) Hides-3. Both intervals were drilled with 8.5 in. diameter drill bits.

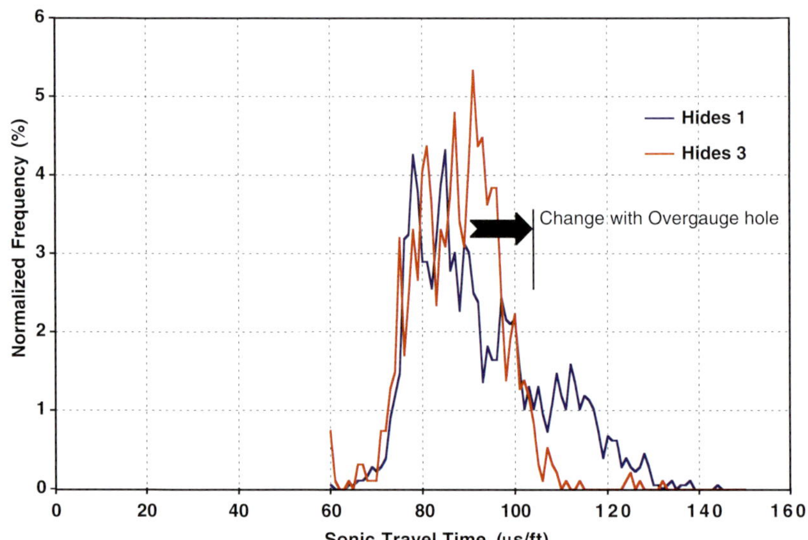

Figure 11. The sonic transit-time frequency distributions for Hides-1 and Hides-3 across the Ieru formation.

sandstone layers (which would have assisted compaction) and is overlain by a massive carbonate sequence. Furthermore, the Ieru formation has been subjected to severe uplift and deformation, which not only radically affects the fluid pressure, but also the porosity and log signature of the sediments. Not surprising, therefore, is that the Ieru formation does not display classic undercompaction anomalies, such as repeatable sonic-velocity reversals on all wells. Discrete reversals in electrical and drilling logs are observed in places, which may reflect some undercompaction, but these are not meaningful in the absence of a compaction trend. They could just as easily be caused by lithological variations, fracturing, or wellbore instability. Under these conditions, undercompaction may be considered a secondary mechanism at best. The previous comments only apply to observations made of the wells in this study; undercompaction may be an important mechanism in geologically recent depositional areas in PNG (e.g., Eastern Gulf of Papua).

Papua New Guinea is a tectonically active area, which is an important consideration in studying formation pressures. Tectonic shearing of low-permeability sediments can result in significant localized overpressures (Yassir and Addis, 2002); unfortunately, these are not commonly associated with porosity anomalies (Yassir and Bell 1996; Yassir and Addis, 2002). Tectonic shearing could be an important overpressure mechanism in PNG, as it seems to be in other overpressured regions, for example, Trinidad, Azerbaijan, and the Gulf of Alaska (Yassir and Addis, 2002). Another component of this mechanism is tectonic overburden loading (Kota et al., 1996). This involves addition of overburden load by overthrusting, which could result in a porosity anomaly.

The volumetric expansion of fluids at depth, related to hydrocarbon generation (or, less likely, smectite dehydration), can result in significant pressuring with some associated porosity increase (Yassir and Bell, 1996). The possibility that oil and gas generation has contributed to the overpressures should not be ignored. A formation can also become overpressured through connection to a pressure source. In PNG, this is most likely to occur by tectonic squeezing of a formation, which causes fluid migration and entrapment into a structure. The shallow pressure gradients associated with hydrocarbon columns result in pressure anomalies at depth, which could be mistaken for overpressure, especially if there is connectivity between the hydrocarbon column in the reservoir and overburden formation pressures. This is the generally accepted interpretation of the high reservoir pressures in the Hides-Angore structure (Grainge, 1993, Eisenberg, 1993; Kotaka, 1996).

Uplift is commonly cited as a potential overpressure mechanism. In theory, a formation with hydrostatic pressures can be uplifted, rendering its pressure out of equilibrium with the shallower depth. This requires the seal to perfectly maintain its volume to prevent pressure equilibration. In practice, however, all geological seals expand and can commonly fracture where they are subjected to a reduction in effective stresses (through uplift in this case). Minimal expansion results in pressure dissipation and even underpressuring in some cases. Uplift as an overpressure mechanism is not proposed for PNG.

CONCLUSIONS

The previous discussion has demonstrated the difficulty of understanding the pressure environment in the Papuan fold belt. It is a geologically complex area with locally compartmentalized pressures. Also, reservoir pressures appear unrelated to the Ieru overburden pressures.

Correct selection of a reference datum, from which to plot any measured data, is crucial to correct interpretation where surface elevations are not constant and where high structural dips and pressure compartmentalization occur. A conventional approach to well planning, which assumes the wellhead to be close to sea level, is not applicable to pressure prediction and evaluation in a mountainous terrain. To make analysis of pressure data from a number of wells meaningful for the shales, the effect of overburden should be considered.

Using different pressure-depth plots, this study has found some pressures to be linked to surface outcrop and others to be compartmentalized. Reservoir pressures of two foreland wells (Ketu-1 and Elevala-1) are shown to be hydrostatic, and two of the fold-belt wells, SE Mananda-1X and SE Mananda-2X, are likely connected to outcrop of reservoir formations high in the fold belt, which explains their mild overpressure. Paua-1X and Kutubu-1X, however, are most likely compartmentalized.

In the course of this study, it was found that standard pressure-prediction techniques using wire-line logs are ineffective in this area. One reason for this is the lack of suitable conditions necessary for reliable log interpretations, specifically, the absence of a relatively uniform and thick shale zone to establish representative compaction trend lines. The use of generic compaction curves is also seen to be of limited use in this area.

Nevertheless, comparison of the sonic-log responses from wells within the same field, wells spatially distant, wells from the active fold belt, and wells from the foreland basin showed little absolute difference in log response. Wellbore instability is likely to be an important influence on log response in these areas and introduces significant errors into the pore-pressure analysis. Another important consideration is the origin of the overpressure, which influences its log response. Undercompaction is unlikely responsible for overpressuring in the area; instead, the large, horizontal stresses acting in this area are likely to be a predominant factor in controlling the pressure. A weak relationship between log response and effective stress was observed for the shaly units, which suggests some geomechanical control on the log response. The data, however, are too sparse to be effectively used for pressure prediction at this stage.

The main conclusion of this study is that in a complex, tectonically active area, such as the Papuan fold belt, normal methods of pressure prediction may not apply, but rigorous analysis and correlation between data can still yield valuable information on the pressure regime.

ACKNOWLEDGMENTS

We would like to thank BP Developments Australia Pty. Ltd. and CSIRO Petroleum for their support of this research. The chapter has also benefited from technical reviews by Philip Caldwell, Mick McWalter, and three anonymous reviewers.

REFERENCES CITED

Addis, M. A., N. R. Barton, S. C. Bandis, and J. P. Henri, 1990, Laboratory studies on the stability of vertical and deviated wellbores: 65th Annual Society of Petroleum Engineers Conference, SPE Paper 20406, p. 19–30.

Addis, M. A., N. C. Last, D. Boulter, L. Ramisa-Roca, and R. A. Plumb, 1993, The quest for borehole stability in the Cusiana field, Colombia: Oilfield Review, v. 5, no. 2/3, p. 33–43.

Addis, M. A., N. Yassir and A. Hennig, 1998, Problems in pore pressure detection: distinguishing between overpressure and wellbore instability, *in* Pressure regimes in sedimentary basins and their prediction: American Association of Drilling Engineers Industry Forum, p. 6.

Aplin, A. C., and Y. Yang, 1995, Assessment of β, the compression coefficient of mudstones and its relationship with detailed lithology: Marine and Petroleum Geology, v. 12, no. 8, p. 955–963.

Baldwin, B., and C. O. Butler, 1985, Compaction curves: AAPG Bulletin, v. 69, no. 4, p. 622–626.

Blakeman, E. R., 1982, A case study of the effect of shale alteration on sonic transit times: 23rd Annual Society of Professional Well Log Analysts Logging Symposium, 14 p.

Boonen, P., C. Bean, R. Tepper, and R. Deady, 1998, Important implications from a comparison of LWD and wireline acoustic data from a Gulf of Mexico well: 39th Annual Society of Professional Well Log Analysts Logging Symposium, 14 p.

Bowers, G. L., 1995, Pore pressure estimation from velocity data: accounting for overpressure mechanisms besides undercompaction: SPE Paper 27488, Society of Petroleum Engineers Drilling and Completion, v. 10, p. 85–95.

Brereton, N. R., and C. J. Evans, 1987, Rock stress orientations in the United Kingdom from borehole breakouts: Report of the Regional Geophysics Research Group, British Geological Survey, no. RG 87/14, 36 p.

Daniels, M. C., 1993, Formation pressure measurements and their use in oil exploration in the Kutubu project, Papua New Guinea: Proceedings of the Second Papua New Guinea Petroleum Convention, p. 579–588.

Eaton, B. A., 1972, The effect of overburden stress on geopressure prediction from well logs: SPE Paper 3719, Journal of Petroleum Technology, v. 24, p. 929–934.

Eisenberg, L. I., 1993, Hydrodynamic character of the Toro Sandstone, Iagifu/Hedinia area, southern Highlands Province, Papua New Guinea: Proceedings of the Second Papua New Guinea Petroleum Convention, p. 447–458.

Eisenberg, L. I., M. V. Langston, and R. E. Fitzmorris, 1994, Reservoir management in an hydrodynamic environment, Iagifu-Hedinia area, southern Highlands, Papua New Guinea: Society of Petroleum Engineers Asia Pacific Oil and Gas Conference, p. 81–90.

Grainge, A., 1993, Recent developments in prospect mapping in the Hides/Karius area of the Papuan fold belt: Proceedings of the Second Papua New Guinea Petroleum Convention, p. 527–537.

Hill, K. C., 1991, Structure of the Papuan fold belt, Papua New Guinea: AAPG Bulletin, v. 75, no. 5, p. 857–872.

Hill, K. C., J. Forwood, C. Rodda, C. Smyth, and G. Whitmore, 1993, Structural styles and hydrocarbon prospectivity around the northern Muller anticline, Papua New Guinea: Proceedings of the Second Papua New Guinea Petroleum Convention, p. 325–334.

Hill, G. S., S. J. Price, M. S. Foster, R. W. Stephenson, D. Ellis, and J. A. Lyslo, 1996, Seismic acquisition in the Papuan fold belt: a new approach: Proceedings of the Third Papua New Guinea Petroleum Convention, p. 445–458.

Holbrook, P. W., 1995, The relationship between porosity, mineralogy and effective stress in granular sedimentary rocks: 36th Annual Society of Professional Well Log Analysts Logging Symposium, 12 p.

Hornby, B. E., and S. K. Chang, 1985, A case study of shale and sandstone alteration using a digital sonic tool: 26th Annual Society of Professional Well Log Analysts Logging Symposium, 11 p.

Hsu, K., G. Minerbo, J. Aron, D. Codazzi, V. Ernst, and T. Lau, 1998, Case studies of time-lapse sonic logging: 39th Annual Society of Professional Well Log Analysts Logging Symposium, 14 p.

Kota, B., D. Trattner, P. Woyengu, and P. K. Webb, 1996, Extreme overpressures and anomalous fluid densities from RFT pressure surveys in Papua New Guinea wells: some results and implications: Proceedings of the Third Papua New Guinea Petroleum Convention, p. 481–493.

Kotaka, T., 1996, Formation water systems in the Papuan basin, Papua New Guinea: Proceedings of the Third Papua New Guinea Petroleum Convention, p. 391–405.

Last, N. C., and M. R. McLean, 1995, Assessing the impact of trajectory on wells drilled in an overthrust region: Society of Petroleum Engineers Annual Meeting, SPE 30465, p. 161–172.

Last, N. C., R. A. Plumb, R. Harkness, P. Charlez, J. Alsen, and M. R. McLean, 1996, Brief: an integrated approach to wellbore stability in the Cusiana field: SPE Paper 30464, Journal of Petroleum Technology, v. 48, p. 245–246.

McLean, M. R., and M. A. Addis, 1990, Wellbore stability: a review of current methods of analysis and their field application: International Association of Drilling Contractors/Society of Petroleum Engineers Drilling Conference, IADC/SPE Paper 19941, p. 261–274.

Mouchet, J. P., and A. Mitchell, 1989, Abnormal pressure while drilling: Manuels Technique Elf Aquitaine 2, 264 p.

Phelps, J. C., and C. N. Denison, 1993, Stratigraphic thickness variations and depositional systems of the Ieru Formation, southern Highlands and western Provinces, Papua New Guinea: Proceedings of the Second Papua New Guinea Petroleum Convention, p. 169–190.

Ramos, G. G., D. E. Mouton, B. S. Wilton, and Leksmono, 1998, Integrating rock mechanics with drilling strategies in a tectonic belt, offshore Bali, Indonesia: Proceedings of the Society of Petroleum Engineers/International Society of Rock Mechanics Eurock '98, SPE/ISRM 47286, p. 69–75.

Skempton, A. W., 1970, The consolidation of clays by gravitational compaction: Quarterly Journal of the Geological Society, v. 125, p. 373–411.

Twynam, A. J., P. A. Caldwell, and K. Meads, 1994, Glycol-enhanced water-based muds: case history to demonstrate improved drilling efficiency in tectonically stressed shales: International Association of Drilling Contractors/ Society of Petroleum Engineers Drilling Conference, IADC/SPE Paper 27451, p. 191–207.

Ward, C. D., K. Coghill, and M. D. Broussard, 1995, Pore- and fracture-pressure determinations: effective-stress approach: SPE Paper 28297 1994, Journal of Petroleum Technology, v. 47, p. 123–124.

Winkler, K. W., 1997, Acoustic evidence of mechanical damage surrounding stressed boreholes: Geophysics, v. 62, no. 1, p. 16–22.

Woodland, D. C., 1990, Borehole instability in the western Canadian overthrust belt: SPE Paper 17508, Society of Petroleum Engineers Drilling Engineering, v. 5, p. 27–33.

Wu, P., D. Scheibner, and W. Borland, 1993, A case of near-borehole shear velocity alteration: 34th Annual Society of Professional Well Log Analysts Logging Symposium, p. 12.

Yassir, N., and M. A. Addis, 2002, Relationships between pore pressure and stress in different tectonic environments, *in* A. R. Huffman and G. L. Bowers, eds., Pressure regimes in sedimentary basins and their prediction: AAPG Memoir 76, p. 79–88.

Yassir, N. A., and J. S. Bell, 1996, Abnormally high fluid pressures and associated porosities and stress regimes in sedimentary basins: Society of Petroleum Engineers Formation Evaluation, v. 48, no. 3, p. 5–10.

10

Geological Controls and Variability in Pore Pressure in the Deep-Water Gulf of Mexico

Michael A. Smith
Minerals Management Service, New Orleans, Louisiana

ABSTRACT

In most areas of the world, pressure-related drilling problems are the leading cause for abandoning a deep-water well or else requiring expensive remedial changes in the drilling and casing programs to reach its targeted reservoir depths. This chapter discusses geological controls and trends in the onset of geopressure in the deep-water Gulf of Mexico, shallow water flow from overpressured sands in the top-hole section, and other pressure-related problems unique to deep water. Pore-pressure prediction has become a subject of intense current interest with several joint industry projects and predictive models now available for government and company participation.

INTRODUCTION

As exploration moves into deeper water in the Gulf of Mexico, pore-pressure prediction and the correct anticipation of overpressured sands becomes more and more critical to the effective evaluation of federal outer continental shelf (OCS) lease blocks. Since 1992, the growth in deep-water activity has been reflected in numerous leasing, drilling, and production statistics. The number of exploratory wells drilled and the number of Exploration Plans filed for deep-water lease blocks have increased by about a factor of 5 since 1994, but many of these leases will expire without being drilled. In addition, many deep-water blocks, initially leased after the OCS Deep Water Royalty Relief (DWRR) Act in 1996 provided economic incentives to develop deep-water fields, will be available by 2006. During the last eight years of the 1990s, the number of deep-water active leases increased from about 1500 to nearly 3900 (Figure 1), about half of the active present-day OCS blocks, including a record number of lease blocks since 1996 in ultradeep water (>5000 ft [1524 m]). Baud et al. (2000) noted that, in the 1990s, the average Gulf of Mexico field size in more than 1500 ft (457 m) of water was 60 million BOE, 12 times the average shallow-water discovery. Deep-water oil now provides more than half of the region's production, and increases in gas production have also offset the shallow-water decline in recent years, with much of new volume coming from subsea completions.

In this chapter, we look at the occurrence of geopressure in about 100 wells in deep water from Viosca Knoll to Alaminos Canyon, most of them drilled in more than 2000 ft (610 m) of water during the last five years. We also analyze shallow water-flow encounters and trends in these areas. As exploratory drilling begins in previously untested geological trends in ultradeep water, new technology and equipment will be needed to control unique pressure-related drilling problems encountered in the exploration and development of hydrocarbon resources in this emerging province.

Smith, Michael A., 2002, Geological Controls and Variability in Pore Pressure in the Deep-Water Gulf of Mexico, in A. R. Huffman and G. L. Bowers, eds., Pressure regimes in sedimentary basins and their prediction: AAPG Memoir 76, p. 107–113.

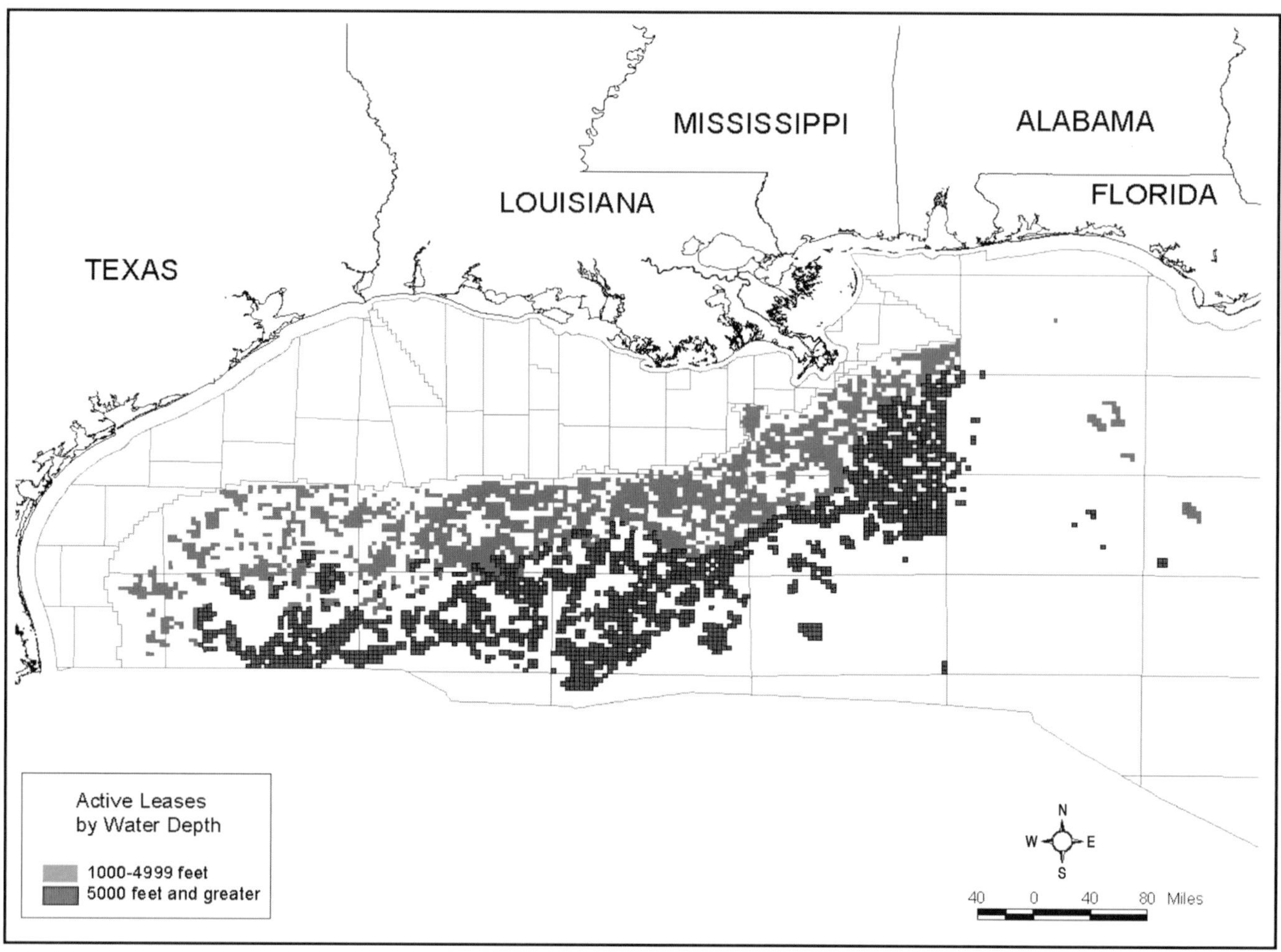

Figure 1. Deep-water (>1000 ft [305 m]) and ultradeep-water (>5000 ft [1524 m]) active leases in the Gulf of Mexico.

PORE-PRESSURE GRADIENTS

Minerals Management Service (MMS) geological reviews of exploration and development plans and applications for permit to drill on Gulf of Mexico OCS leases include a discussion of possible abnormal pressure zones. Geopressure is defined as the situation where pore fluid pressure exceeds normal hydrostatic pressure (Fertl, 1976; Dutta, 1987). This onset of moderate overpressure in continental shelf deltaic sediment occurs where pore pressures are equivalent to 12.5 pound per gallon (ppg) mud weights. In deep water, however, the fracture gradient and shallow casing shoe tests are lower, and the onset of even mild overpressures of 9.5 to 12.0 ppg contributes to many drilling problems such as shallow water flow. Burial rates, geothermal gradients, compaction, and diagenetic reactions are the primary factors affecting the occurrence of geopressure (Law et al., 1998). In deep-water wells, the large seawater column also results in greater depths to abnormal pressure, so depths below the mud line (bml) or sea floor were used in this study in place of vertical subsea depths. Geological factors that control the deposition of turbidite systems, sequence stratigraphy, major faults, unconformities, and salt also affect pore pressure. In complexly faulted structures, formation pressures may be compartmentalized and may vary between different sands.

We analyzed predicted and actual pore pressures, sedimentation rates, and formation temperatures in the deep-water Gulf of Mexico and prepared trend maps of the occurrence of geopressure for this province. The top of geopressure was defined as the depth at which pore-pressure equivalent mud weights, referenced to kelly bushing elevation, exceeded 12.5 ppg. The wells in this study are located in four deep-water sections that include, from east to west, Viosca Knoll/Mississippi Canyon/Atwater Valley, Green Canyon, Garden Banks, and East Breaks/Alaminos Canyon. The upper slope (less than 1000 m of water) in Mississippi Canyon has a thicker Pliocene section with a shallower top of geopressure, an average of about 6950 ft (2118 m) bml, than the deeper water parts of this area. In deeper water, the average top of geopressure occurs in the Miocene at about 10,700 ft (3261 m) bml. In the younger Pliocene–Pleistocene section to the west in Green Canyon, Garden Banks, and East Breaks, the average top of geopressure occurs at about 8700 ft (2652 m) bml. In the deeper water sections in Green Canyon, Garden Banks, and Alaminos Canyon to the south and southeast, however, the top of geopressure occurs in the Miocene at an average depth of about 11,200 ft (3414 m) bml. Throughout the deep-water Gulf of Mexico, as shown in Figure 2, it appears that older and more compacted strata have a deeper top of geopressure than occurs in younger strata.

Except for the northeastern corner of Mississippi Canyon, the thermal gradient in the eastern study area is lower than that of deep-water areas to the west, generally about 1.05°F/100 ft (0.58°C/30.5 m). The thermal gradient falls from an average of 1.25°F/100 ft (0.69°C/30.5 m) in East Breaks to about 1.0°F/100 ft (0.555°C/30.5 m) in Garden Banks, and in Green Canyon the temperature gradient appears to decrease from 1.3 to 0.8°F/100 ft (from 0.72 to 0.44°C/30.5 m) to the southeast with greater water depths. These observations suggest that lower thermal gradients may correspond to a deeper top of geopressure.

Salt domes and ridges that form the boundaries of salt-withdrawal minibasins cause increased pore pressure in the surrounding sediment. This fact results in anomalously high pore pressures in wells drilled on the flanks of a salt dome relative to wells drilled through equivalent strata toward the center of the basin. Pore-pressure ramps or steep increases also occur adjacent to salt masses, and some deep-water exploratory wells have had to be abandoned during attempts to drill through overpressured fractured shale associated with a salt diapir before the reservoir interval was reached. Below tabular salt sheets, formations can be

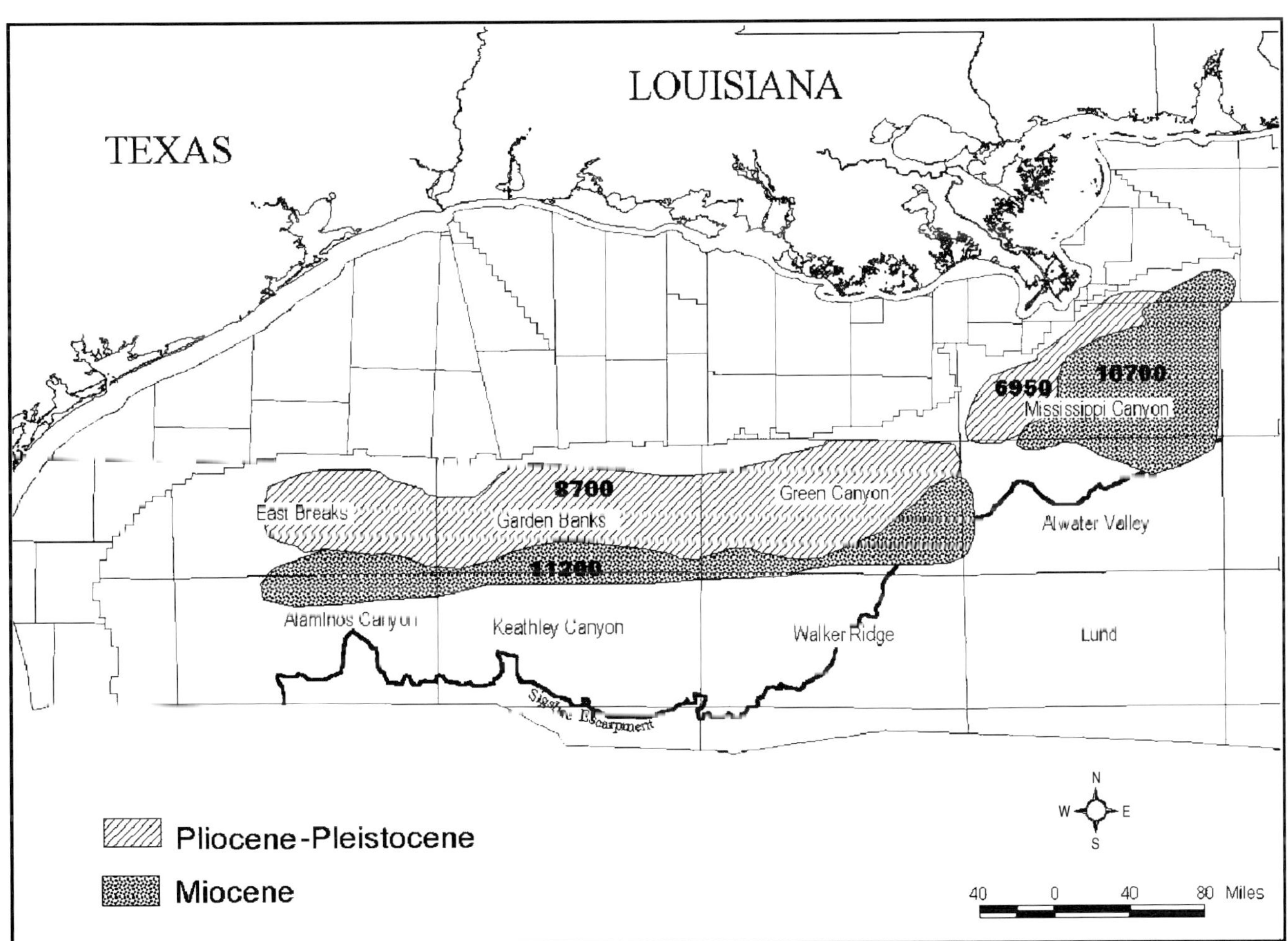

Figure 2. Average depth and stratigraphic interval for the occurrence of moderate overpressures (12.5 ppg pore pressure), deep-water Gulf of Mexico.

overpressured because of an effective seal, and in some subsalt wells a pressure kick has been encountered in the rubble zone below salt. In general, however, the top of subsalt geopressure occurs at greater depths and deeper in the stratigraphic section than in wells without salt.

SHALLOW WATER-FLOW SANDS

Water flow from an overpressured shallow aquifer occurring above the first pressure-containing casing string can significantly impact drilling and cementing practices in addition to the setting depth and number of shallow casing points. This shallow subsurface geohazard may even cause an operator to change a surface location or lose a well. Shallow water-flow sands were deposited as continental slope/fan sequences during upper Pleistocene progradation, the building out of prodelta sandy zones. Since 1984, shallow water-flow occurrences have been reported in about 70 Gulf of Mexico lease blocks covering 55 oil and gas fields or prospects. With a few exceptions, water-flow incidents occur at water depths exceeding 1700 ft (518 m) with a mean value at about 3000 ft (914 m) of water. Water-flow problem sands also typically occur from 950 to 2000 ft (290–610 m) bml but have been reported from 450 to 3500 ft (137–1067 m) below the sea floor. Individual channel-sand units display slumping zones or debris flows with a chaotic seismic character and, in some cases, tilted and rotated slump blocks. In the Mississippi Canyon and southern Viosca Knoll areas, some of the shallowest channel sands can be identified as part of a particular distributary system such as the old Timbalier Channel, Southwest Pass Canyon, or Einstein levee/channel system. High-sedimentation rates and an impermeable mud or clay seal from a condensed section are the main factors contributing to overpressures in shallow water-flow sands (Alberty et al., 1997). These sands occur in several depositional subbasins that are generally bounded by salt ridges or walls. No significant water-flow occurrence, however, is found over tabular salt sills that are 1000 to 10,000 ft (305–3048 m) below the sea floor in some areas. This fact may suggest that communication with the deeper stratigraphic section contributes to abnormal pressures in shallow sands or that the salt forms a positive sea floor topographic feature, preventing sediment loading that might contribute to the generation of overpressures.

The integration of high-resolution multichannel and reprocessed conventional two-dimensional (2-D) and three-dimensional (3-D) seismic data for the top-hole section, further refined by seismic facies analysis, can identify sand bodies with moderate or high shallow water-flow potential. In assessing shallow water-flow risk, information from surrounding wells and shallow borehole tests also provides important data for drilling program design. The MMS Notice to Lessees and Operators (NTL) on shallow hazards requirements for the Gulf of Mexico OCS, NTL 98–20, is currently undergoing extensive revisions (Stauffer et al., 1999). The updated NTL will accommodate the shifting focus of drilling into deeper water and the improved technology and data now available to mitigate deep-water geohazards such as shallow water flow.

Mitigating approaches that have been used in the drilling of shallow water-flow areas include measurement while drilling (MWD) logging plus an annular pressure measurement while drilling (PWD) tool, monitoring and confirming shallow water-flow occurrences with remotely operated vehicles (ROV), and drilling the shallow section as a pilot hole. Additional casing strings and quick-setting foam cements, borehole tests to 1500 to 5000 ft (457–1524 m) bml before development drilling, and other geophysical and engineering techniques that are currently under development have also been employed. The loss of integrity plus buckling or collapse of shallow casing strings in development wells has caused serious economic loss in several cases. Establishing a database of known shallow water-flow occurrences and the most effective methods for controlling them will greatly advance the partnership between the MMS and offshore operators in containing this critical deep-water hazard (Smith, 1999).

OVERPRESSURED SANDS IN ULTRADEEP WATER

In low-margin deep-water drilling areas with abruptly increasing pore pressures and weak fracture gradients, extra casing strings are needed to maintain control in the shallower part of the well. A conventional single-gradient mud system and marine riser maintain bottom-hole pressure with a single mud density from the rig to the bottom of the well, which may require extra casing strings to prevent weaker formations from fracturing. In addition, loop currents or other strong deep-water currents might limit drilling at times because of high riser loads. With a dual-gradient system, however, mud is diverted to separate riser return lines with the effect of replacing the mud from the drilling riser with seawater and referencing pressure gradients relative to the sea floor (Smith and Gault, 2002). The larger hole size maintained at total depth with this technology also allows more completion and production options for deep-water reservoirs.

The northern Gulf of Mexico Basin can be divided into various arcuate tectonic provinces that parallel the shelf/slope break (Diegel et al., 1995; Karlo and Shoup, 1999). Salt-withdrawal minibasins on the continental slope, such as those in the Green Canyon and Garden Banks areas, are bounded by salt walls and filled with the ponded turbidite sands that provide reservoirs for most of the earlier deep-water Gulf of Mexico discoveries. A tabular salt canopy tectonic province occurs in a basinward direction in Walker Ridge and Keathley Canyon, and the Sigsbee Escarpment defines its extent. The middle to lower continental slope contains fold/thrust belts with large prospective geological structures that are the focus of current deep-water drilling and include several recent discoveries (Peel, 1999; Rowan et al., 2000). Figure 3 shows the distribution of hydrocarbon plays in the deep-water Gulf of Mexico, including untested plays in ultradeep water.

In the centroid concept, pore pressure in a reservoir sand at the crest of a high-relief overpressured structure can exceed pore pressure in the bounding shale. Deep-water areas with extensive shallow faulting are particularly vulnerable to low-margin drilling conditions that require extra casing strings. The top of a large, high-relief fold or anticlinal structure at various depths in an exploratory well may contain fluid pressures that approach the fracture gradient in adjacent shale (Traugott, 1997). The mud log from a 1996 ultradeep-water well (Figure 4) provides an example of substantial pore-pressure increases that required closely spaced additional casing strings in the shallow section. This exploratory well was abandoned less than

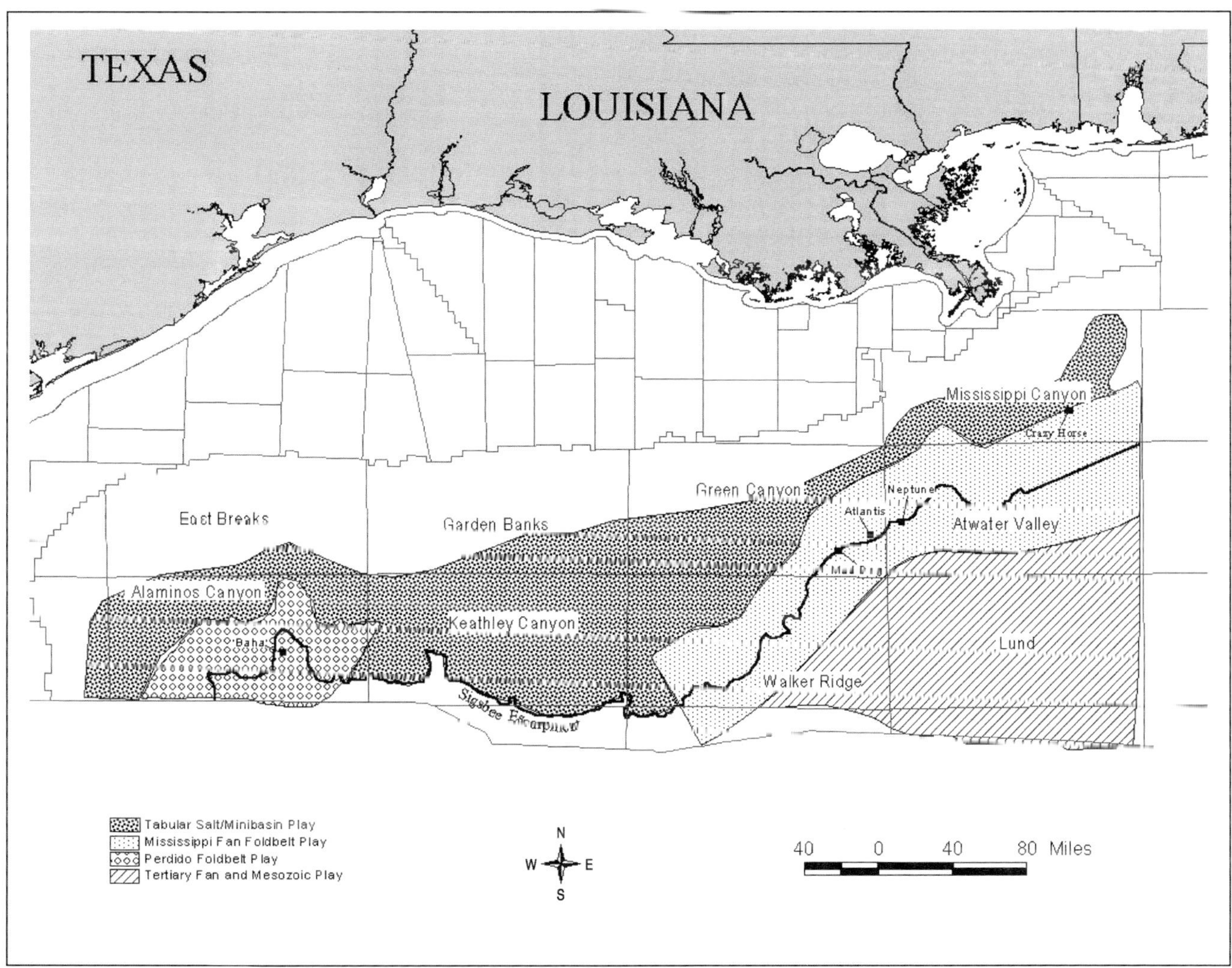

Figure 3. Established and frontier deep-water Gulf of Mexico hydrocarbon plays.

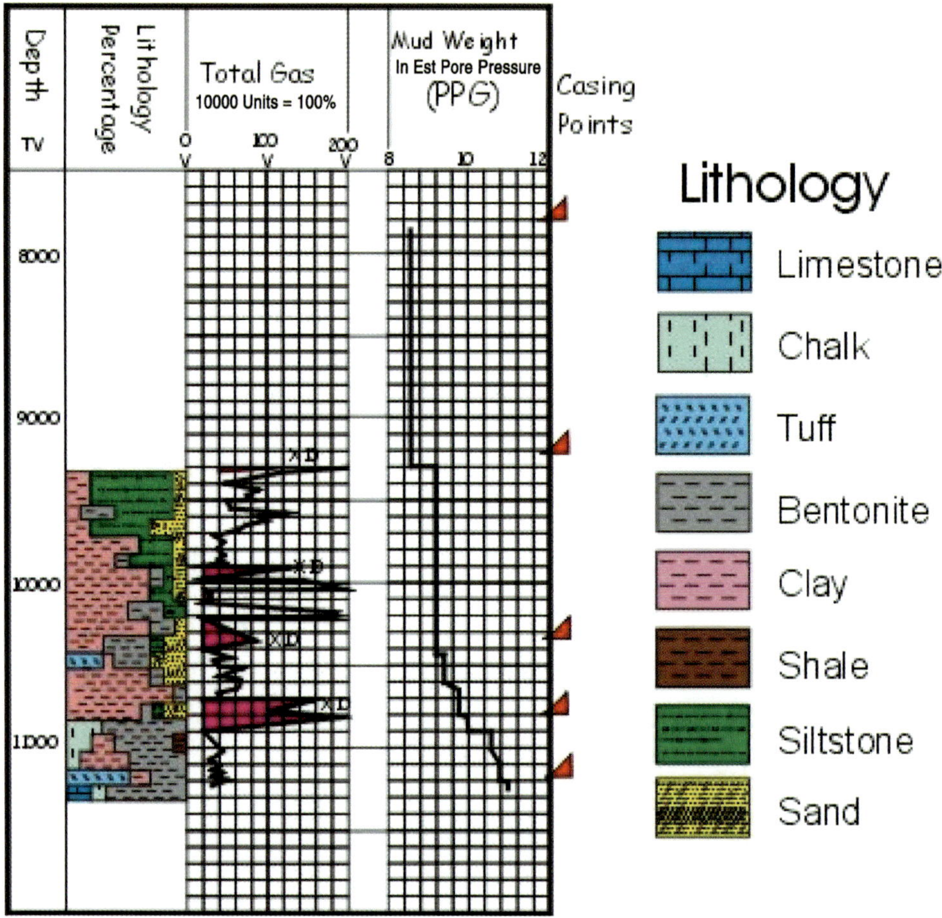

Figure 4. Ultradeep-water exploratory well that encountered rapid pore-pressure buildup requiring extra shallow casing strings. Higher values in the Total Gas track are marked with an x.

3000 ft (914 m) bml because of the narrow margin between pore-pressure and fracture gradient in addition to its small hole size well above the prospective target interval. The use of a dual-gradient/riserless drilling approach and other innovative casing and diverter systems that are under development, however, may contribute the new technologies required for successful exploration in the deepest Gulf of Mexico leases.

CONCLUSIONS

Many of the serious and costly drilling problems in deep water are related to the pore-pressure/fracture-gradient relationship. Other pressure-related hazards, such as shallow water flow, require better predrill identification and quantification of overpressured problem sands. In many Gulf of Mexico frontier deep-water areas, a lack of offset wells mandates better pressure models that incorporate all available geological data. Operations geologists and geophysicists in the MMS are working with deep-water operators to establish databases and methodologies that will improve industry's success in dealing with deep-water geohazards well into the new millennium.

ACKNOWLEDGMENTS

This project was initiated as a result of excellent presentations at the 1998 American Association of Drilling Engineers Industry Forum on Pressure Regimes in Sedimentary Basins and their Prediction. Preliminary results were presented at the 1998 MMS Information Transfer Meeting and the 1999 AAPG International Conference in Birmingham, England. I thank two anonymous reviewers and, particularly, James C. Niemann for their insightful comments, which greatly improved this chapter. Some of the ideas presented here were clarified by discussions with Jim Bridges, Matt Czerniak, Nader Dutta, Pete Harrison, Alan Huffman, Bob Peterson, Paul Post, and Selim Shaker. Finally, I am grateful to the MMS Gulf of Mexico Region, especially to Darcel Waguespack, Fred Times, and Wayne Plaisance, for help and support in the preparation of this chapter.

REFERENCES CITED

Alberty, M. W., M. E. Hafle, J. C. Minge, and T. M. Byrd, 1997, Mechanisms of shallow waterflows and drilling practices for intervention: Offshore Technology Conference Proceedings Paper OTC 8301, p. 241–247.

Baud, R. D., R. H. Peterson, C. Doyle, and G. E. Richardson, 2000, Deepwater Gulf of Mexico: America's emerging frontier: Minerals Management Service Outer Continental Shelf Report 2000-022, 89 p.

Diegel, F. A., J. F. Karlo, D. C. Schuster, R. C. Shoup, and P. R. Tauvers, 1995, Cenozoic structural evolution and tectono-stratigraphic framework of the northern Gulf Coast continental margin: AAPG Memoir 65, p. 109–151.

Dutta, N. C., ed., 1987, Geopressure: Society of Exploration Geophysicists Reprint Series 7, 365 p.

Fertl, H. W., 1976, Abnormal formation pressures: Amsterdam, Elsevier, 382 p.

Karlo, J. F., and R. C. Shoup, 1999, Large patterns become predictive tools to define trends, reduce exploration risk: Offshore, v. 59, no. 7, p. 94–95, 156.

Law, B. E., G. F. Ulmishek, and V. I. Slavin, eds., 1998, Abnormal pressures in hydrocarbon environments: AAPG Memoir 70, 264 p.

Peel, F., 1999, Structural styles of traps in deepwater fold/thrust belts of the northern Gulf of Mexico (abs.): AAPG International Conference, extended abstracts volume, p. 392.

Rowan, M. G., B. D. Trudgill, and J. C. Fiduk, 2000, Deepwater, salt-cored foldbelts: lessons from the Mississippi Fan and Perdido foldbelts, northern Gulf of Mexico: American Geophysical Union Monograph 115, p. 173–191.

Smith, M. A., 1999, MMS regulatory approach to shallow water flow mitigation: Proceedings of the 1999 International Forum on Shallow Water Flows, paper 15, unpaginated.

Smith, K. L., and A. D. Gault, 2002, Subsea mudlift drilling: a new technology for ultradeep-water environments, in A. R. Huffman and G. L. Bowers, eds., Pressure regimes in sedimentary basins and their prediction: AAPG Memoir 76, p. 171–175.

Stauffer, K. E., A. Ahmed, R. C. Kuzela, and M. A. Smith, 1999, Revised MMS regulations on shallow geohazards in the Gulf of Mexico: Offshore Technology Conference Proceedings Paper OTC 10728, v. 1, p. 79–81.

Traugott, M., 1997, Pore/fracture pressure determinations in deep water: World Oil, v. 218, no. 8, p. 68–70.

11

An Easily Derived Overburden Model Is Essential for the Prediction of Pore-Pressure Gradients and Fracture Gradients for Wildcat Exploration in the Gulf of Mexico

Fred R. Holasek
Diamond Offshore Team Solutions, Inc., Houston, Texas

ABSTRACT

Analysis of overburden and Poisson's ratio profiles, in addition to being essential to pore-pressure and fracture-gradient analysis, is critical to calculation of maximum pore pressure where using a microbasin fluid-migration model to predict the pore pressure for a proposed well. A review of the sensitivity of pore-pressure analysis to a range of overburden and Poisson's ratio profiles has been made. A model is presented that is currently being successfully used in the Gulf of Mexico.

INTRODUCTION

Exploration for oil and gas has led our industry into new areas including deep water. This has given our industry new challenges and made us revisit neglected theories.

Two immediate challenges that greet us in new areas of the Gulf of Mexico including deep water are that the pore-pressure and fracture-gradient prediction techniques are not as reliable as desired. The purpose of this chapter is to suggest a method to estimate the overburden and Poisson's ratio profiles for an area of interest. This method is currently being used with a microbasin fluid-migration model to successfully predict pore pressure and fracture gradient for proposed wildcat wells.

Holasek, Fred R., 2002, An Easily Derived Overburden Model Is Essential for the Prediction of Pore-Pressure Gradients and Fracture Gradients for Wildcat Exploration in the Gulf of Mexico, in A. R. Huffman and G. L. Bowers, eds., Pressure regimes in sedimentary basins and their prediction: AAPG Memoir 76, p. 115–124.

BACKGROUND

In the past, exploration for oil and gas has led our industry to drill deeper and to move offshore. When pore-pressure and fracture-gradient prediction gained increased importance, methods based on sound theory were developed to meet these needs for specific areas. For many years, these methods worked reasonably well where applied to specific areas. When our industry moved to different areas, including deep-water areas, not surprisingly, the methods did not work as well as desired.

Two primary reasons exist for the failure of these techniques. One reason for failure is that the tolerance for error is greatly reduced in new frontiers of exploration of deeper wells and deeper water. The second reason is that the existing techniques did not present a reliable method of predicting pore pressure prior to the drilling of a proposed wildcat well in a way that honored geological events.

Some work has been done with seismic stacking velocities. Working closely with many major exploration

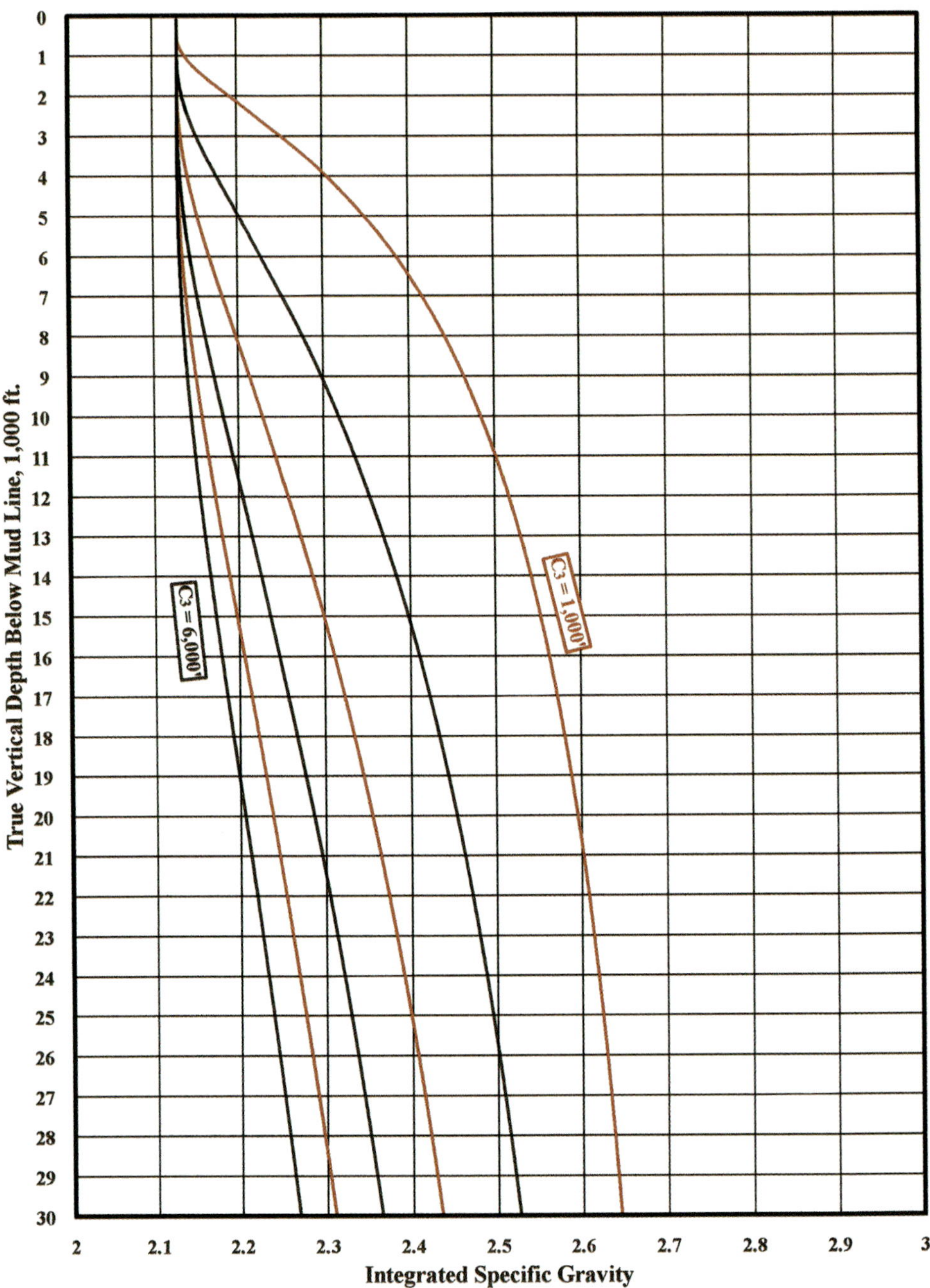

Figure 1. Integrated specific gravity vs. depth.

corporations, I have noted a general disappointment with the results. Using the refined overburden model and honoring all geological events dramatically improves the results of stacking-velocity analysis.

I have noted in the past four years that many of the major exploration corporations are using and perfecting microbasin fluid-migration models to estimate the pore-pressure profile for proposed wells. The concept of active fluid migration from the deeply buried Mesozoic carbonate source rocks in the Gulf of Mexico appears to be becoming widely accepted (Sassen, 1993).

To get the desired results from the models, all geological events must be honored including the pressure at which active fluid migration occurs. This requires a good estimate of overburden at the proposed drilling location.

The objective of the chapter is to present a general overburden model that is currently being successfully

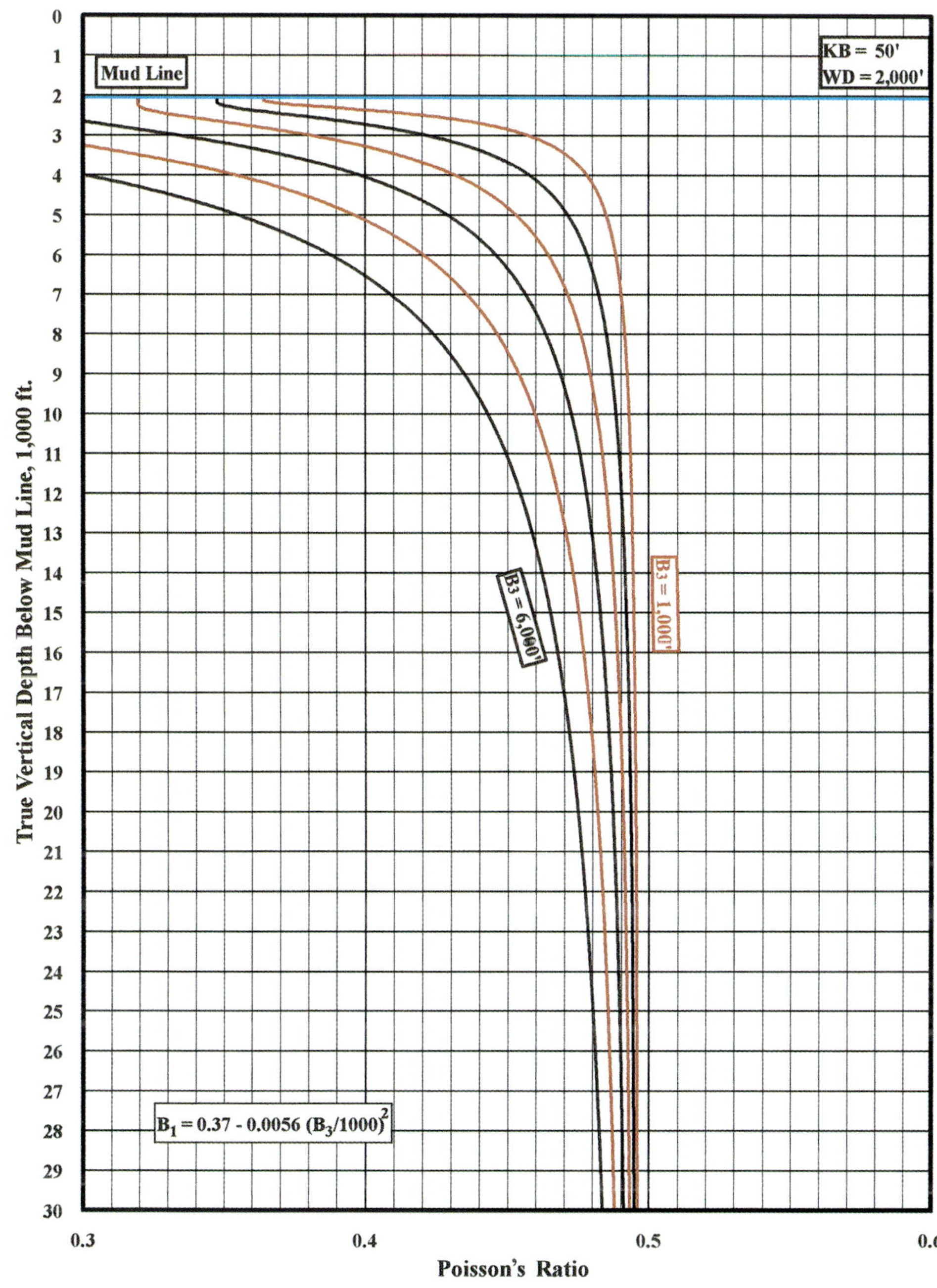

Figure 2. Poisson's ratio vs. depth. KB = kelly bushing; WD = water depth.

used in the Gulf of Mexico to estimate overburden in conjunction with estimation of pore pressure for rank wildcat drilling locations.

THEORY

One of the most popular theories for fracture-gradient prediction was developed by Eaton. Eaton's equation (restated in Eaton and Eaton, 1997) is

$$F/D = [\nu/(1 - \nu)]\,[(S/D) - (p/D)] + (p/D) \quad (1)$$

where F = fracture extension pressure (psi); p = pore-pressure (psi); S = overburden pressure (psi); D = true vertical depth; and ν = Poisson's ratio (dimensionless). The use of this equation to produce a fracture gradient vs. true vertical depth requires

1. Pore-pressure gradient (p/D) vs. true vertical depth
2. Overburden pressure gradient (S/D) vs. true vertical depth
3. Poisson's ratio vs. true vertical depth

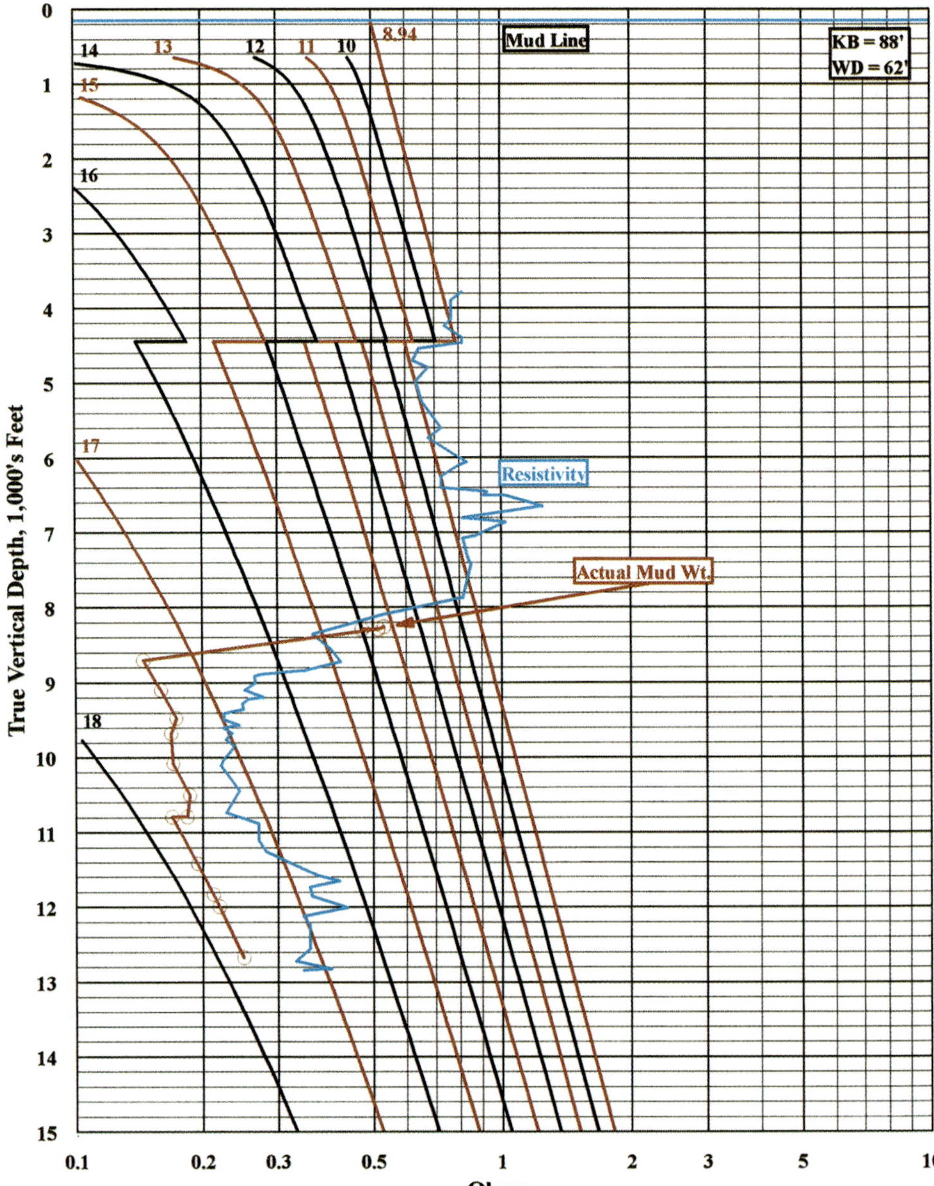

Figure 3. Resistivity vs. pore-pressure equivalent mud weight. Company: confidential; well: confidential; location: confidential, offshore Texas. KB = kelly bushing; WD = water depth. Reasons for the curve jumps are addressed in Holasek (2002).

One of the most popular theories for deriving formation pore pressure from well logs was developed by Eaton (1975). Eaton's equations for formation pore-pressure gradient prediction using shale resistivity (R_o), conductivity (C_o), and interval transit time Δt_o are

$$p/D = (S/D) - [(S/D) - (p_n/D)_n]\,[R_o/R_n]^a \quad (2)$$

$$p/D = (S/D) - [(S/D) - (p_n/D)]\,[C_n/C_o]^a \quad (3)$$

$$p/D = (S/D) - [(S/D) - (p_n/D)]\,[\Delta t_n/\Delta t_o]^b \quad (4)$$

where p/D = formation pore-pressure gradient (psi/ft); S/D = overburden pressure gradient (psi/ft); p_n/D = normal pore-pressure gradient (0.465) psi/ft; R_o = shale resistivity (Ω m); R_n = normal compaction-line shale resistivity (Ω m); C_o = shale conductivity from well log (mmho/m); C_n = normal shale conductivity (mΩ/m); Δt_o = interval transit time through shale (μs/ft); Δt_n = normal interval transit time through shale (μs/ft); a is given as a constant with a value of 1.2; b is given as a constant with a value of 3.0.

DISCUSSION

Close examination of the equations for pore-pressure equations suggest that they would be very sensitive to

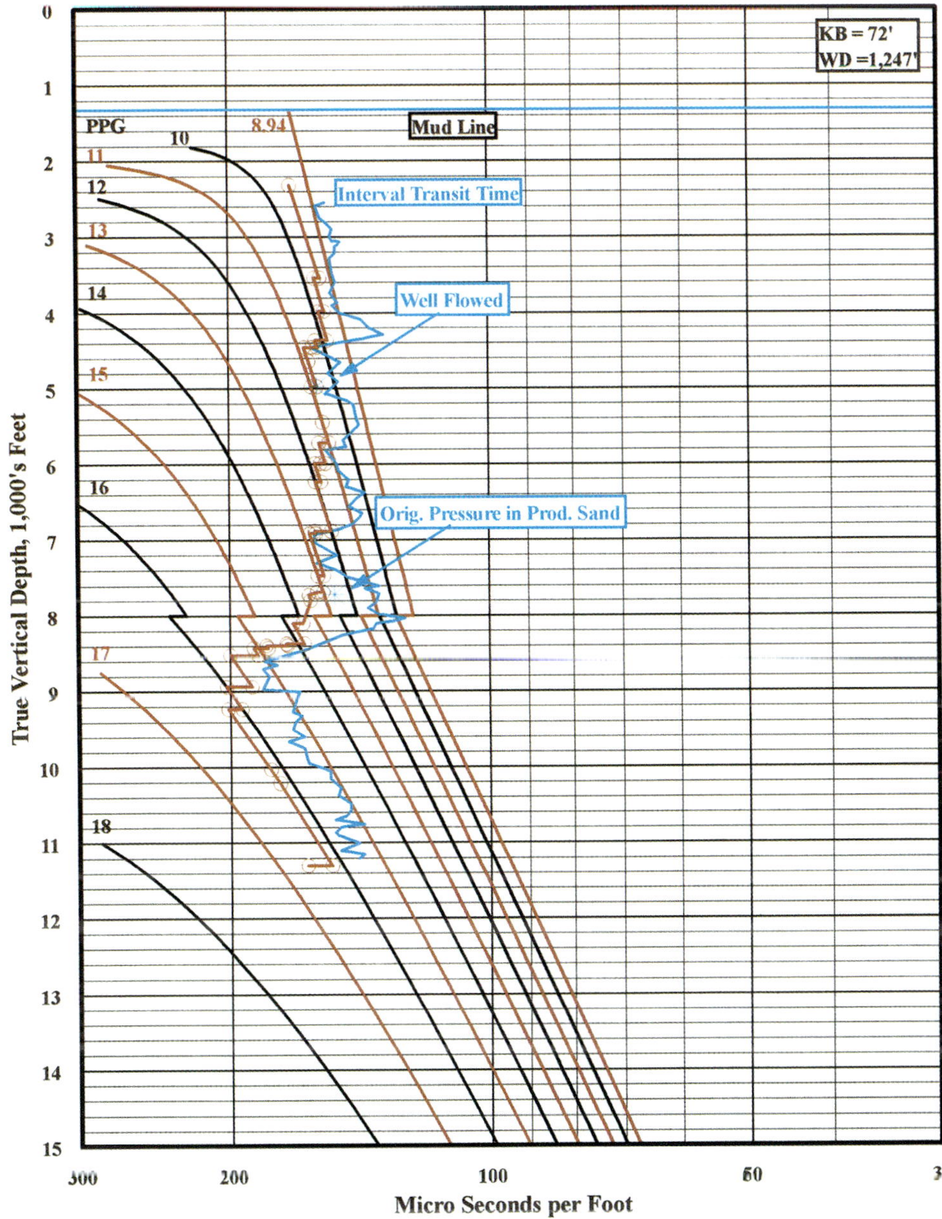

Figure 4. Interval transit time vs. pore-pressure equivalent mud weight. Company: confidential; well: confidential; location: confidential, offshore Louisiana. KB = kelly bushing; WD = water depth. Reasons for the curve jumps are addressed in Holasek (2002).

changes in overburden profile. Examination of the fracture-gradient equation suggests that it would be very sensitive to changes in overburden and to a lesser degree, changes in Poisson's ratio profile.

Immediately, five questions arise:

1. How much does the overburden profile change in the Gulf of Mexico?
2. How much does the Poisson's ratio profile change in the Gulf of Mexico?
3. How much do the observed changes in overburden profile change the resistivity vs. pore-pressure relationship?
4. How much do the observed changes in overburden profile change the interval transit time vs. pore-pressure relationship?
5. How much do the observed changes in overburden profile and Poisson's ratio profile change the fracture-gradient profile?

GENERAL CONCEPT

Ideally, hundreds of high-quality density logs from mud line to more than 20,000 ft (>6096 m) should be analyzed, and the resulting overburden profiles should

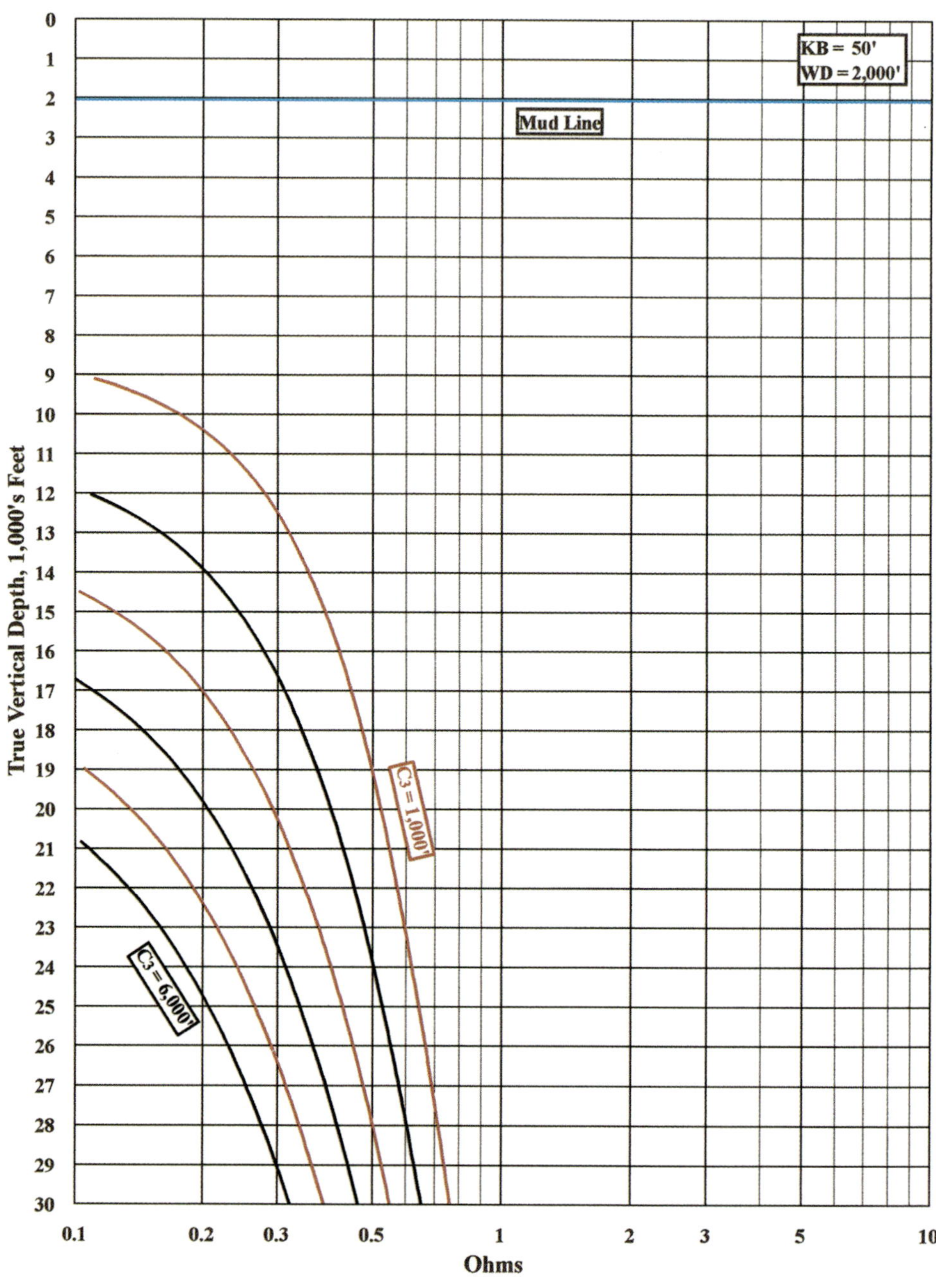

Figure 5. Resistivity vs. pore-pressure equivalent mud weight. 17 ppg curves vs. comparison factor C_3. KB = kelly bushing; WD = water depth.

be related. Very few high-quality density logs from mud line to more than 20,000 ft (>6096 m) exist.

Another approach was taken. A general overburden model was developed from the few available and usable density logs. Also, a general Poisson's ratio model was developed using the associated drilling data. The general overburden model and the pore-pressure equations were used to develop a real-time interactive graphical program to allow matching of actual data. The general overburden model, the general Poisson's ratio model, and the fracture-gradient equation were combined in a program to allow matching of actual data.

The general concept was to assume an overburden profile and analyze the logs and well data to produce a pore-pressure profile. Then the pore-pressure profile was used in the fracture-gradient model and the overburden and Poisson's ratio models were modified to match actual data. The new overburden profile was used to reanalyze the log and well data to produce a second pore-pressure profile. This process was re-

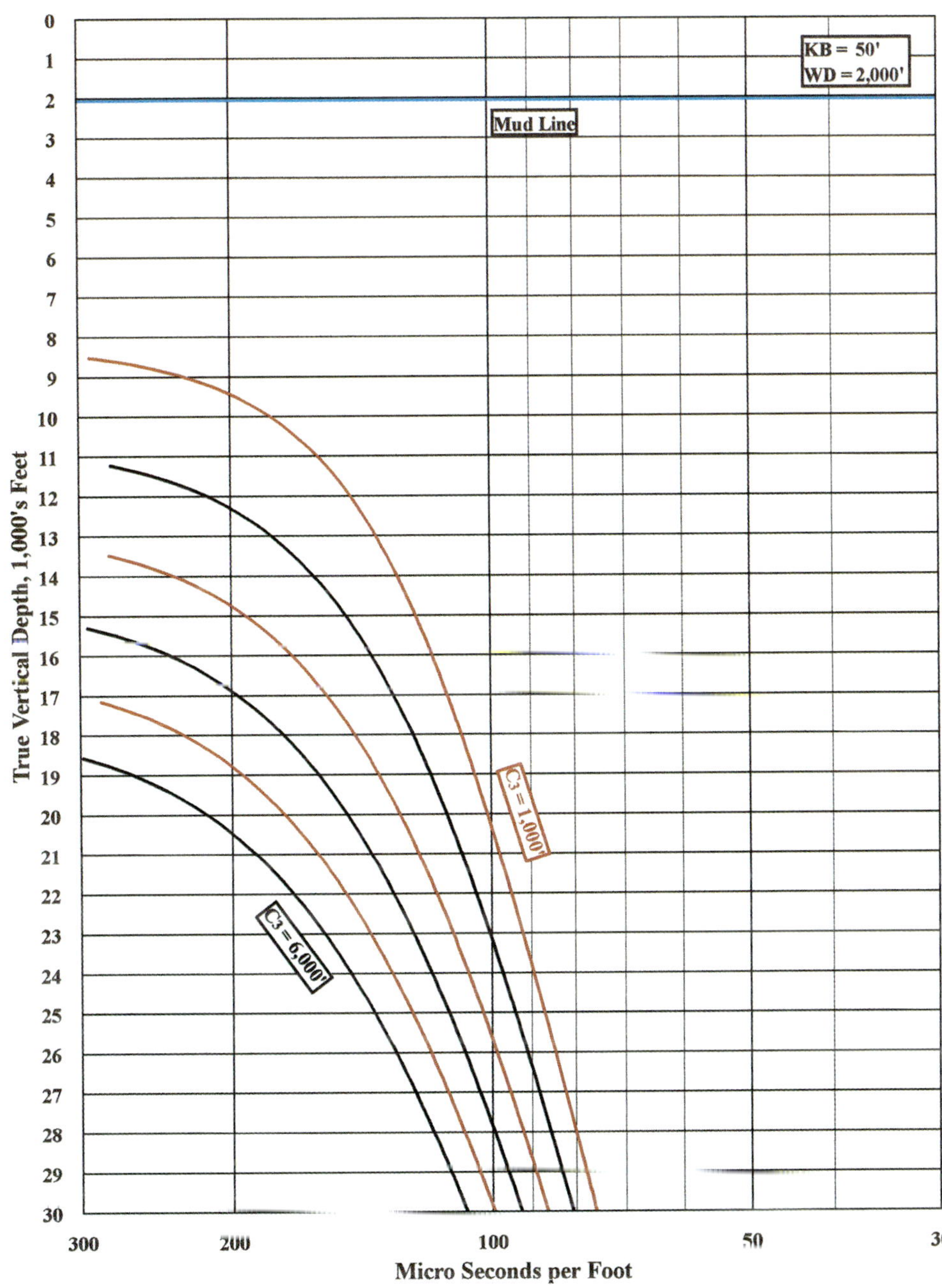

Figure 6. Interval transit time vs. pore-pressure equivalent mud weight. 17 ppg curves vs. comparison factor C_3. KB = kelly bushing; WD = water depth.

peated to yield a unique solution for the overburden profile and Poisson's ratio profile.

GENERAL OVERBURDEN MODEL

Briefly reviewing the basics, the overburden for any given point is the sum of weight of air from kelly bushing (KB) to mean sea level, plus the weight of seawater from mean sea level to mud line, plus the weight of the sediments from mud line to the point of interest. Where used in the gradient form, the reference point is the KB. The general equation for overburden is

$$OB = \int_{TVD_{MSL}}^{TVD_{mud\ line}} f(Spg_{sw})d(TVD) + \int_{TVD_{mud\ line}}^{TVD_0} f(Spg_{sed})d(TVD) \quad (5)$$

where OB = integrated overburden specific gravity; Spg_{sw} = specific gravity of sea water; Spg_{sed} = specific gravity of the sediment at TVD; TVD_{MSL} = true vertical depth from KB to mean sea level; $TVD_{mud\ line}$ = true vertical depth from KB to mud line; and TVD_o = true vertical depth at point of interest.

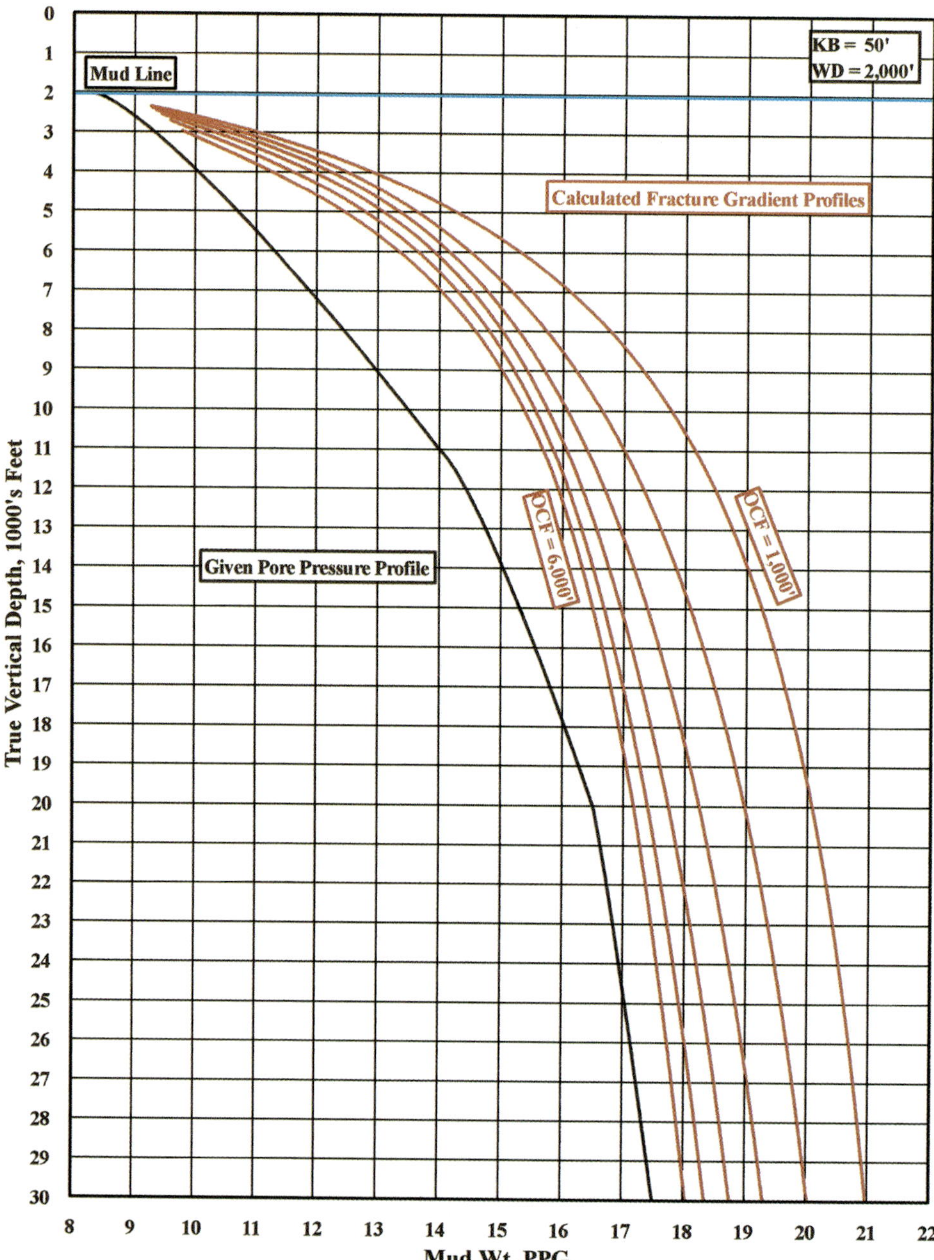

Figure 7. Fracture gradient vs. overburden comparison factor (OCF) C_3. KB = kelly bushing; WD = water depth.

Please note that the overburden is presented using specific gravity. I have found that the advantages of using specific gravity to relate to closely associated disciplines outweigh any minor inconveniences. Depth is presented in feet, which is the convention in the Gulf of Mexico (metric values in parentheses).

The overburden of air from KB to sea level is assumed to be negligible. The calculation of contribution to overburden of seawater is assumed to approach the linear function of seawater specific gravity in the area of interest. The remaining part of the overburden is estimating the specific gravity of the sedimentary rocks to the point of interest.

Define $\mathrm{Spg_{intg}}$ as

$$\mathrm{Spg_{intg}} = \left(\int_{\mathrm{TVD_{mud\ line}}}^{\mathrm{TVD_0}} f(\mathrm{Spg_{sed}}) d(\mathrm{TVD}) \right) / \mathrm{TVDBML_0} \quad (6)$$

$$Spg_{intg} \cong \left(\sum_{TVDBML_{mud\ line}}^{TBDBML_0} (Spg_{sed})\Delta(TVDBML)\right) / TVDBML_0 \quad (7)$$

where Spg_{intg} = integrated specific gravity of formation sediment; TVDBML = true vertical depth below mud line; $TVDBML_{mud\ line}$ = 0 ft; $TVDBML_o$ = true vertical depth below mud line at the point of interest. The overburden stress can then be written as

$$S = 0.4335\ (Spg_{SW}WD + Spg_{intg}TVDBML_o) \quad (8)$$

where WD is water depth (ft), and 0.4335 is a conversion factor from specific gravity to psi/ft.

Formation density logs in digital form were evaluated for Spg_{intg}. I noted that the profile varied from area to area and block to block. I also noted that Spg_{intg} could be matched with the following generalized equation.

$$Spg_{intg} = C_1 + (C_2 - C_1)/(\exp(C_3/(TVDBML)^{(0.8)(n)}) \quad (9)$$

where $C_1 = Spg_{intg}$ at the mud line. In the evaluation of many wells, this factor did not deviate significantly from a value of 2.13. C_2 equals Spg_{intg} at infinity. In the evaluation of many wells, this factor has rarely deviated from a value of 2.8. Nearby salt (halite) can reduce the value. C_3 is the comparison factor, and TVDBML is measured in feet. In the evaluation of many wells, this factor has ranged from as small as 1000 ft (305 m) to as much as 6000 ft (1829 m). The variable n is the exponential modifier. In the evaluation of many wells, this factor did not deviate from the value of 1.0 with the exception of below a salt raft. Note that integration of the overburden function must honor discontinuities such as halite.

Question No. 1–How Much Does the Overburden Profile Change in the Gulf of Mexico?

A plot of integrated specific gravity vs. depth for C_3 comparison factors ranging from 1000 ft (305 m) to more than 6000 ft (>1829 m) was prepared and is presented in Figure 1. In general, wells with a C_3 comparison factor greater than 4000 ft (1219 m) were in areas of higher depositional rates. Wells with a C_3 comparison factor less than 3000 ft (914 m) were in areas of low depositional rates, that is, older strata nearer the mud line.

The majority of the wells drilled in the Gulf of Mexico should be contained within the plotted limits of the C_3 comparison factor. To put the range of Spg_{intg} curves into perspective, a constant overburden gradient of 1 psi/ft below mud line would plot as a vertical line with a Spg_{intg} of 2.3.

General Poisson's Ratio Model

The Poisson's ratio model was developed from wells analyzed with a useful density log to derive a unique overburden equation. The Poisson's ratio model was developed to match leak-off tests. The Poisson's ratio model is represented by the following general equation:

$$\nu = B_1 + (B_2 - B_1)/(\exp(B_3/(TVDBML)^{(1.16)(n)}) \quad (10)$$

where $B_1 = \nu$ at the mud line. In the evaluation of many wells, this factor generally could be estimated using the following relationship:

$$B_1 = 0.37 + 0.0056(C_3/1000)^2 \quad (11)$$

where $B_2 = \nu$ at infinity. In the evaluation of many wells, this factor did not deviate from the value of 0.497. B_3 is the comparison factor, and TVDBML is measured in feet. In the evaluation of many wells, this factor generally is equal to C_3. The variable n is the exponential modifier. In the evaluation of many wells, this factor rarely deviated from the value of 1.0.

Question No. 2–How Much Does the Poisson's Ratio Profile Change in the Gulf of Mexico?

A plot of Poisson's ratio vs. depth for B_3 comparison factors ranging from 1000 ft to 6000 ft (305–1829 m) is presented in Figure 2.

Pore-Pressure Analysis

Two interactive graphic programs were developed for pore-pressure analysis. One was based on Eaton's resistivity vs. pore-pressure equation, and the other was based on the Eaton's interval transit time vs. pore-pressure equation.

Each model has a provision for three different base lines, sometimes referred to as normal pressure (0.465 psi/ft = 8.94 ppg) lines. Each base line is an equation for a straight line on a semi-log base, with intercept at mud line and slope available for modification. The equation for each base line is:

$$\text{base line} = a(10^{(b)(depth)}) \quad (12)$$

where a = the intercept of the base line at the mud line, and b = the slope of the base line.

Eaton's equations were solved for a series of constant pore-pressure gradient curves from 10–20 ppg, incorporating the base line and general overburden equation, and allowing for the exponent within Eaton's equation to be modified.

The interactive graphic program also plotted actual mud weights used and known pressure data such as repeat formation tests, bottom-hole pressure survey results, or estimated bottom-hole pressure from well-kick data.

An example of the resistivity vs. pore-pressure plot is shown in Figure 3. An example of the interval transit time vs. pore-pressure plot is shown in Figure 4. Reviewing Figures 3 and 4, it is readily apparent that a base trend is not necessary for pore-pressure analysis.

Most of the wells encountered multiple geological units with different base-line slopes and intercepts. The geological units appeared to have different and complex depositional and fluid-migration histories.

Question No. 3–How Much Do the Observed Changes in Overburden Profile Change the Resistivity vs. Pore-Pressure Relationship?

A resistivity vs. pore-pressure plot was prepared showing only the 17 ppg pore-pressure curve for a fixed set of variables. Without changing any other variables, a 17 ppg curve was prepared for different C_3 comparison factors from 1000 ft to 6000 ft (305–1829 m) (Figure 5). A significant difference exists between the curves. Constant pressure curves less than 17 ppg had progressively less difference. Conversely, constant pressure curves greater than 17 ppg had progressively greater differences. Clearly, resistivity vs. pore pressure equations are very sensitive to the changes in the overburden profile seen in the Gulf of Mexico.

Question No. 4–How Much Do the Observed Changes in Overburden Profile Change the Interval Transit Time vs. Pore-Pressure Relationship?

A plot of interval transit time vs. pore pressure was prepared showing only the 17 ppg pore-pressure curve for a fixed set of variables. Without changing any other variables, a 17 ppg curve was prepared for different C_3 comparison factors from 1000 ft to 6000 ft (305–1829 m) (Figure 6) .

A significant difference exists between the curves. Constant pressure curves less than 17 ppg had progressively less difference. Conversely, constant pressure curves greater than 17 ppg had progressively greater differences. Clearly, interval transit time vs. pore-pressure equations are very sensitive to the changes in the overburden profile seen the Gulf of Mexico.

Question No. 5–How Much Do the Observed Changes in Overburden Profile and Poisson's Ratio Profile Change the Fracture-Gradient Profile?

A set of fracture-gradient profiles was calculated for a given pore-pressure profile for different C_3 comparison factors from 1000 ft to 6000 ft (305–1829 m). The noted relationships for B_1 and B_3 were used for this plot.

For this chapter, C_3 has been defined as the "overburden comparison factor" (Figure 7). There is a significant difference between the curves. Clearly, fracture-gradient equations are very sensitive to changes in the overburden profile seen in the Gulf of Mexico.

CONCLUSIONS

The sensitivity of Eaton's pore-pressure and fracture-gradient equations can easily be used to refine an overburden profile for a given area. Honoring all geological events, an overburden profile can be easily derived for a proposed drilling location using conventional integration techniques. Using the derived overburden profile and a fluid-migration model, a pore-pressure profile can be derived for the proposed well. Using the derived overburden profile and a derived interval velocity profile, an analysis of the seismic stacking velocity for pore pressure can be made.

REFERENCES CITED

Eaton, B. A., 1975, The equation for geopressured prediction from well logs: Society of Petroleum Engineers Paper 5544, 5 p.

Eaton, B. A., and T. L. Eaton, 1997, Fracture gradient prediction for the new generation: World Oil, v. 218, no. 10 (October), p. 93–100.

Holasek, F., 2002, Evaluation of deepwater drilling prospects using new concepts to identify, quantify, and mitigate (IQM) risks for well design: Society of Professional Engineers Paper 74489.

Sassen, R., 1993, Association of oil seeps and chemosynthetic communities with oil discoveries upper continental slope, Gulf of Mexico, *in* N. C. Rosen, ed., Transactions of the 43rd annual convention of the Gulf Coast Association of Geological Societies: GCAGS, v. 43, p. 349–355.

12

Fracture-Gradient Predictions in Depleted Sands in the Gulf Coast Sedimentary Basin

Baldeo Singh
Unocal, Indonesia, Balikpapan, Indonesia

Nelson Emery
Unocal, Lafayette, Louisiana

ABSTRACT

The primary objective of this work was to estimate fracture gradients in depleted sands. This information is necessary in predicting lost-circulation zones and optimizing drilling plans. Large reserves are known to exist in deeper horizons underneath mature fields in several basins in the United States, such as the Gulf Coast, the Permian basin, and the West Coast.

Historically, fracture-gradient predictions are made for shales to enable selection of casing points. In this chapter, the shale fracture gradients are related to the fracture gradient in sands (virgin conditions) using a lithology factor. The lithology factor is a function of Poisson's ratio, a physical property considered to be the most significant rock-mechanical property in lithology determination. For ease of application, Poisson's ratio is related to the shaliness of the sand. The fracture gradient so computed is then discounted to account for pressure depletion using a field correlation from the literature.

Although the method of fracture-gradient prediction has only been tested on Gulf Coast wells, the formulation is general and may have application in other basins. In its partial form, this procedure may also be used to predict contrast between shale and sand fracture gradients to design stimulation jobs.

FRACTURE GRADIENTS IN THE EARTH

Several techniques have been suggested for estimating fracture gradients in the earth (Matthews and Kelly, 1967; Pennebaker, 1968, Eaton, 1969; Christman, 1973). They all attempt to relate the overburden gradient and pore pressure to the horizontal stress gradient (or fracture gradient). For a gravity-dominated system, such as the Gulf Coast, the basic equation is written as

$$[G_f - G_p] = K[G_{ob} - G_p] \qquad (1)$$

where G_f is the fracture gradient of shale, G_{ob} is the overburden gradient, G_p is the pore-pressure gradient of the formation, and K is the stress factor. The proportionality factor, K (stress factor), relates the vertical effective stress to the horizontal effective stress being generated due to compaction. The stress factor is essentially an empirical function that accounts for the complexity of lithology and diagenesis processes as the rock is buried in the ground. Pilkington (1978) reviewed the published data on fracture gradients (Matthews and Kelly, 1967; Pennebaker, 1968; Eaton, 1969; Christman, 1973). He found that all of these data could be normalized to produce a single stress-factor curve. This observation has been used widely in making predictions for fracture gradients.

Singh, Baldeo, and Nelson Emery, 2002, Fracture-Gradient Predictions in Depleted Sands in the Gulf Coast Sedimentary Basin, in A. R. Huffman and G. L. Bowers, eds., Pressure regimes in sedimentary basins and their prediction: AAPG Memoir 76, p. 125–129.

Table 1. Summary of Important Parameters in Predicting Lost Circulation through Depleted Sands, Gulf Coast, United States

Field Name	Gamma-Ray Log Value for Sand	Pore Pressure (ppg)		Fracture Gradients (ppg)			Mud Weight (ppg)	Remarks: Lost Circulation?
		Virgin Sand	Depleted Sand	Shale	Virgin Sand	Depleted Sand		
Ship Shoal–208(E–14)	30	12.6	1.9	17.1	15.2	11.1	12.7	Yes
North Fresh Water Bayou(F–20)	30	14.9	9.5	18.5	17.0	14.5	15.7	Yes
North Fresh Water Bayou(E–9)	30	13.8	9.4	18.5	16.5	14.5	17.0	Yes
West Cameron–196(A–4)	30	9.72	6.98	16.7	13.7	12.7	12.0	No
West Cameron–280(A–4 ST)	30	8.2	4.2	16.7	13.1	11.6	11.2	No
West Cameron–280(A–4 ST)	30	9.7	3.4	16.7	13.7	11.4	11.2	No
Vermillion–39(12)	30	8.8	0.72	16.3	13.1	10.3	10.0	No
South Marsh Island–48(C–2)	40	10.2	2.6	17.6	15.0	12.0	11.0	No
Lake Pagie(B–17)	30	11.0	2.4	17.9	15.0	11.6	14.5	Yes

PROPOSED MODIFICATION

In the proposed model, the stress factor, K, is subdivided into two factors: one to account for lithology, and the other to account for the nonlinearity of the stress pattern found in the earth. The lithology factor is based on the plane-strain solution from elasticity theory for horizontal effective stress as a function of vertical effective stress and is a function of Poisson's ratio. It is equal to $\nu/(1 - \nu)$, where, in this chapter, ν is Poisson's ratio for shale. Poisson's ratio is considered to be the most significant rock-mechanical property in lithology determination. All other nonlinear effects due to factors such as diagenesis and compaction are lumped in the second stress-correction term, K_c. This correction term, K_c, is expected to be area specific and represents the characteristics of that particular region. The resulting equation may now be written as

$$[G_f - G_p] = K_c[\nu/(1 - \nu)][G_{ob} - G_p] \quad (2)$$

By doing so, we have incorporated the effects of lithology explicitly, through Poisson's ratio. Although the previous formulation was written for shale, it is general in nature and can be easily modified to represent sand by using the Poisson's ratio of sand. Also note that the correction term, K_c, is independent of lithology.

We write, explicitly, an equation for the fracture gradient of a sand adjacent to the shale:

$$[G_f^{sd} - G_p] = K_c[\nu^{sd} / (1 - \nu^{sd})][G_{ob} - G_p] \quad (3)$$

where G_f^{sd} is the fracture gradient in the sand, and ν^{sd} is Poisson's ratio of the xsand.

Dividing equation 2 by equation 3, we get

$$[G_f - G_p]/[G_f^{sd} - G_p] = [\nu/(1 - \nu)][(1 - \nu^{sd})/\nu^{sd}] \quad (4)$$

By this manipulation, we are able to relate the fracture gradients in shale to the fracture gradient in sand using Poisson's ratio. This is a very useful result and is used in this chapter in computing fracture gradients in sands at virgin pore-pressure conditions where the fracture gradient in an adjacent shale is known. Historically, good correlations exist in the literature to predict fracture gradients in shales, such as Eaton's relation (1969) for the Gulf Coast. Alternatively, these may be obtained from well-characterized leak-off tests.

ESTIMATION OF POISSON'S RATIO USING GAMMA-RAY-LOG VALUES

The ultimate objective here is to relate Poisson's ratio to gamma-ray–log values, which, unlike Poisson's ratio, are readily available. A linear model is used to relate the two.

Typical formation sand is a mixture of numerous minerals, such as clays, quartz, and dolomite. For simplicity of modeling, however, we have assumed that the two dominant constituents comprising sands are pure shales (primarily clays) and pure sand (primarily quartz). For a given shaliness, f_{sh}, Poisson's ratio of the formation sand is then computed as follows:

$$\nu^{sd} = (1 - f_{sh})\nu^{qz} + (f_{sh})\nu^{s} \quad (5)$$

where ν^{qz} and ν^{s} are Poisson's ratio for quartz and clean shale, respectively. These are assumed constant

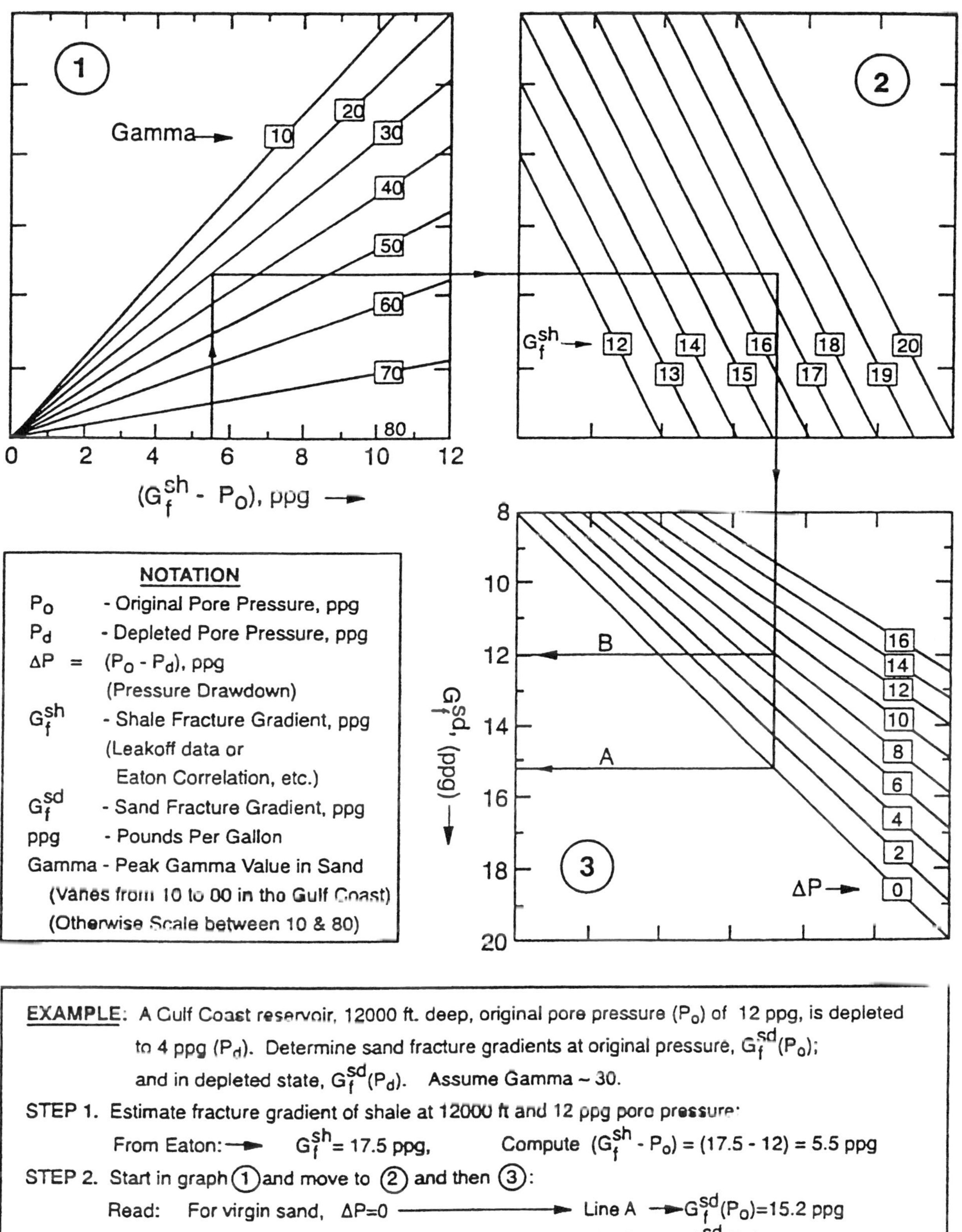

Figure 1. Nomogram to compute fracture gradients in sand (Gulf Coast sedimentary basin). Figure used with permission of Unocal.

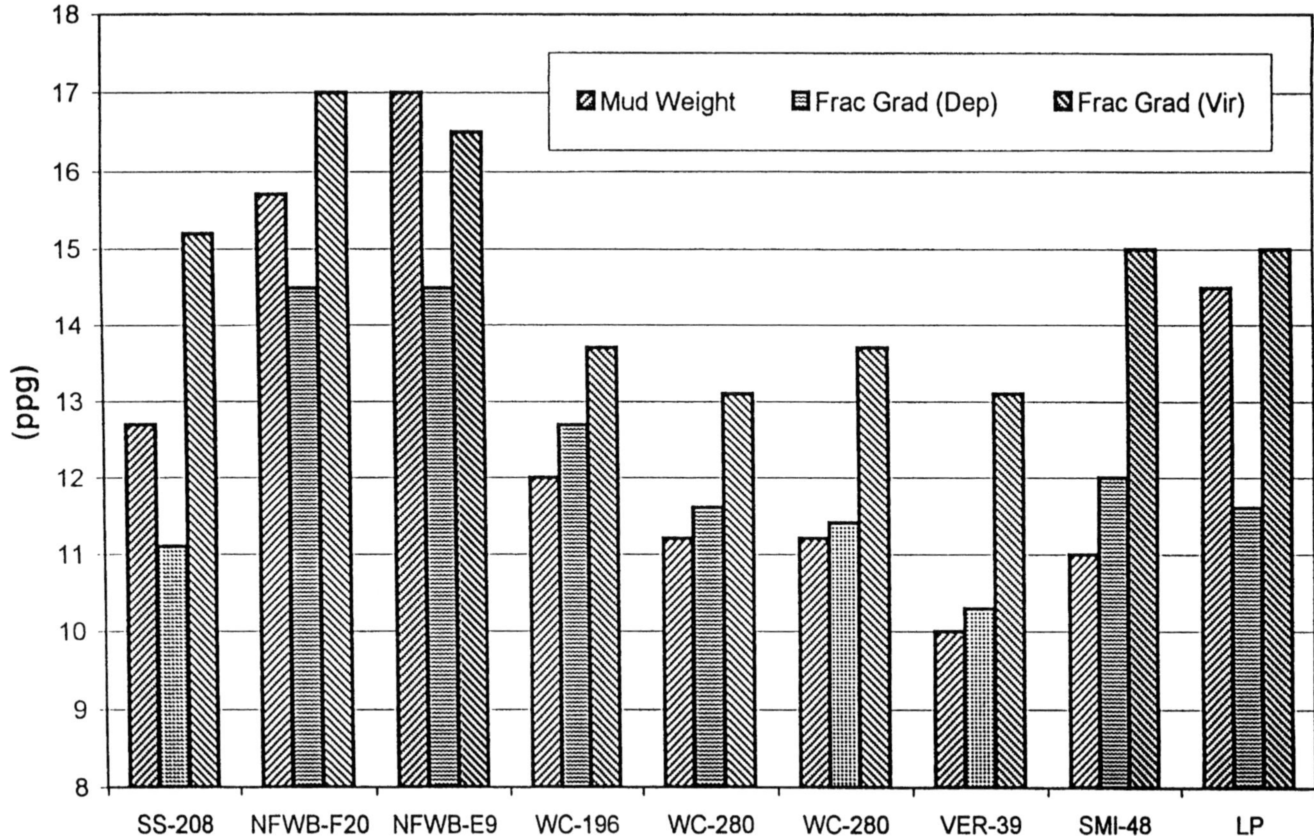

Figure 2. Comparison of fracture-gradient prediction in depleted reservoir and mud weight used. Losses occured for SS-208, NFWB-F20, NFWB-E9, LP. No losses occured for WC-196, WC-280, VER-39, SMI-48. Dep = depleted; Vir = virgin; SS = Ship Shoal; NFWB = North Fresh Water Bayou; WC = West Cameron; VER = Vermillion; SMI = South Marsh Island; LP = Lake Pagie.

and are chosen as 0.125 and 0.25 for sand and shale, respectively. Note that these values may vary and are region specific. The shaliness of a sand, f_{sh}, reflects the relative amount of contamination of the sand by shale and is estimated using linear interpolation as follows:

$$f_{sh} = (GR - GR_{qz})/(GR_{sh} - GR_{qz}) \quad (6)$$

where GR is the gamma-ray–log count in the sand under consideration, GR_{sh} is the gamma-ray–log count in clean shale, and GR_{qz} is the gamma-ray–log count in clean sand; GR_{sh} and GR_{qz} are considered fixed and are area specific. These values are obtained as described in the following paragraphs.

First, base lines are established for clean shale and clean sand for a specific region. Only thick sections of shales (representative of the clean shale found in the formation) are used. Because thick sections of shales are abundant in fields, these data are readily available. Based on the logs from the Gulf Coast, a value of 80 API units is the gamma-ray–log value typically used for GR_{sh}.

But the task of obtaining a base-line gamma value for pure (uncontaminated) sand is not a trivial one. Most sand sections are contaminated to some degree. Ideally, a sand unit made out of quartz only has a gamma-ray–log value close to 0. But because of inherent presence of minute amounts of contamination, along with logging instrument resolution, a value of 10 API units is selected to represent a gamma value of pure sand (made out of quartz), GR_{qz}. The results are, however, found to be rather insensitive to minor variation in these values.

FRACTURE GRADIENTS IN DEPLETED SANDS

Salz (1977) measured fracture gradients in several reservoirs (including depleted reservoirs) in the Vicksburg Formation at the McAllen Ranch area, south Texas. To start with, some of these reservoirs were highly abnormally pressured to a pressure gradient of about 0.96 psi/ft. The reservoirs were depleted to pressures as low as 0.2 ppg. Salz measured in-situ stress by conducting well-characterized fracturing tests. The fracture gradients were measured before and after the depletion. Based on more than 70 fracture treatments, he developed a correlation that could predict fracture gradient of a depleted reservoir if the fracture gradient in the virgin formation was known.

Salz's correlation is used in this chapter for predicting fracture gradient in the depleted sand. The correlation can be written as follows:

$$G_f^{sd}\ [\mathrm{Dep}] = G_f^{sd}[\mathrm{Vir}] \times \exp[-0.57(P_i - P_f)] \quad (7)$$

where G_f^{sd} [Dep] is the fracture gradient in depleted sand (ppg), G_f^{sd}[Vir] is the fracture gradient in virgin sand (ppg), P_i is initial formation pressure gradient (psi/ft), and P_f is final formation pressure gradient (psi/ft).

CASE HISTORIES—FRACTURE-GRADIENT PREDICTION IN DEPLETED SANDS IN THE GULF COAST RESERVOIRS

Table 1 presents lost-circulation data in depleted sands from seven lost-circulation–prone Unocal fields in the Gulf Coast. Fracture-gradient predictions in nine depleted sands from eight wells are compared with the mud weights that were used while drilling.

First, fracture gradients in adjacent shales were estimated using Eaton's (1969) correlation. Then, equations 4, 5, and 6 were used to calculate the fracture gradient for a virgin sand using gamma values for the sand. This estimate was then corrected for depletion using Salz's correlation, equation 7.

For simplicity and ease of application, a nomogram was developed incorporating the previous steps. The nomogram is shown in Figure 1. An example, included in the nomogram, illustrates how to use it.

RESULTS AND DISCUSSION

Excellent agreement was found between the predicted vs. observed lost-circulation cases for the depleted field data in Table 1 (nine sands from seven Gulf Coast fields). Lost circulation was observed in all of the cases where the predicted fracture gradient was less than the mud weight. The results are shown graphically in Figure 2. Note also the change in the fracture gradients of the depleted sands as compared with the virgin sands. In its partial form, this fracture-gradient prediction procedure may also be used to predict the contrast between the shale and the sand fracture gradients and thus improve the design of stimulation jobs.

REFERENCES CITED

Christman, S. A., 1973, Offshore fracture gradients: Journal of Petroleum Technology, v. 25, p. 910–914.

Eaton, B. A., 1969, Fracture gradient prediction and its application in oilfield operations: Journal of Petroleum Technology, v. 21, p. 1353–1360.

Matthews, M. K., and J. Kelly, 1967, How to predict formation pressure and fracture gradient: Oil & Gas Journal, v. 65, no. 8, p. 92–106.

Pennebaker, E. S., 1968, An engineering interpretation of seismic data: Society of Petroleum Engineers 43rd Annual Fall Meeting, SPE 2165, 12 p.

Pilkington, P. E., 1978, Fracture gradient estimates in Tertiary basins: Petroleum Engineer International, v. 50, no. 5, p. 138–148.

Salz, L. B., 1977, Relationship between fracture propagation pressure and pore pressure: Society of Petroleum Engineers 52nd Annual Fall Meeting, SPE 6870, 8 p.

13

Consolidation State, Permeability, and Stress Ratio as Determined from Uniaxial Strain Experiments on Mudstone Samples from the Eugene Island 330 Area, Offshore Louisiana

Beth B. Stump
Texaco Worldwide Exploration and Production, New Orleans, Louisiana

Peter B. Flemings
Pennsylvania State University, University Park, Pennsylvania

ABSTRACT

Uniaxial strain experiments conducted on mudstone cores from overpressured horizons in Eugene Island Block 330 (Gulf of Mexico) reveal information about consolidation state, compaction behavior, and permeability. Maximum past effective stresses for two mudstone samples were experimentally derived and are within 200 psi of porosity-based estimates of in-situ stress. Laboratory measurements of stress ratio ($K_0 = 0.85$) compare well with in-situ measurements made during leak-off and stress tests ($K_0 = 0.84 - 0.91$). The high K_0 values suggest that the sediment deformation is primarily plastic at in-situ levels of effective stress. A slope change on the stress-strain curve supports the observation of primarily plastic deformation following yield. Direct measurements of mudstone permeability at estimated in-situ levels of effective stress reveal layer-parallel and layer-perpendicular permeability of 5.32×10^{-4} md (5.25×10^{-19} m^2) and 1.17×10^{-4} md (1.15×10^{-19} m^2), respectively.

INTRODUCTION

Strain in a geologic basin is typically assumed to occur uniaxially (Roegiers, 1989). We use this uniaxial strain assumption to replicate geologic deformation in the laboratory (Karig and Hou, 1992; Karig, 1996). Deformation behavior of an uncemented sample depends on stress history and physical properties. Uniaxial strain experiments provide an estimate of the maximum past stress of an undisturbed sample. Using this estimate of maximum past stress, we assess the consolidation state and stress history of the sample.

Laboratory tests on undisturbed samples provide further insight into compaction behavior and elastic properties by providing continuous measurements of stress ratio, K_0, defined here as the ratio of change in horizontal effective stress to change in vertical effective stress (Karig and Morgan, 1994). We compare experimental values of K_0 with stress ratios calculated from leak-off tests and fracture completions in this area. Nomenclature and symbols used in this chapter are presented in Table 1.

Stump, Beth B., and Peter B. Flemings, 2002, Consolidation State, Permeability, and Stress Ratio as Determined from Uniaxial Strain Experiments on Mudstone Samples from the Eugene Island 330 Area, Offshore Louisiana, in A. R. Huffman and G. L. Bowers, eds., Pressure regimes in sedimentary basins and their prediction: AAPG Memoir 76, p. 131–144.

Table 1. Nomenclature

Variable	Property	Units*
A	sample cross sectional area	L^2
e	void ratio	L^3/L^3
f	acoustic formation factor	dimensionless
g	gravitational acceleration	L/T^2
h	hydraulic head	L
k	permeability	L^2
K	hydraulic conductivity	L/T
K_0	stress ratio	dimensionless
l	sample length	L
p'	mean effective stress	M/LT^2
P_f	fluid pressure	M/LT^2
q	differential stress	M/LT^2
Q	volumetric flow rate	L^3/T
S_v	vertical stress	M/LT^2
v_p	compressional-wave velocity	L/T
δP_f	P_f difference across the sample	M/LT^2
δt	wire-line–measured transit time	T/L
δt_{ma}	matrix transit time	T/L
ε	strain	L^3/L^3
ϕ	porosity	L^3/L^3
ϕ_i	initial sample porosity	L^3/L^3
μ	fluid viscosity	M/LT
ν	Poisson's ratio	dimensionless
ρ_f	fluid density	M/L^3
σ_c	preconsolidation stress	M/LT^2
σ_h	horizontal effective stress	M/LT^2
σ_v	vertical effective stress	M/LT^2

*L = length; T = time; M = mass.

We describe our methodologies for uniaxial strain experiments and permeability tests. We then present our experimental results and draw comparisons to in-situ estimates and previous studies. Finally, we discuss factors that affect laboratory and in-situ measurements and close with implications of our findings.

METHODS

Sample Description

Conventional core taken from two wells in the Eugene Island 330 field (Figure 1) provided mudstone samples for the deformation experiments. The Pennzoil 330 A-20ST2 (Pathfinder) well was cored in 1993; 343 ft (104.5 m) of core was recovered (95% recovery). In 1994, 43.2 ft (13.1 m) of core was recovered from the Pennzoil 316 A-12 well (34% recovery). All of the 4 in. (101.6 mm) diameter cores were cut into 3 ft (.914 m) sections and then stabilized in the core barrels using a quick-hardening epoxy resin. Cores were then slabbed into one-third and two-thirds parts. Laboratory samples were taken from the two-thirds part. Slabs were archived in cold storage at Core Labs (Houston) and then moved to Penn State University, where they are stored in a humid, chilled room. The A-20ST2 core was sealed in wax to preserve moisture.

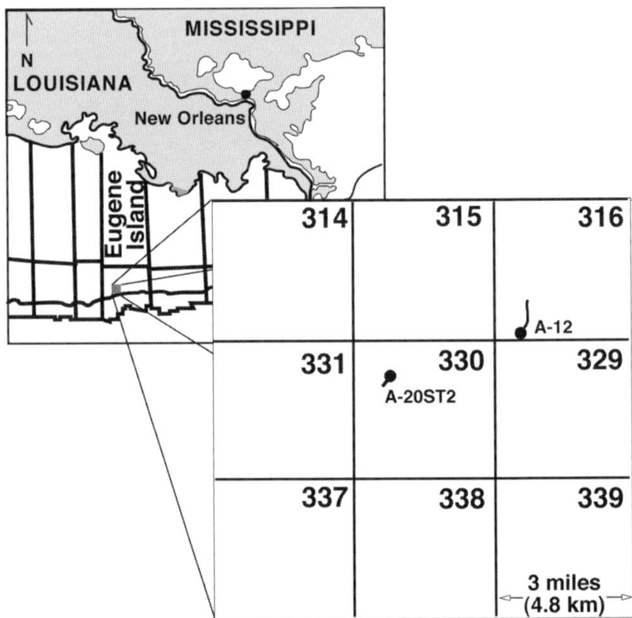

Figure 1. Basemap locates cored wells from which laboratory samples were taken. Circles indicate bottom-hole location.

The pressure regime (Figure 2) of the Eugene Island 330 area consists of a shallow, sand-rich, hydrostatically pressured zone overlying an interbedded transition zone of moderate overpressure (12 lb/gal equivalent-mud weight [EMW]) and a severely overpressured section (>14 lb/gal EMW) at depth. The samples for the deformation experiments were taken from the severely overpressured outer neritic mudstone section adjacent to the Lentic 1 sand (Figure 3). Alexander and Flemings (1995) present a detailed description of the geologic evolution of this field. Table 2 provides a summary of estimated in-situ fluid pressures and effective stress, as well as the composition of the core samples.

The A-12 core was subsampled between the Lentic 1 upper and lower sands (Figure 3a). The Lentic 1 sand has been interpreted as a turbidite/distal shelf deposit (Alexander and Flemings, 1995). At the onset of the K_0 experiment, the A-12 mudstone (T96) sample was 63.5 mm long and 31.01 mm in diameter, with an initial porosity of 0.39. The sample was taken from core barrel 1 and contains 38% quartz, 39% clay, and small fractions of potassium feldspar, plagioclase, and calcite (Core Laboratories, 1994). See Table 2 for a detailed

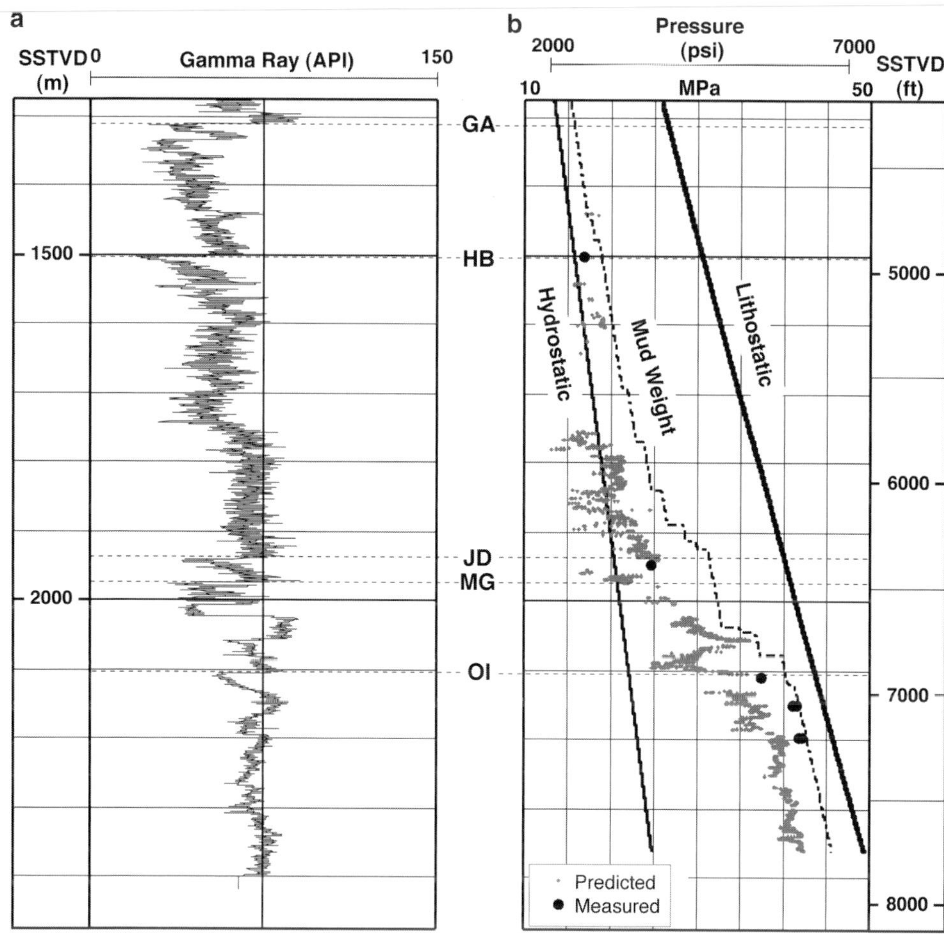

Figure 2. A pressure regime typical of the Eugene Island 330 area is shown for the A-20ST2 well. (a) Gamma ray shows depths of key sands. (b) Pressure track is bounded by hydrostatic (0.465 psi/ft) and lithostatic gradients. Dashed line represents pressure calculated from drilling mud weights, gray dots indicate fluid pressure calculated from porosity in the mudstones, and circles are measured fluid pressures in adjacent sands.

composition description. Laser particle size analysis revealed 2.5% fine-grained, 15.4% very fine grained, 62.1% silt-size, and 20.0% clay-size particles.

Fluid pressure (P_f) in the mudstone overlying the Lentic 1 sand in the A-12 well is estimated as 35.0 MPa (5080 psi; 14.6 lb/gal EMW) from a porosity-effective stress method (Hart et al., 1995). Vertical (overburden) stress (S_v) is calculated by integrating the bulk density log and is 42.1 MPa (6105 psi). The estimated in-situ vertical effective stress (σ_v) is 7.1 MPa (1025 psi), as defined by equation 1 (Terzaghi, 1925).

$$\sigma_v = S_v - P_f \qquad (1)$$

The A-20ST2 (T77) sample (Figure 3b) was taken between the Cris S flooding surface and the Lentic 1 sand (Alexander and Flemings, 1995). Losh (1998) provides a discussion of the structural analysis of the A-20ST2 core. The A-20ST2 sample had an initial length of 52.73 mm, a diameter of 30.94 mm, and an initial porosity of 0.37. The sample contains 35% quartz and 54% clay (Table 2) (Losh et al., 1994). Estimated fluid pressure in this mudstone interval is 39.5 MPa (5730 psi; 15.0 lb/gal EMW). Calculated S_v is 46.8 MPa (6782 psi). The estimated in-situ vertical effective stress in the A-20ST2 at 2240 m (7350 ft) is 7.3 MPa (1052 psi).

Experiment Description

K_0 Tests

K_0 tests, defined here as consolidation experiments conducted under uniaxial strain conditions, were conducted in a triaxial servohydraulic load frame controlled by computers at Cornell University (see Karig [1996] for a photograph of the configuration). An initial isotropic stress state at the onset of the experiment, at effective-stress levels well below anticipated sample yield, was used to minimize preshearing of the sample (Mesri and Hayat, 1993). After the sample was stabilized in a period of 24–48 hr, the experiment proceeded under uniaxial strain conditions. Experiment duration was determined, in part, by the length of time necessary to reach yield. The durations of the T96 (A-12) and T77 (A-20ST2) experiments were 8 and 13 days, respectively.

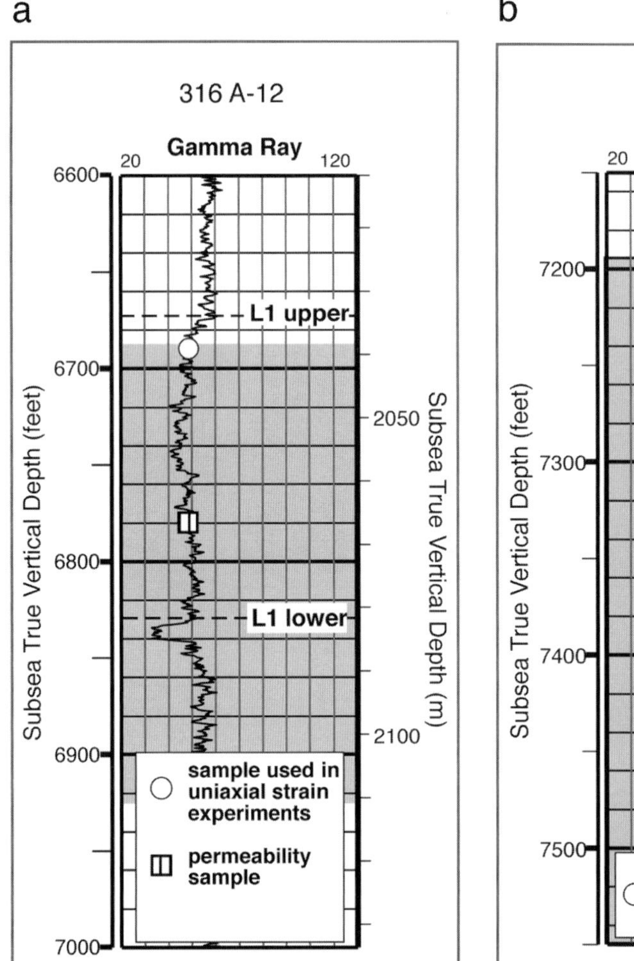

Figure 3. Cylindrical laboratory samples were drilled perpendicular to bedding at locations denoted with circles. Gray area represents entire cored interval. (a) 316 A-12; (b) 330 A-20ST2.

Table 2. Sample Descriptions

Well, Sample	Subsea True Vertical Depth	Composition	Sonic-Derived In-Situ P_f (psi)	Estimated In-Situ σ_v (psi)	Estimated In-Situ K_0
316 A-12 T96	6690 ft (2039 m)	38% quartz, 5% potassium feldspar, 9% plagioclase, 3% calcite, 1% dolomite, and 39% clay (48% illite, 38% smectite, 7% kaolinite, 8% chlorite)	5080 (35.0 MPa)	1025 (7.1 MPa)	0.91 (6690 ft)
316 A-12 P01, P03	6781 ft (2067 m)	40% quartz, 1% potassium feldspar, 9% plagioclase, 2% pyrite, and 48% clays (45% illite, 35% smectite, 7% kaolinite, 12% chlorite)	5166 (35.6 MPa)	1045 (7.2 MPa)	0.43 (6798 ft)
330 A-20ST2 T77	7350 ft (2240 m)	35% quartz, , 8% potassium feldspar, 3% plagioclase, and 54% clay (52% illite, 32% smectite, 15% kaolinite, 1% chlorite)	5730 (39.5 MPa)	1052 (7.3 MPa)	>0.84 (7277 ft)

All experiments were run under controlled axial load at a rate of 0.1 psi/min, which approximates an initial strain rate of 1×10^{-7} s^{-1}. A controlled load experiment was used to allow fluid dissipation in these low-permeability samples. As consolidation increases with increasing axial load and the sample compresses, sample compressibility decreases, which in turn causes the strain rate to decrease. Within the apparatus, sintered titanium disks with 2 μm pore spaces allow fluid drainage from the top and bottom of the cell during consolidation. Linear variable-displacement transducers (LVDT) measure axial strain. Karig (1996) presented a full description of the apparatus used in these experiments.

Jacob's (1949) relation provides a method for calculating porosity changes from strain, ε (compression positive), during these experiments:

$$\varepsilon = \frac{-d\phi}{1-\phi} \qquad (2)$$

By substituting $d\phi = \phi - \phi_i$ and rearranging, we get an expression for porosity as a function of initial porosity (ϕ_i) and axial strain:

$$\phi = \frac{\phi_i - \varepsilon}{1-\varepsilon} \qquad (3)$$

Initial porosity (ϕ_i) is calculated in the laboratory using initial bulk density, pycnometer-derived grain density, and an assumed fluid (brine) density of 1.07 g/cm^3.

Void ratio (e), used to graphically determine consolidation state, is a function of porosity:

$$e = \frac{\phi}{1-\phi} \qquad (4)$$

Permeability Tests

Mudstones in the Eugene Island 330 area act as seals to pressure and hydrocarbon migration. Permeability measurements in mudstone allow us to evaluate seal integrity and estimate time scales of overpressure dissipation. Permeability is determined directly in the laboratory from the rate of fluid flow through a sample of known length, using a constant-head test.

During this experiment, constant fluid pressure is held at one end of the sample; the outflow end is kept at atmospheric pressure. We calculate hydraulic conductivity, K, from Darcy's law.

$$\frac{Q}{A} = -K\left(\frac{\Delta h}{\Delta l}\right) \qquad (5)$$

Using the definition of hydraulic head, h, and substituting the expression for hydraulic conductivity (equation 6) into equation 5, we get an expression for permeability, k (equation 7).

$$K = \frac{k\rho_f g}{\mu} \qquad (6)$$

$$k = \frac{Q\mu}{A}\frac{\Delta l}{\Delta P_f} \qquad (7)$$

Laboratory measurements of permeability in low-permeability sediments are difficult. The identification and elimination of potential errors are the keys to accurate direct-permeability measurements (Tavenas et al., 1983). Leakage through the external fitting is perhaps the largest and most unavoidable source of error. We attempt to quantify this error by running leak-rate tests in the cell containing no sample. Stump (1998) detailed methods used to decrease error in these consolidation experiments. Such methods include surrounding the sample with silicone oil to reduce osmotic effects and maintaining a confining pressure on the sample to minimize preferred flow between the latex sleeve and the sample.

Our experiments were run using an inflow fluid pressure ranging from 49 to 150 psi to redissolve any gas bubbles that exsolved during core retrieval. We also used saline brine (35 ppt) to minimize osmotic effects that can decrease apparent permeability (Neuzil, 1986). Last, to account for the transient nature of the flow at the beginning of the test (Olsen et al., 1985), we allow the test to equilibrate and reach steady-state flow before measuring flow rate.

RESULTS

Determination of Maximum Past Stress

Preconsolidation stress (σ_c), defined as the maximum past effective stress, was determined using the graphical Casagrande method (1936) (Figure 4a). If the laboratory experiment replicates the burial-induced deformation path, and the sample is relatively undisturbed and uncemented, the observed yield stress corresponds to the sample's preconsolidation stress (Karig and Morgan, 1994). Sample yield is defined as the break in slope of the stress-strain curve, which

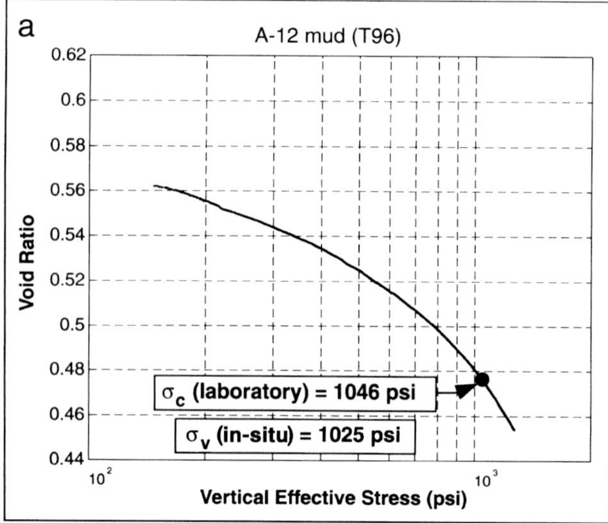

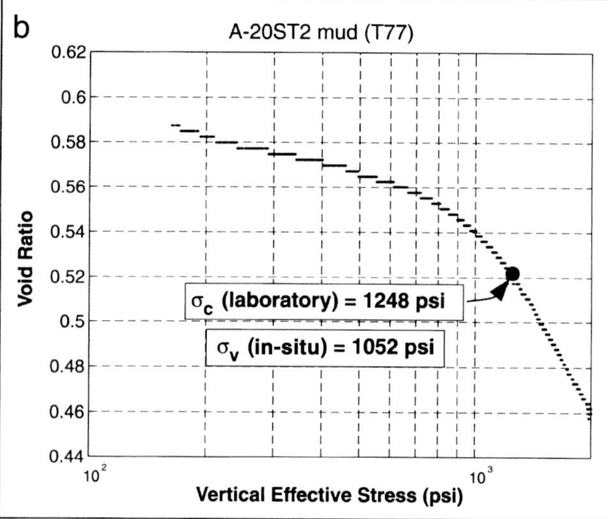

Figure 4. Maximum past stress (preconsolidation stress, σ_c), is determined using Casagrande's (1936) graphical method. Estimated in-situ σ_v is calculated from integrated bulk density and fluid pressure estimated by a porosity-effective stress method. (a) For the A-12 (T96), σ_c is within 0.2 MPa (25 psi) of estimated in-situ σ_v. (b) In the A-20ST2 (T77), experimental σ_c is approximately 1.4 MPa (200 psi) greater than estimated in-situ σ_v.

denotes the transition from elastic reconsolidation to first-time consolidation (Karig and Morgan, 1994).

The preconsolidation stress for the A-12 mud (T96) sample was experimentally determined as 7.2 MPa (1046 psi) (Figure 4a), which is within 0.1 MPa (21 psi) of the estimated in-situ σ_v of 7.1 MPa (1025 psi), calculated from integrated bulk density and pressure derived from a porosity-effective stress analysis. For the A-20ST2 sample (T77), the experimentally derived σ_c is 8.6 MPa (1248 psi) (Figure 4b), which is 1.3 MPa (196 psi) greater than estimated in-situ σ_v of 7.3 MPa (1052 psi), based on porosity-effective stress method. Table 3 presents a summary of results from the K_0 experiments.

Stress Ratio, K_0

The stress ratio, K_0, referred to as the coefficient of earth pressure at rest (Brooker and Ireland, 1965), is the ratio between horizontal effective stress and vertical effective stress under uniaxial strain conditions (Jones, 1994). We calculate K_0 as the slope of the $\sigma_h - \sigma_v$ curve (equation 8).

$$K_0 = \frac{\Delta\sigma_h}{\Delta\sigma_n} \tag{8}$$

K_0 increased in the A-12 (T96) and A-20ST2 (T77) experiments following sample yield (Figure 5). In the A-12 (T96) sample (Figure 5a), at vertical effective stresses less than σ_c, K_0 was 0.52. At larger values of vertical effective stress, K_0 was 0.86. Similarly, in the A-20ST2 (T77), K_0 increased from 0.63 to 0.85 following σ_c (Figure 5b).

Permeability

Constant-head permeability tests on samples from the A-12 well provided estimates of layer-parallel (P01) and layer-perpendicular (P03) mudstone permeability. Flow rates through the layer-parallel (P01) sample were measured using two pressure differences (ΔP_f) across the length of the sample: 0.34 and 0.66 MPa (49 and 96 psi, respectively) (Figure 6a). Confining pressure remained at 6.9 MPa (1000 psi) for the duration

Table 3. Summary of Deformation Experiment Results

Well	Sample	Subsea True Vertical Depth	σ_c (psi)	Preyield K_0	Postyield K_0
316 A-12	T96	6690 ft (2039 m)	1046 (7.2 MPa)	0.52 ν = 0.34	0.86 ν = 0.46
330 A-20ST2	T77	7350 ft (2240 m)	1248 (8.6 MPa)	0.63 ν = 0.39	0.85 ν = 0.46

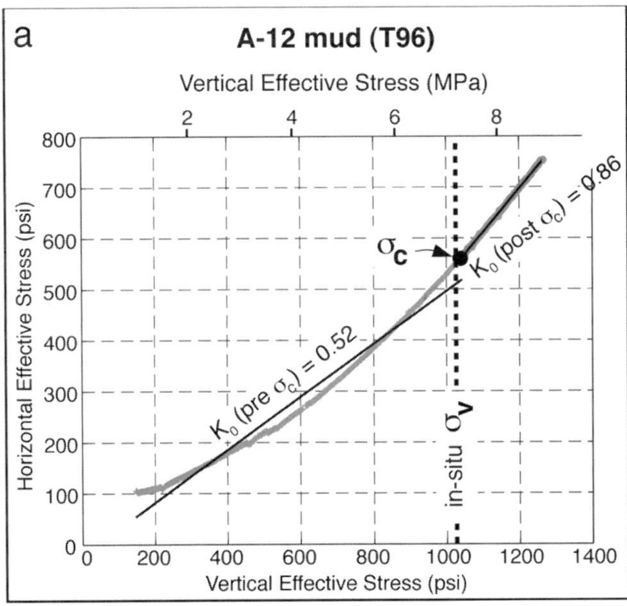

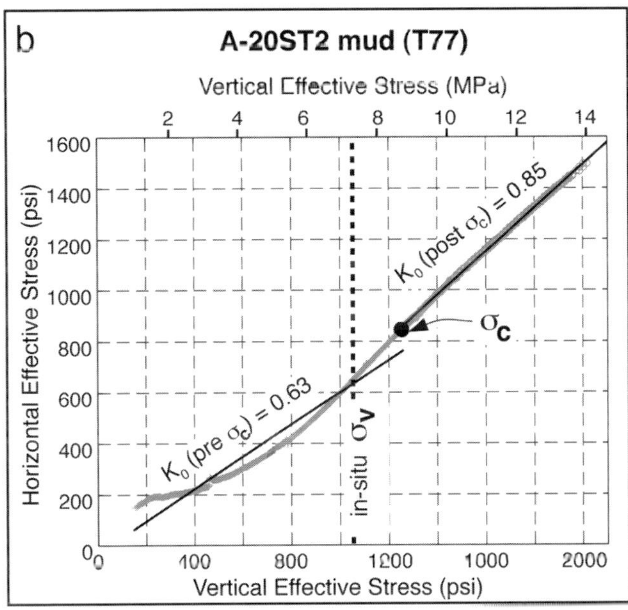

Figure 5. Stress ratio K_0 is calculated throughout the deformation experiments as the slope of horizontal vs. vertical effective stress. (a) In the A-12 (T96), pre-σ_c K_0 is 0.52; post-σ_c K_0 is 0.86. The dotted line denotes in-situ σ_v and the circle represents experimental σ_c. (b) In the A-20ST2 (T77) sample, K_0 increased from 0.63 to 0.85.

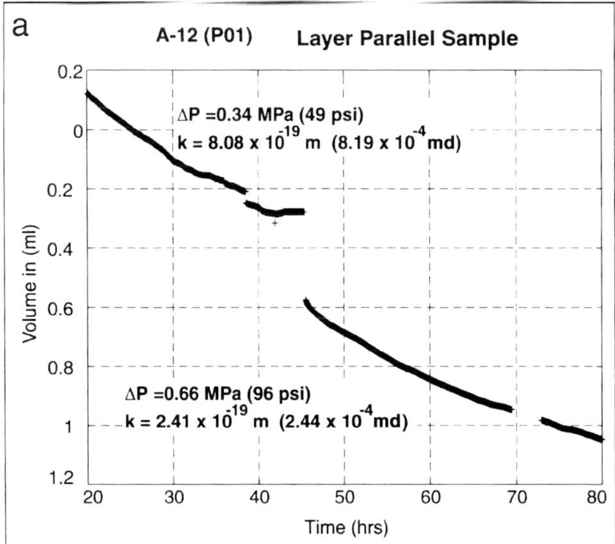

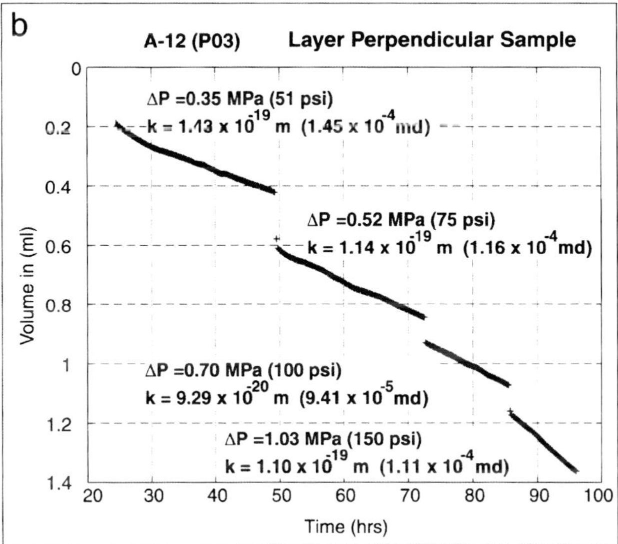

Figure 6. (a) Direct measurements of layer-parallel (P01) permeability yield an average permeability of 5.25 × 10^{-19} m^2 (5.32 × 10^{-4} md). Initial porosity of P01 sample is 0.40; final porosity was difficult to determine because sample was misshapen upon removal from cell. (b) Average layer-perpendicular permeability (P03) is 1.15 × 10^{-19} m^2 (1.17 × 10^{-4} md). During the P03 experiment, porosity decreased from 0.48 to 0.38.

of the experiment. Average layer-parallel permeability at the two pressure differences is 5.25 × 10^{-19} m^2 (5.32 × 10^{-4} md). The layer-perpendicular sample, P03, was tested at four pressure differences of 0.35, 0.52, 0.70, and 1.03 MPa (51, 75, 100, 150 psi, respectively) (Figure 6b). Confining pressure for the P03 experiment was 800 psi. Layer-perpendicular permeability measurements averaged 1.15 × 10^{-19} m^2 (1.17 × 10^{-4} md), ranging from 9.29 × 10^{-20} m^2 to 1.43 × 10^{-19} m^2.

Two leak-rate tests, run at a variety of ΔP_f, identified leak rates of 0.042 and 0.048 mL/hr. Correcting for the leak rate of the system, the values for layer-parallel and layer-perpendicular permeability are 3.82 × 10^{-19} m^2 and 6.47 × 10^{-20} m^2 (3.87 × 10^{-4} and 6.56 × 10^{-5} md), respectively.

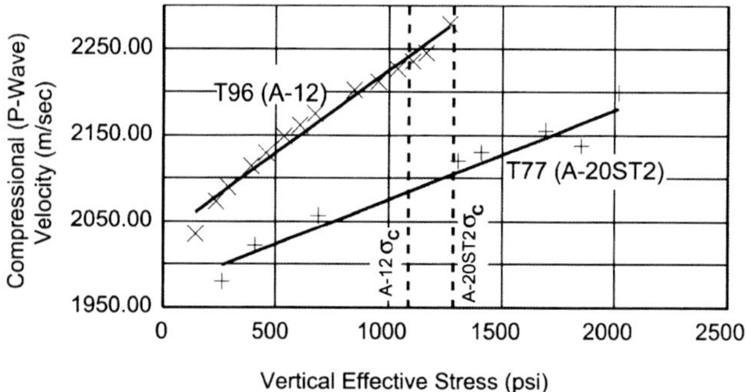

Figure 7. Compressional velocity increases with increasing vertical effective stress during all three K_0 tests. Symbols represent velocity measurements taken during T96 ($\times$) and T77 ($+$) experiments. The change in velocity with vertical effective stress (i.e., the slope of linear regressions through these data) is 1.91×10^{-2} for the T96 (A-12) and 1.04×10^{-2} for T77 (A-20ST2). Vertical lines are shown at the experimental preconsolidation stresses for the A-12 and A-20ST2 samples (1046 and 1248 psi, respectively).

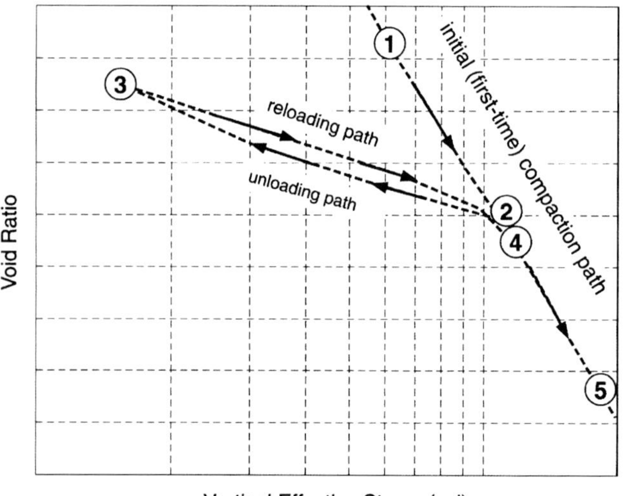

Figure 8. A generalized mudstone deformation path includes primary compaction (1–2). As effective stress decreases (2–3), sample experiences some porosity rebound, but does not decompact along the original deformation path. Upon reloading (3–4), sample follows similar path until it reaches maximum past stress. Deformation at higher stresses (4–5) tracks along a primary compaction path.

Velocity

Compressional-wave velocity (v_p) was measured during K_0 experiments on the A-12 and A-20ST2 mud samples. Compressional waves at a frequency of 400 kHz were generated along the core axis (i.e., perpendicular to bedding). Velocity measurements have an accuracy of ± 0.02 km/s (Karig, 1996). As vertical effective stress increases and the sediment compacts, v_p increases (Figure 7). At σ_c, compressional velocity in the A-12 (T96) sample was 2234 m/s (delta transit time [DT] = 136 μs/ft). The A-20ST2 (T77) velocity at σ_c was 2101 m/s (DT = 145 μs/ft). By comparison, in-situ wire-line sonic traveltime measurements at core sample depths were 155 μs/ft in the A-12 and 149 μs/ft in the A-20ST2.

DISCUSSION

Assessment of Consolidation State

A normally consolidated sediment has never been subjected to a higher stress than its current stress (Jones, 1994). An overconsolidated sediment is one that has a maximum past effective stress that is greater than the current effective stress. We determined the maximum past stress experimentally by observing the change in deformation behavior during the uniaxial strain experiment. Mudstone compaction is largely irreversible because deformation is primarily plastic with a small elastic component. When a sample is brought from depth to the surface, it experiences a decrease in effective stress and a consequent rebound in void ratio (point 3 on Figure 8). This rebound results from elastic expansion and opening of microcracks (Karig and Hou, 1992). As the sample is reloaded in the laboratory, the deformation path follows a slope similar to, but not identical with, the unloading path until the stress reaches the maximum past stress (point 4 in Figure 8). Mesri and Choi (1985) demonstrated the effectiveness of the void ratio–effective stress relationship in determining maximum past stress experimentally. As the vertical effective stress increases beyond the maximum past effective stress, the slope of the void ratio–effective stress curve changes, reflecting a change from primarily elastic to primarily plastic deformation (Turcotte and Schubert, 1982; Atkinson, 1993). On a traditional stress-strain plot the elastic to plastic transition is manifested by a change from linear to nonlinear behavior.

Our experimentally derived σ_c values are within 0.2 and 1.4 MPa (25 and 200 psi) of our prediction of the in-situ stresses based on a porosity-effective stress method. The agreement between these different approaches suggests that both approaches are imaging the preconsolidation stress of the sample. Unfortunately, neither approach taken independently can determine if these sediments are overconsolidated. Bowers (1994) refers to overconsolidation, or a late-stage decrease in effective stress, as "unloading."

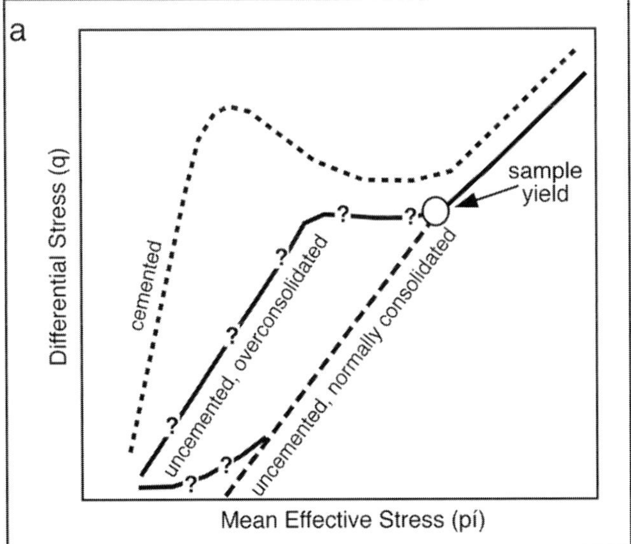

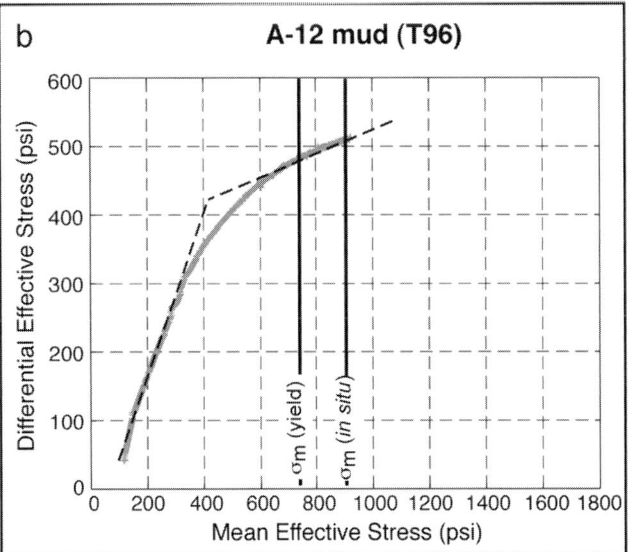

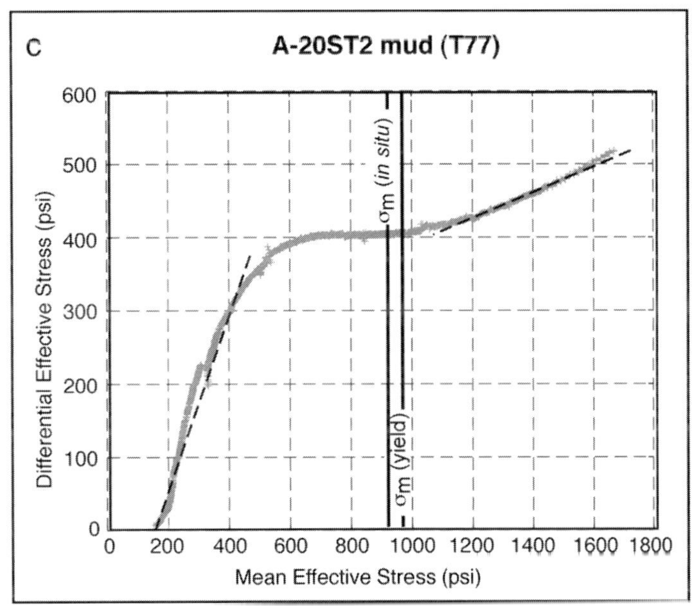

Figure 9. Crossplots of mean effective stress (p') with differential stress (q) can illuminate the difference between cementation and a state of overconsolidation. (a) A generalized figure (adapted from Karig, 1996) shows the p'-q signatures for cemented, uncemented normally consolidated, and uncemented overconsolidated sediments. (b) A-12 (T96) sample shows a bilinear deformation path, indicative of uncemented, normally consolidated sediment. Vertical lines denote mean effective stress at experimental yield ($\sigma_m = 727$; $\sigma_v = 1046$, $\sigma_h = 567$ psi) and from in-situ estimates ($\sigma_m = 889$; $\sigma_v = 1025$, $\sigma_h = 820$ psi). (c) The A-20ST2 (T77) p'-q signature suggests that this sample may be overconsolidated. Mean effective stress at yield in the T77 experiment was 977 psi ($\sigma_v = 1248$, $\sigma_h = 842$ psi); in-situ estimate of mean effective stress is 940 psi ($\sigma_v = 1052$, $\sigma_h = 884$ psi).

Cementation vs. Overconsolidation

One possible way to determine if the sediment is overconsolidated is illustrated on plots of mean effective stress ($p' = (2\sigma_h + \sigma_v)/3$) vs. differential stress ($q = \sigma_v - \sigma_h$). A normally consolidated sample exhibits a bilinear behavior on a p'-q plot, whereas both overconsolidated and cemented samples exhibit more complicated behavior (Figure 9a, adapted from Karig [1996]). One interpretation of the A-12 sample behavior is that this sediment is normally consolidated, whereas the A-20ST2 sediment is overconsolidated

Table 4. Comparison of Postyield K_0 Values with Previous Work

Sediment Type	Composition	K_0
A-12 mud (this study)	38% quartz, 39% clay, some potassium feldspar, plagioclase, calcite	0.86
A-20ST2 mud (this study)	35% quartz, 54% clay	0.85
Ottawa sand mixture (Karig and Hou, 1992)	90% quartz, 10% clay	0.53
silty clay (Karig and Hou, 1992)	50% silica powder (including quartz, potassium feldspar), 50% clay	0.62
Boston blue clay (Mesri and Hayat, 1993)	35% quartz, 30% clay, 23% plagioclase, 8% potassium feldspar	0.56
St. Alban clay (Mesri and Hayat, 1993)	25% quartz, 26% clay, 33% plagioclase, 11% potassium feldspar	0.49
Bearpaw Shale (Brooker and Ireland, 1965)	30% quartz, 65% clay, 5% potassium feldspar	0.70
London clay (Brooker and Ireland, 1965)	15% quartz, 85% clay	0.67
Weald clay (Brooker and Ireland, 1965)	30% quartz, 70% clay	0.54
Goose Lake flour (Brooker and Ireland, 1965)	25% quartz, 75% clay	0.51

(Figure 9b, c). Neither sample demonstrated a decrease in differential stress indicative of cement breakdown. This observation is consistent with the lack of typical cements like calcite in these samples (Core Laboratories, 1994; Losh et al., 1994).

Although we used the Casagrande method, Bryant et al. (1986a) suggested that this method may underpredict the maximum past stress by as much as 35%. They proposed an alternative method of computing regressions through the reloading and first-time compaction curves. In this method the intersection of the two lines identifies the maximum past stress. Blum et al. (1996), however, observed no systematic difference between the Casagrande and Bryant methods in determining the preconsolidation stress of their samples.

A few potential sources of error are inherent in our calculations because of our assumptions. First, we assume that geologic deformation is approximately uniaxial. We then consider uniaxial strain experiments to be a replication of sediment burial in a geologic basin and therefore presume the experimental yield stress to be indicative of the maximum past stress. In an extensional basin such as the Eugene Island 330 area the assumption of pure uniaxial strain may cause the maximum past vertical effective stress observed in the laboratory to differ from the in-situ maximum past vertical effective stress.

Second, we assume that the deformation experiments are run under drained conditions. That is, we assume that strain rates are sufficiently slow to allow excess fluid pressure to dissipate, such that the fluid pressure in the sample is constant. If this assumption is invalid and the excess fluid is not drained from the sediments during the experiments, we are overestimating effective stress for a given porosity. Preliminary calculations indicate that for an average strain rate of $1 \times 10^{-7}\ s^{-1}$ in a sample with permeability of $1 \times 10^{-19}\ m^2$ (1×10^{-4} md) the accumulated excess pressure is negligible ($P_f/S_v < 0.05$). As the A-12 mud-permeability measurements are in the range of $1 \times 10^{-19}\ m^2$, we consider all of these tests to be representative of drained behavior and therefore that no overpressure was induced in the samples during the experiments.

Stress Ratio, K_0

Experimentally derived K_0 values at vertical stresses above the preconsolidation stress for the A-12 (T96) and A-20ST2 (T77) samples agree well with in-situ stress ratios calculated from leak-off test and stress-test data (Stump, 1998). A stress test in the A-20ST2 well provided an in-situ estimate of horizontal effective stress, σ_h (Flemings et al., 1994). Vertical effective stress was calculated using bulk density and sonic-derived estimates of fluid pressure (Stump and Flemings, 2000). The calculated in-situ stress ratio in the A-20ST2 well is 0.84, which is nearly identical with the postyield K_0 value (0.85). A leak-off test close to the A-12 well provided an in-situ stress ratio at sample depth of 0.91 (Finkbeiner, 1998; Finkbeiner et al., 2001). This in-situ stress ratio is slightly higher than the experimental postyield K_0 of 0.86. The mudstone K_0 values measured in our experiments are slightly higher than values of K_0 presented in previous studies (Table 4).

Both the A-12 and the A-20ST2 mud samples showed an increase in K_0 following yield. The sharp increase following yield results from a change in deformation. Prior to yield, during the reloaded phase, deformation is recoverable (elastic). Following yield, as the sample consolidates along its first-time compaction path, deformation is mostly plastic. Karig and Hou (1992) measured K_0 values of 0.35 and 0.62 for the elastic and first-time consolidation phases of deformation in silty clays, respectively.

In isotropic sediments, during elastic deformation under uniaxial strain conditions, the stress ratio is a

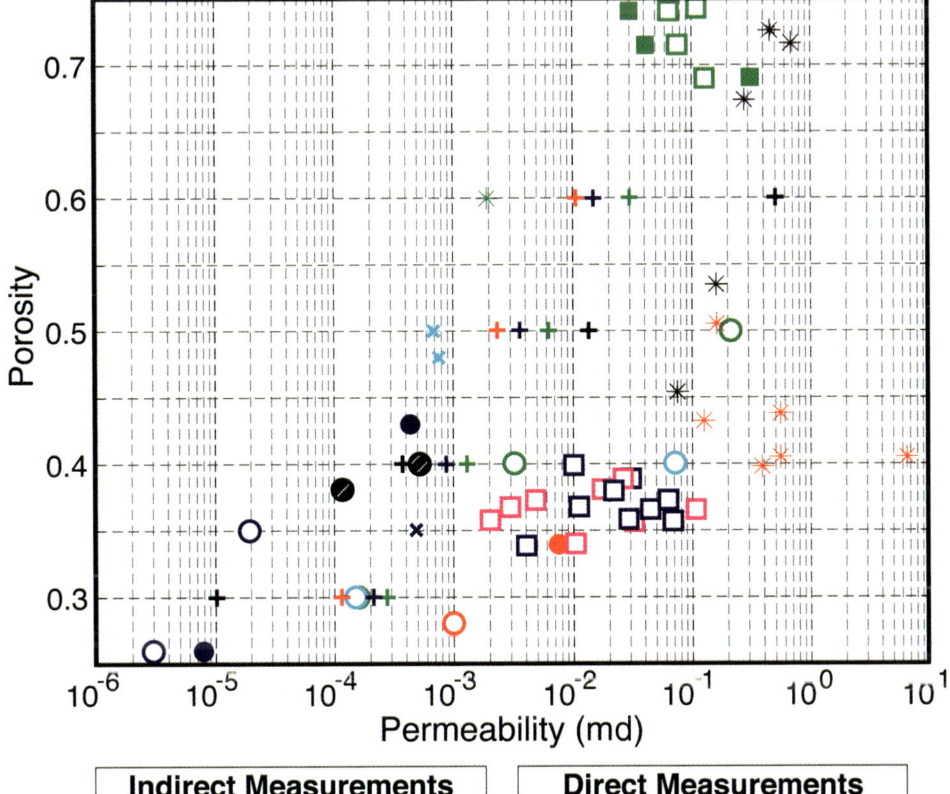

Figure 10. Comparison plot of porosity-permeability data published by previous authors shows that our permeability measurements (black circles) compare well with previous measurements made in similar sediments. Plus symbols (+) represent modeled results from Bryant et al. (1986a) and Gordon and Flemings (1998) for Gulf of Mexico mudstones. Indirect permeability measurements, made during consolidation tests, are shown as asterisks (*), ×s, and filled symbols. Empty symbols denote direct measurements of permeability (constant-head and flow-pump tests).

direct function of Poisson's ratio. We recognize the anisotropy of clays, but estimate an average Poisson's ratio using the experimental K_0. Equation 9 is derived from Hooke's Law.

$$K_0 = \frac{\Delta\sigma_h}{\Delta\sigma_n} = \frac{\nu}{1-\nu} \quad (9)$$

By rearranging, we calculate the Poisson's ratio of these samples from the experimental K_0 measured during the reload (elastic) phase of consolidation.

$$\nu = \frac{K_0}{1+K_0} \quad (10)$$

In the A-12 sample the K_0 of the reloaded phase is 0.52, corresponding to a Poisson's ratio of 0.34. Dynamic Poisson's ratio, calculated from wire-line dipole sonic measurements at the depth of the A-12 mud sample, is 0.39. The relationship used to calculate dynamic Poisson's ratio from wire-line logs is $\nu = [(\Delta t_s/\Delta t_c)^2 - 2\,]/[2(\Delta t_s/\Delta t_c)^2 - 2]$, where Δt_s is shear-wave traveltime and Δt_c is compressional-wave traveltime.

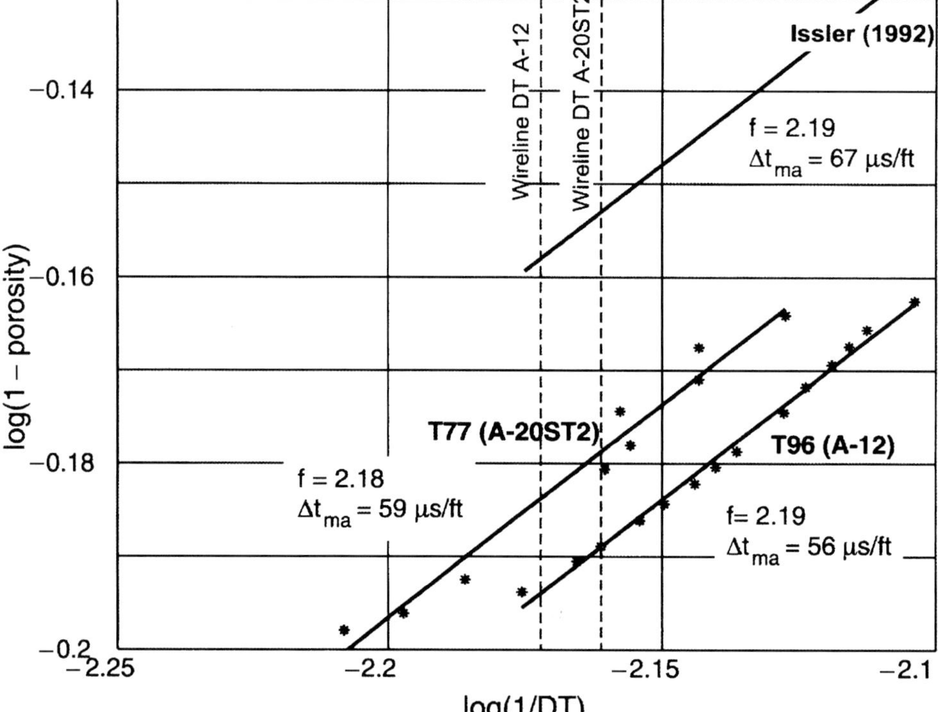

Figure 11. Crossplot of porosity and compressional velocity data is used to determine acoustic formation factor (f) and matrix velocity (Δt_{ma}) (equation 11). Specifically, the slope of the regression line is $1/f$. The y intercept of the line is equal to $(1/f)\log(\Delta t_{ma})$. The correlation coefficients (R^2) for T77 and T96 were 0.9850 and 0.9962, respectively. For reference, Issler's (1992) relationship is shown.

In the A-20ST2, the pre-σ_c K_0 of 0.63 corresponds to a Poisson's ratio of 0.39. A Poisson's ratio measured during a uniaxial stress test on a mudstone sample from the A-20ST2 well (D. Karig, 1998, personal communication) was 0.35.

Permeability

Permeability calculations (5.32×10^{-4} md, 1.17×10^{-4} md) from constant-head tests on samples from the A-12 well fall in the range of measurements made in previous studies (Figure 10). Dewhurst et al. (1998) measured an average permeability of 7.5×10^{-3} md (7.4×10^{-18} m^2) for a silt-rich (40% clay) sample with 34% porosity. Measurements of permeability during consolidation tests on deep-water core from the Pigmy Basin, Gulf of Mexico, yielded an average value of 3×10^{-4} md (3×10^{-19} m^2) (Bryant et al., 1986b). Wetzel (1990) measured an average permeability of 8.6×10^{-3} md (8.5×10^{-18} m^2) for turbidites taken from several hundred feet below the sea floor. Bryant et al. (1975) evaluated permeability from consolidation tests on various Gulf of Mexico sediments. For samples with composition and porosity similar to our samples, permeability ranged from 6.6×10^{-4} md (6.5×10^{-19} m^2) to 9.6×10^{-4} md (9.5×10^{-19} m^2) (Bryant et al., 1975).

The ratio of layer-parallel to layer-perpendicular permeability, a ratio of 4.5 for our data, is due to the anisotropy of the mudstone. Vasseur et al. (1995) showed that the difference between layer-parallel and layer-perpendicular permeabilities increases for increasing levels of compaction. Taylor and Fisher (1993) also observed anisotropy in their permeability measurements of sediments from the Nankai accretionary prism.

The permeability of a mudstone layer in a basin may exceed laboratory estimates of permeability because of the influence of fractures. Neuzil (1994) recognized the differences between laboratory and regional estimates for the Pierre shale and some clay till, but also observed that several muds showed very similar values for both laboratory and regional permeability.

Velocity

The porosity-velocity relationships from these deformation experiments (Figure 11) correlate well with a relationship developed by Raymer et al. (1980) and enhanced by Raiga-Clemenceau et al. (1986).

$$\phi = 1 - \left(\frac{\Delta t_{ma}}{\Delta t}\right)^{1/f} \qquad (11)$$

Issler (1992) calculated an acoustic formation factor, f, of 2.19 and a Δt_{ma} of 67 μs/ft for noncalcareous, low total organic carbon shale. As shown in Figure 11, values calculated during our K_0 tests on the A-12 and

A-20ST2 mudstone samples correlate well with Issler's (1992) values. Linear regression of the T77 (A-20ST2) data reveals $f = 2.18$ and $\Delta t_{ma} = 59$ μs/ft. T96 (A-12) data yields an f value of 2.19 and a Δt_{ma} value of 56 μs/ft. The observations that both the experimentally derived porosity-velocity relationship and the acoustic formation factor values agree with Raiga-Clemenceau et al. (1986) and Issler (1992), respectively, is significant because the in-situ fluid pressures in the mudstones were calculated from sonic-derived porosities, using equation 11 and Issler's (1992) values.

CONCLUSIONS

Deformation experiments conducted on two mudstone samples from the Eugene Island 330 area are relatively compatible with estimated in-situ behavior. Experimentally observed yield stresses agree with porosity-derived estimates of in-situ vertical effective stresses. Experimental stress ratios following sample yield correlate well with in-situ measurements. Constant head tests reveal mudstone-permeability estimates of 5.32×10^{-4} and 1.17×10^{-4} md.

ACKNOWLEDGMENTS

This research was supported by the Gas Research Institute (Contract 5095-260-3558) and the Penn State GeoFluids Consortium. We thank Pennzoil for donating the core used in these experiments. The A-20ST2 (Pathfinder) core was retrieved by the Global Basins Research Network (funded by DOE and industry partners). We especially thank Dan Karig for both conducting these experiments and for critically reviewing this chapter. We thank Alan Huffman for providing valuable feedback on this chapter. We also thank Heather Johnson for assistance with figures.

REFERENCES CITED

Alexander, L. L., and P. B. Flemings, 1995, Geologic evolution of a Plio-Pleistocene salt-withdrawal minibasin: Eugene Island Block 330, offshore Louisiana: AAPG Bulletin, v. 79, p. 1737–1756.

Atkinson, J., 1993, An introduction to the mechanics of soils and foundations: New York, McGraw Hill, 337 p.

Blum, P., J. Xu, and S. Donthireddy, 1996, Geotechnical properties of Pleistocene sediments from the New Jersey upper continental slope, *in* G. S. Mountain, K. G. Miller, P. Blum, C. W. Poag, and D. C. Twichell, eds., Proceedings of the Ocean Drilling Program, scientific results, v. 150: College Station, Texas, Ocean Drilling Program, p. 377–384.

Bowers, G. L., 1994, Pore pressure estimation from velocity data: accounting for overpressure mechanisms besides undercompaction: Society of Petroleum Engineers, SPE 27488, p. 515–529.

Brooker, E. W., and H. O. Ireland, 1965, Earth pressures at rest related to stress history: Canadian Geotechnical Journal, v. 2, p. 1–15.

Bryant, W. R., W. Hottman, and P. Trabant, 1975, Permeability of unconsolidated and consolidated marine sediments, Gulf of Mexico: Marine Geotechnology, v. 1, p. 1–14.

Bryant, W., A. Wetzel, E. Taylor, and W. Sweet, 1986a, Consolidation characteristics and permeability of Mississippi fan sediments, *in* A. H. Bouma et al., eds., Initial reports of the Deep Sea Drilling Project, Leg 96: Washington, D.C., U.S. Government Printing Office, p. 797–809.

Bryant, W., A. Wetzel, and W. Sweet, 1986b, Geotechnical properties of intraslope sediments, Gulf of Mexico, Deep Sea Drilling Project Leg 96, Site 619, *in* A. H. Bouma et al., eds., Initial reports of the Deep Sea Drilling Project, Leg 96: Washington D.C., Government Printing Office, p. 819–824.

Casagrande, A., 1936, The determination of the pre-consolidation load and its practical significance: 1st International Conference on Soil Mechanics, v. 1, p. 60–64.

Core Laboratories, 1994, Sedimentology and petrology, conventional core, no. A12 well, Eugene Island Block 316: Reservoir Geology File no. 194117, unpaginated.

Dewhurst, D. N., A. C. Aplin, J-P. Sarda, and Y. Yang, 1998, Compaction-driven evolution of porosity and permeability in natural mudstones: Journal of Geophysical Research, v. 103, p. 651–661.

Finkbeiner, T., 1998, In situ stress, pore pressure, and hydrocarbon migration and accumulation in sedimentary basins: Ph.D. dissertation, Stanford University, Stanford, California, 193 p.

Finkbeiner, T., M. Zoback, P. Flemings, and B. Stump, 2001, Stress, pore pressure, and dynamically constrained hydrocarbon columns in the South Eugene Island 330 field, northern Gulf of Mexico: AAPG Bulletin, v. 85, no. 6, p. 1007–1031.

Flemings, P., M. D. Zoback, B. A. Bishop, and R. N. Anderson, 1994, State of stress in the Pathfinder well, *in* R. N. Anderson, ed., Global Basins Research Network data volume: Global Basins Research Network, p. 548–586.

Gordon, D. S., and P. B. Flemings, 1998, Generation of overpressure and compaction-driven fluid flow in a Plio-Pleistocene growth-faulted basin, Eugene Island 330, offshore Louisiana: Basin Research, v. 10, p. 177–196.

Hamilton, E. L., 1964, Consolidation characteristics and related properties of sediments from experimental Mohole (Guadalupe site): Journal of Geophysical Research, v. 69, p. 4257–4269.

Hart, B. S., P. B. Flemings, and A. Deshpande, 1995, Porosity and pressure: role of compaction disequilibrium in the development of geopressures in a Gulf Coast Pleistocene basin: Geology, v. 23, p. 45–48.

Issler, D. R., 1992, A new approach to shale compaction and stratigraphic restoration, Beaufort-Mackenzie basin and Mackenzie Corridor, northern Canada: AAPG Bulletin, v. 76, p. 1170–1189.

Jacob, C. E., 1949, Flow of ground water, *in* H. Rouse, ed., Engineering hydraulics: New York, John Wiley and Sons, p. 321–386.

Jones, M., 1994, Mechanical principles of sediment deformation, *in* A. Maltman, ed., The geological deformation of sediments: New York, Chapman and Hall, p. 37–71.

Karig, D. E., 1996, Uniaxial reconsolidation tests on porous sediments: mudstones from Site 897, *in* R. B. Whitmarsh, D. S. Sawyer, A. Klaus, and D. G. Masson, eds., Proceedings of the Ocean Drilling Program, scientific results, v. 149: College Station, Texas, Ocean Drilling Program, p. 363–373.

Karig, D. E., and G. Hou, 1992, High-stress consolidation experiments and their geologic implications: Journal of Geophysical Research, v. 97, p. 289–300.

Karig, D. E., and J. Morgan, 1994, Tectonic deformation: stress paths and strain histories, *in* A. Maltman, ed., The geological deformation of sediments: New York, Chapman and Hall, p. 167–204.

Losh, S. L., 1998, Oil migration in a major growth fault: structural analysis of the Pathfinder core, South Eugene Island 330, offshore Louisiana: AAPG Bulletin, v. 82, p. 1694–1710.

Losh, S., L. Eglinton, and J. Wood, 1994, Coring and inorganic geochemistry in the Pathfinder Well, *in* R. N. Anderson, ed., Global Basins Research Network data volume: Global Basins Research Network, p. 183–194.

Marine Geotechnical Consortium, 1985, Geotechnical properties of northwest Pacific pelagic clays, *in* G. R. Heath, et al., eds., Initial Reports of the Deep Sea Drilling Project, Leg 86, Hole 576A: Washington, D.C., U.S. Government Printing Office, p. 723–758.

Marlow, M. S., H. J. Lee, and A. W. Wright, 1984, Physical properties of sediments from Lesser Antilles margin along the Barbados Ridge, *in* B. Biju-Duval et al., eds., Initial reports of the Deep Sea Drilling Project, Leg 78A: Washington, D.C., U.S. Government Printing Office, p. 549–558.

Mesri, G., and Y. K. Choi, 1985, The uniqueness of the end-of-primary (EOP) void-ratio effective stress relationship: Proceedings of the 11th International Conference on Soil Mechanics and Foundation Engineering, v. 2, p. 587–590.

Mesri, G., and T. M. Hayat, 1993, The coefficient of earth pressure at rest: Canadian Geotechnical Journal, v. 30, p. 647–666.

Neuzil, C. E., 1986, Groundwater flow in low-permeability environments: Water Resources Research, v. 22, p. 1163–1195.

Neuzil, C. E., 1994, How permeable are clays and shales?: Water Resources Research, v. 30, p. 145–150.

Olsen, H. W., R. W. Nichols, and T. L. Rice, 1985, Low gradient permeability measurements in a triaxial system: Geotechnique, v. 35, p. 145–157.

Raiga-Clemenceau, J., J. P. Martin, and S. Nicoletis, 1986, The concept of acoustic formation factor for more accurate porosity determination from sonic transit time data: Society of Professional Well Log Analysts 27th Annual Logging Symposium Transactions, paper G, unpaginated.

Raymer, L. L., E. R. Hunt, and J. S. Gardner, 1980, An improved sonic transit time-to-porosity transform: Society of Professional Well Log Analysts 21st Annual Logging Symposium Transactions, paper P, unpaginated.

Roegiers, J-C., 1989, Elements of rock mechanics, *in* M. J. Economides, and K. G. Nolte, eds., Reservoir stimulation: Englewood Cliffs, New Jersey, Prentice Hall, p. 2-1, 2-22.

Stump, B. B., 1998, Illuminating basinal fluid flow in Eugene Island 330 (Gulf of Mexico) through *in situ* observations, deformation experiments, and hydrodynamic modeling: Master's thesis, Pennsylvania State University, University Park, Pennsylvania, 121 p.

Stump, B. B., and P. B. Flemings, 2000, Overpressure and fluid flow in dipping structures of the offshore Gulf of Mexico (E.I. 330 field): Journal of Geochemical Exploration, v. 69–70, p. 23–28.

Tavenas, F., P. Leblond, P. Jean, and S. Leroueil, 1983, The permeability of soft clays, part I: methods of laboratory measurement: Canadian Geotechnical Journal, v. 20, p. 645–660.

Taylor, E., and A. Fisher, 1993, Sediment permeability at the Nankai accretionary prism, site 808, *in* I. A. Hill et al., eds., Proceedings of the Ocean Drilling Program, scientific results, v. 131: College Station, Texas, Ocean Drilling Program, p. 235–243.

Terzaghi, K., 1925, Principles of soil mechanics, II—compressive strength of clay: Engineering News Record, v. 95, p. 796.

Turcotte, D. L., and G. Schubert, 1982, Geodynamics: applications of continuum physics to geological problems: New York, John Wiley and Sons, 450 p.

Vasseur, G., I. Djeran-Maigre, D. Grunberger, G. Rousset, D. Tessier, and B. Veide, 1995, Evolution of structural and physical parameters during experimental compaction: Marine and Petroleum Geology, v. 12, p. 941–954.

Wetzel, A., 1990, Consolidation characteristics and permeability of Bengal fan sediments drilled during Leg 116, *in* J. R. Cochran et al., eds., Proceedings of the Ocean Drilling Program, scientific results, v. 116: College Station, Texas, Ocean Drilling Program, p. 363–368.

14

Method for Determining Regional Force-Balanced Loading and Unloading Pore-Pressure Regimes and Applying Them in Well Planning and Real-Time Drilling

Phil Holbrook
Force Balanced Petrophysics, Houston, Texas

ABSTRACT

Fluid-expansion effective-stress unloading requires a very low net permeability seal and a relatively high (>25°C/km) regional geothermal gradient. Transition from loading-limb to unloading-limb pore-pressure calibration depends upon the recognition of the regional seal in each well. A regional unloading-limb stress-strain coefficient is calibrated below the regional sealing surface.

The depth to fluid-expansion unloading in 16 worldwide basins ranges from 2 to 6 km. The onset of fluid-expansion unloading occurs between the 90 and 120°C isotherms in these basins. The occurrence of a regional seal to contain fluid expansion is determined from a regional overburden, high fracture-propagation pressure correspondence. These factors control net (intergranular + fracture) permeability that is required to form an effective fluid-expansion pressure seal.

In each well within a region, pore pressures above this seal can be determined from a force-balanced loading-limb stress-strain relationship. The loading-limb stress-strain coefficients sigma max (σ_{max}) and alpha exponent (α) are dependent only on mineralogic constants. These coefficients are composite physical properties of minerals that are independent of depth or location. Loading-limb pore fluid pressure could be considerably above hydrostatic pressure before encountering the unloading pore-pressure regime.

The optimum regional unloading-limb in-situ stress-strain relationship is determined from an appropriate correspondence between measured pore pressures with respect to petrophysically measured strain. A common regional unloading-limb stress-strain exponent (α_{offset}) is determined with respect to the fluid-expansion seal in each well.

Each petrophysical sensor has a different nonlinear response that is not readily transformed to porosity through single-mineral chart book functions. Petrophysical sensor to porosity transforms depend upon the physical properties of the mineral grains. Each petrophysical sensor has its own borehole environmental problems. If sensor- and mineral-specific porosities are different for a given foot, the discrepancy should be resolved before the (strain = 1.0 − porosity) determination is made. The details of the sensor- and mineral-specific nonlinear porosity transforms for resistivity, γ-γ density, and transit time for water-saturated sedimentary rocks are described.

Holbrook, Phil, 2002, Method for Determining Regional Force-Balanced Loading and Unloading Pore-Pressure Regimes and Applying Them in Well Planning and Real-Time Drilling, in A. R. Huffman and G. L. Bowers, eds., Pressure regimes in sedimentary basins and their prediction: AAPG Memoir 76, p. 145–157.

The regional unloading-limb stress-strain exponent (α_{offset}) joins the loading limb (α) at the regionally extensive seal. This stress-strain fitting approach maintains a mechanically sensible fluid pressure to strain proportionality throughout the regionally confined unloading-limb pressure regime. Detailed fluid-pressure comparisons in many stacked unloading-limb pressure compartments indicate that this is commonly the case.

Field experience has shown that there is no unique unloading-limb exponent (α_{offset}). Differing regional temperature gradients and complex hydrocarbon richness exert strong influences on in-situ hydrocarbon cracking. The liquid- to gas-phase change accompanying complex hydrocarbon cracking is the dominant fluid-expansion mechanism.

Mineral grain dissolution can return unloaded sedimentary rocks to the gravitational loading limb, but not beyond. The overlying Newtonian gravitational loading-limb stress-strain coefficients (σ_{max} and α) are mineralogically general throughout approximately biaxial normal fault regime basins. The underlying unloading-limb stress-strain exponent (α_{offset}), however, is region specific and needs to be determined empirically as described previously.

PORE PRESSURE AND FRACTURE PRESSURE DERIVED SIMULTANEOUSLY FROM FORCE-BALANCED STRESS-STRAIN PHYSICS

Pore pressure is the fluid load-sharing element in the subsurface. Solid mineral grains bear the remaining load. The entire load is both generated and borne by the Earth's solid and fluid matter. The effective stress theorem is the force-balanced physical-mathematical expression for porous granular solids that compose the Earth's sedimentary crust. In this rigorous physical expression, the fluid scalar pore pressure (P_P) is the difference between the two solid element scalars, average confining load (S_{ave}), and average effective stress (σ_{ave}), that is, $P_P = S_{ave} - \sigma_{ave}$.

The Earth's sedimentary crust is a continuous closed-form solid-fluid mechanical system. The (S_{ave}) and (σ_{ave}) terms of the effective stress theorem are interdependent within the Earth through constant mineral and fluid coefficients. Holbrook (2002), in chapter 3 of this volume, presents and explains a Newtonian force-balanced in-situ strain-linked system of equations that applies to both loading and unloading pressure regimes in the subsurface. Both loading and unloading pore-pressure regimes depend on overburden, mineralogy, and in-situ strain. The only difference between loading-limb and unloading-limb pressure regimes is their different stress-strain coefficients (α) or (α_{offset}).

There are dynamic interactions within the Earth's closed mechanical system that regulate pore-fluid pressure in both the loading and unloading pore-fluid pressure regimes. On a geologic time scale, fluids are the continuous pore-pressure transmission system. The containing solid granular matrix and fracture system regulates pore-pressure profiles in the subsurface (Holbrook, 1998).

A regional stratigraphic column can be characterized as a series of moderate-permeability pressure compartments with interlayered low-permeability pressure-regulating seals. On a geologic time scale, fluids within a continuous moderately permeable lithostratigraphic body reach a seal relative hydrostatic pressure gradient. The overriding pressure regulation dynamics are that pore pressure at a partially sealing cap rock's minimum work leak point can be no greater than the fracture propagation pressure of the overlying cap rock at that leak point. Intergranular Darcy flow regulates pore pressures where fluid pressures are below the force-balanced fracture-propagation pressure limit.

The keys to understanding and quantitatively predicting loading vs. unloading pore-pressure regimes are the subsurface dynamic interactions of these force – balanced terms. The first key to unloading-limb pore-pressure determination is the recognition of a regional caprock seal. This is the tie point of the loading and unloading limbs in stress-strain space. The second key requirement is a regionally calibrated unloading-limb stress-strain exponent (α_{offset}) below that seal. These two key factors are interdependent. They vary from region to region and must be calibrated from offset wells on that scale.

STRESS-STRAIN LOADING AND UNLOADING HYSTERESIS FOR GRANULAR SOLIDS

The loading limb defines the relationships between stress and strain for sediments that are currently at their maximum state of compaction. The loading-limb pore-pressure regime contains hydrostatic and disequilibrium compaction suprahydrostatic fluid pressures. Minerals are the discrete solid load-bearing elements of the Earth. The loading-limb stress-strain coefficients (σ_{max} and α) are global in nature dependent principally upon mineralogic composition (Holbrook, 1995). Each mineral has only two volumetric plastic stress-strain coefficients that are operative during granular consolidation under effective-stress loading.

Each mineralogic end member has a unique volumetric elastic stress-strain coefficient (K = bulk modulus). A porous sedimentary rock has a proportionally lower bulk modulus (Holbrook et al., 1999). Stress-strain unloading-reloading hysteresis loops are centered about the porous rock or sediment bulk modulus. Hysteresis loops are wider at faster unloading-reloading rates (Holbrook, 1996). The unloading-reloading stress paths approach singularity at geologic strain rates (see figure 1 in Holbrook, 1996).

Geologic loading rate volumetric effective-stress compaction is assumed to be power-law proportional to volumetric in-situ strain ($1.0 - \phi$) (Holbrook, 2002). The (σ_{max}) mineralogic coefficients are related to the average bond strength and hardness of the mineral's crystalline lattice (Holbrook, 2002). The reversible thermal and elastic stress-strain properties of minerals were measured in laboratories decades ago (Carmichael, 1982). The elastic properties of sedimentary rocks are related to and limited by the elastic properties of the minerals of which they are composed.

Figure 1 is a power-law linear effective stress–strain diagram. The plastic loading-limb coefficients (α and σ_{max}) of the first fundamental in-situ effective stress–strain relationship are shown. The plastic and elastic stress-strain limits join at the peak loading-limb point. The geologic unloading-limb angular offset (α_{offset}) falls between the elastic and plastic limits with respect to the peak loading-limb point.

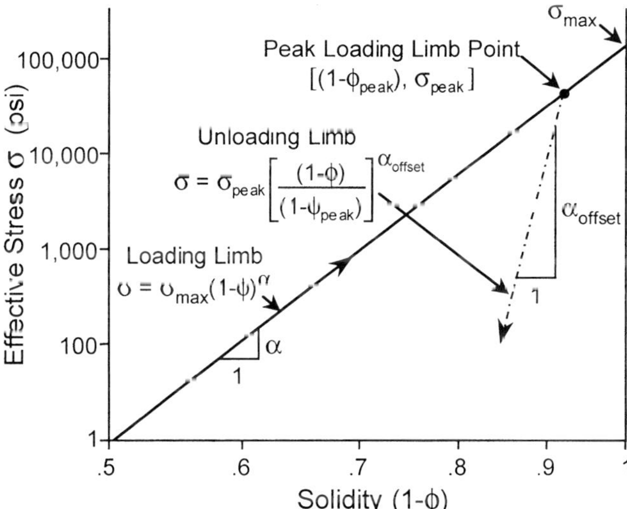

Figure 1. Regional in-situ loading and unloading effective stress–strain angular relationships. The relative in-situ unloading-limb angle effective stress–strain response is power-law linear within plastic and elastic limits. Adapted from Holbrook, 2001.

During natural loading, sedimentary grains are brought closer together, and contact area between grains is increased. The solid element load is borne at these grain contacts and through the mineral lattice to the neighboring grains. Under increasing loads, elastic energy is accumulated in the mineral lattice in proportion to strain. Elastic strain is a miniscule fraction of total strain, which is dominantly plastic.

Also during natural loading, the grain contact area is increased irreversibly following a plastic stress-strain relationship (Holbrook, 2002). The limit of plastic granular consolidation is where all fluid-filled porosity is gone and the rock is totally solid. Solidity (1.0 − porosity) is a direct measure of volumetric in-situ strain. Plastic compaction of granular solids ends where solidity = 1.0. The volumetric strain of zero porosity rocks involves only thermal and elastic coefficients.

The grain contact area necessary to support the average effective-stress load over geologic time is also a function of the average mineralogic crystalline lattice bond strength. Weaker mineral ionic bonds at grain contacts require proportionally more area to bear the same load and vice versa. Grain external contacts and internal mineral lattices bear loads from all directions. Grain mechanical and pressure solution adjustment to force balance is volumetric. Volumetric power-law stress-strain coefficients represent the composite intergrain contact area plus volumetric intragrain force balance in the solid. Pore pressure bears the remaining load.

The two grain matrix mechanical limits in the subsurface are (1) the plastic compactional loading limb; and (2) the elastic unloading limb. The two crucial elements of regional pore-pressure calibration are (1) the location of the peak loading point in the subsurface and (2) the slope of the geologic unloading-limb effective stress–strain exponent (α_{offset}) below the continuous, regional, peak loading-limb surface.

Log and Crossplot Example

The unloading limb joins the peak loading limb in a region that is a continuous surface. Individual wells within a region define points on this surface. Figure 2 is a Gulf Coast example demonstrating how both the peak loading-limb point and the regional unloading-limb stress-strain exponent (α_{offset}) are related. The entire database represents about 14,200 1-ft (0.3 m) samples of effective-stress and strain data. The upper loading-limb segment is about 9700 ft (2957 m) and the lower unloading-limb segment is about 4500 ft (1372 m).

The first panels show the complete stress-strain data set. For clarity, the loading-limb and unloading-limb

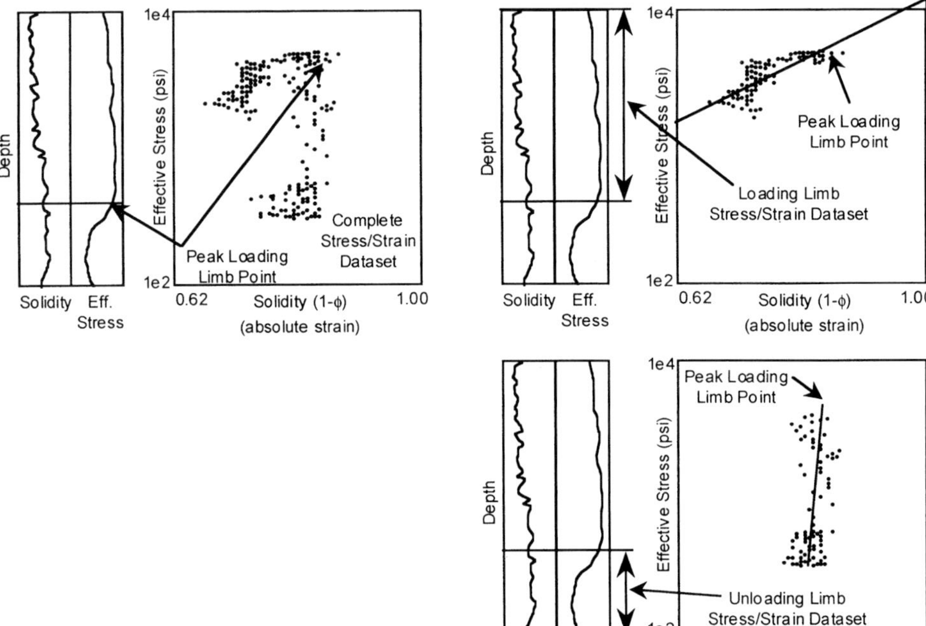

Figure 2. Raw in-situ effective stress–strain relationships. Loading and unloading stress-strain relationships from a Gulf Coast well showing the peak loading-limb point. Most of the scatter in the data is caused by grouping sedimentary rocks of different average mineralogic composition with correspondingly different stress-strain coefficients. Taken from Holbrook, 2001.

segments are shown separately as panels. The loading-limb stress-strain relationship is shown in the top two panels. The regional unloading-limb stress-strain relationship joins the loading limb at the peak loading-limb point. This point is shown on the logs and crossplots of all three panels.

The loading-limb exponent (α) and unloading-limb stress-strain exponent (α_{offset}) are distinctly different. The average mineralogic compactional loading limb angle $\tan^{-1}(\alpha)$ is more than 80° ($\alpha > 5.7$) and is plastic (see Holbrook, 2002). The average unloading limb angle, $\tan^{-1}(\alpha_{offset})$, is 89.2° which corresponds to $\alpha_{offset} = 71.615$. The unloading exponent (α_{offset}) is very close to the bulk modulus mineralogic elastic stress-strain limit. A slight but measurable increase in porosity exists due to in-situ elastic mineral contraction in response to fluid-expansion unloading. Allowable force-balance–dependent pore-fluid pressure measuring relationships must fall between the plastic loading-limb and elastic unloading-limb limits. The data shown in Figure 2 represent an extreme but commonly encountered case of fluid-expansion unloading that is near the elastic mineralogic limit.

The location of the peak loading-limb point is clearly discernible on both the stress-strain crossplots and the individual stress and strain logs shown to the side. The peak loading-limb point is the point of highest solidity in this well. This peak solidity point intersects the regionally continuous fluid-expansion unloading pore-pressure seal.

Regional seals are known to cross stratigraphic boundaries in the North Sea Central Graben as described by Ward et al. (1995, figure 3). High solidity is the common interregional fluid-expansion seal recognition criteria in the North Sea, Gulf Coast, and Far East. All three regions have approximately biaxial normal fault regime stress fields. The same force-balanced physical relationships regulate pore-fluid pressure in all three places. The horizontal/vertical stress ratio in biaxial normal fault regime basins is apparently equal to solidity irrespective of depth and mineralogy (Holbrook, 1996). Solidity (strain) is proportional to the average and principal stresses in the Earth's closed form force balance (Holbrook, 1999).

Force balance determines the upper limit pore-fluid pressure sealing capacity of all cap rocks (Holbrook, 2001). Cap rocks above continuous permeable reservoirs can hold no more pore-fluid pressure than the minimum principal stress at the cap rock's minimum work leak point (where the reservoir pressure is highest with respect to the sum of the cap rock's fracture propagation and capillary entry pressures). The minimum work leak point is on the same surface as the peak loading-limb point in a particular well that penetrates a particular pressure compartment. A regional caprock seal operates in accordance with solid-fluid force balance and can be recognized by its regionally continuous high solidity. Sealing caprock mechanics are the same in both loading and unloading pressure regimes.

FACTORS AFFECTING THE CALCULATION OF POROSITY FROM PETROPHYSICAL SENSORS

The calculation of very accurate porosity (ϕ) from petrophysical sensor data is equivalent to the calculation of very accurate in-situ strain ($1 - \phi$) for further rock-mechanics calculations. For calculation purposes there are several mineral coefficients that are essentially constant ($\pm$ 0.04%) under natural in-situ PV/T conditions. Clay mineral grain density, however, varies by as much as 16% under natural subsurface conditions (Grim, 1968).

Most subsurface sedimentary rocks contain sodium chloride brine under different PV/T x-salinity conditions. Under the known range of geothermal gradients in sedimentary basins, (1) the velocity of sodium chloride brine varies by up to 27%, (2) the density of sodium chloride brine ranges from 0.92 to 1.26 g/cm^3 (also 27%), and (3) the compressibility of sodium chloride brines varies by as much as 61% under known in situ conditions (Holbrook et al., 1999). The electrical conductivity of sodium chloride brines (C_w) varies by five orders of magnitude. All this natural subsurface variability should be considered if one expects to calculate porosity accurately from any petrophysical sensor.

CLAYSTONE DIAGENETIC EFFECTS ON DENSITY, CALCULATED POROSITY, AND OVERBURDEN

During burial diagenesis clay minerals are transformed from low-density highly disordered weathering products into well-crystallized metamorphic micas and chlorites. In moderate to high geothermal gradient areas this complete diagenetic transformation occurs in the upper 5000 m of the Earth's crust. The average grain density of clay minerals in mudstones increases from about 2.64 to 3.15 g/cm^3 as the minerals are transformed primarily through increasing temperature (Aja and Rosenberg, 1992).

Huang et al. (1991) successfully modeled smectite to illite clay mineral transformation as a thermokinetic process in many basins with different geothermal gradients. Aja and Rosenberg (1992) went further, showing data supporting thermodynamic equilibrium between clay minerals during this primarily temperature-controlled diagenetic transformation.

Within a given region, the zero porosity average clay mineral grain matrix (ρ_{clay}) density/depth gradient is estimated. Quartz and calcite have essentially constant densities of 2.65 and 2.71 g/cm^3 in subsurface sedimentary rocks over all known geothermal gradients. Whole rock grain density changes for each sample interval in proportion to the volume-weighted average of clay and nonclay minerals present. Where applied, this procedure reconciles most of the differences between resistivity sensor and γ-γ density sensor-calculated porosities. Accounting for the regional ρ_{clay}/depth density gradient, results in greatly improved porosity, overburden, effective stress, and pore-pressure calculations no matter what petrophysical sensor is used for the primary porosity determination.

POROSITY FROM γ-γ DENSITY SENSOR INPUT

Both average mineral grain density (ρ_{min}) and average downhole fluid density (ρ_{fluid}) must be used to obtain accurate porosity (ϕ) calculations from bulk density (ρ_{bulk}) sensor input. Average clay volume of the solid fraction commonly varies considerably with depth. A borehole attenuation-corrected natural gamma-ray signal from the formation is used to estimate clay volume (V_{clay}) as a fraction of solid for each foot (Holbrook, 1989). Depending upon the stratigraphic sequence type, average mineral grain matrix density is calculated in one of two ways. The average grain matrix density (ρ_{matrix}) used in quartz sand–claystone stratigraphic sequences is

$$(\rho_{matrix}) = [(1.0 - V_{clay}) \times 2.65] + (V_{clay} \times \rho_{clay}) \quad (1)$$

In a calcite–claystone stratigraphic sequence, calcite density (2.71 g/cm^3) is substituted for quartz density (2.65 g/cm^3) in the previous equation. The average claystone grain matrix density (ρ_{clay}) increases gradually with depth determined from the regional geothermal (ρ_{clay}/depth) profile.

Fluid density (ρ_{fluid}) is estimated from a regional PV/T x-sodium chloride salinity profile. The density and bulk modulus coefficients of sodium chloride brines were extracted from voluminous measured density and velocity data by Archer (1992). Archer's equation-of-state thermodynamic molecular interaction coefficients were recast as third-order functions of sodium chloride brine density, pressure, temperature, and molality. The third-order PV/T x-sodium chloride regression of the NaCl brine equation-of-state provides very accurate physically consistent fluid coefficients for porosity (ϕ) from γ-γ bulk density (ρ_{bulk}) measurements estimation and for Gassmann equation forward and inverse modeling (Holbrook et al., 1999). Using the appropriate downhole density coefficients, porosity is calculated from bulk density using the equation

$$\phi = (\rho_{bulk} - \rho_{matrix})/(\rho_{fluid} - \rho_{matrix}) \quad (2)$$

This procedure incorporates the best knowledge we have of grain matrix and fluid densities into the porosity from bulk-density calculation.

POROSITY FROM RESISTIVITY SENSOR INPUT

The common sedimentary mineral grains are infinitely resistive to electric current. Sedimentary rocks conduct electricity primarily through the motion of charged Na^+ and Cl^- ions in the dominantly sodium chloride brine–filled pore space. The formation resistivity factor (F) relates formation water conductivity (C_w) to measured true rock conductivity (C_t). Where porosity (ϕ) equals 1.0, F equals 1.0, and C_w equals C_t.

$$F = C_w/C_t \quad (3)$$

Archie (1941) found a power-law relationship between porosity and formation factor for many sedimentary rocks. The combined intergranular tortuosity-cementation exponent (m) varies with porosity and mineral grain shape in the Archie equation:

$$\phi = F^{(-1.0/m)} \quad (4)$$

Most subsurface brines have salinities equal to or above normal salinity seawater. At these salinities, almost all the total conductivity (C_t) is through the charged Na^+ and Cl^- ions in the water phase (C_w). Variability in the tortuosity-cementation exponent (m) is the pore geometric factor in estimating porosity from resistivity. The power law exponent (m) is a measure of the electrical length/actual length of an insulating porous granular solid. It is a complex function of intergranular pore volume, continuity, and shape.

Figure 3 shows a set of three measured formation factor vs. porosity relationships for the three most common sedimentary minerals. Implicit in each of these curves is the natural grain shape (aspect ratio) of that insulating mineral as it occurs naturally. Grain aspect ratio ranges from 1:1 for perfectly rounded quartz grains to more than 500:1 for an average sedimentary clay. The electrical path though a claystone is many times longer than the electrical path through a quartz grainstone. Considering mineralogically variable tortuosity-cementation (m), the formation factor for a 10% porosity grainstone is 50. The formation factor for a 10% porosity claystone is 240. The inappropriate use of a quartz sandstone formation factor to estimate porosity in a claystone would result in (15/10 PU) or 50% overestimate of claystone porosity.

Mao et al. (1995) determined the quartz grainstone Archie function from laboratory measurements on 155 quartz grainstone core samples. Mao et al.'s data set is many times larger than the earlier Archie or Humble data sets. The value of m increases nonlinearly with decreasing porosity in all three single-mineral empirical formation-factor relationships.

Borai (1987) developed an empirical formation-factor vs. porosity relationship for an equally large core sample data set for pure limestones in Abu Dhubi. The formation factor trend in Borai's data set is offset to higher m values than for quartz grainstones. Mineralogically pure limestones are composed of more platy organic particles. Their average aspect ratio is commonly higher than the generally equant quartz grains. The increased tortuosity-cementation exponent m is evident over the entire range of observed porosities between the rounded quartz and platy limestone data sets (Figure 3).

Holbrook (1996, unpublished data) calculated end-member claystone formation factors from density and resistivity logs on five wells containing only quartz grainstones and claystones. The depth range was from 1000 to 20,000 ft (305–6096 m) covering a wide range of quartz and claystone porosities. The density logs showed unequivocally that the claystones had much lower porosities than the approximately depth equivalent quartz grainstones. A formation-factor ratio ($F_{claystone}/F_{quartz\ grainstone}$) was developed from this data set. The claystone formation factors were leveraged from the well-determined quartz grainstone formation factors. The result is the uppermost m variable claystone formation-factor relationship shown on Figure 3.

All three of the mineralogic end-member formation-factor vs. porosity relationships were determined from large modern data sets. All three relationships are tortuosity-cementation m variable in the same sense. All three are in relative m agreement considering the intergranular tortuosity expected from their different insulating mineral grain aspect ratios. Further details on how porosity is calculated from resistivity measurements can be found in Holbrook et al. (1995).

POROSITY FROM ACOUSTIC TRANSIT-TIME SENSOR INPUT

Sonic logs are commonly used to estimate porosity from transit-time measurements. Generally, there is

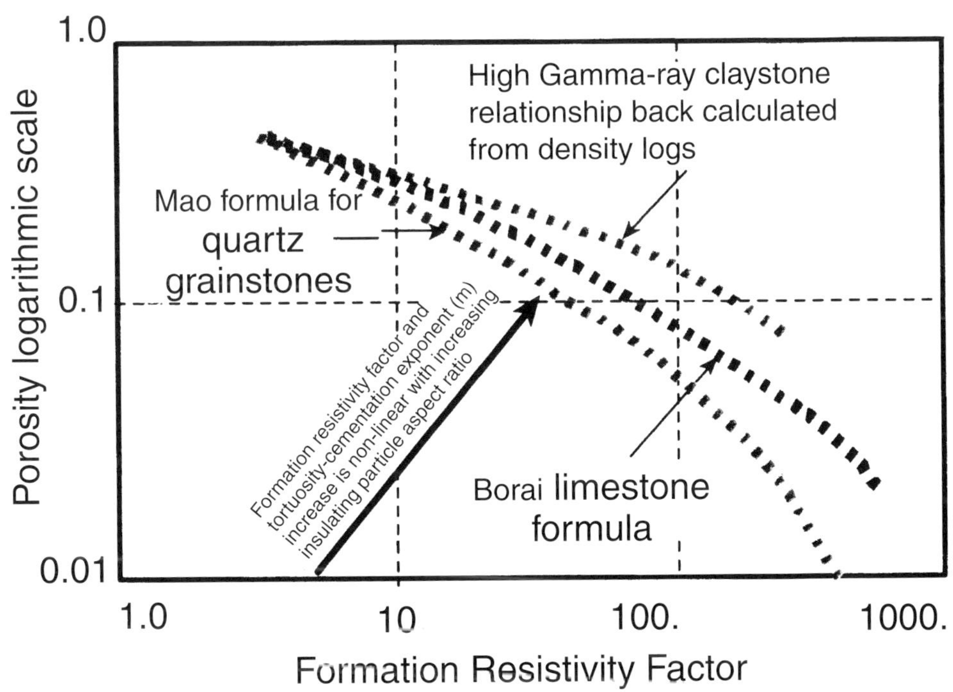

Figure 3. Formation resistivity factor vs. porosity for mineral particles of different aspect ratios. Laboratory sample data was used by Borai (1987) and Mao (1995) to define the *m* variable functions. The claystone *m* variable function was back calculated from in-situ log data using quartz grainstone density log porosity data as a formation-factor reference.

agreement between the velocity-porosity measurements made on laboratory cores at ultrasonic frequencies and that observed with downhole-logging sondes. Almost every study, however, shows a large mineralogic effect, particularly with respect to clay minerals, on measured acoustic velocities (Holbrook et al., 1999).

The propagation of compressional and shear waves through unfractured sedimentary rocks closely follows the extended elastic equations. The Gassmann (1951) equations, Woods equation, and Hashin-Schtrikman (1963) equations, and Archer's (1992) sodium chloride brine relationships are all forms of Hooke's law (Holbrook et al., 1999).

Figure 4 shows the composite elastic coefficients mineralogy–porosity–$V_p^2 - V_s^2$ relationships for the common sedimentary rocks. The V_p^2 and V_s^2 axes of the plot correspond to the bulk modulus (K), shear modulus (μ), and bulk density (ρ_{bulk}) terms in accordance with Gassmann's (1951) granular solid elastic equations.

In-situ claystone elastic coefficients and velocities have been poorly understood. To fill this information gap Goldberg and Gurevich (1998) performed a series of Hashin-Schtrikman inversions on mixed mineralogy V_p^2 and V_s^2 full waveform log data sets. Their average water-wet end-member (zero porosity) claystone elastic coefficients (velocities) fall into a very narrow range shown in Figure 4. This is the same $V_p^2 - V_s^2$ region where electrostatically neutral grainstones pass from a slurry suspension into an amalgamated granular solid.

Though a different physics is involved, there is a correspondence between the convergent formation-factor region (37 ± 3% porosity) in Figure 3 with that of Wood's equation to Gassmann's equation transition region in Figure 4. In both Figures 3 and 4 the single-mineral curves converge as sediments of any mineralogy consolidate from a separate particle slurry to a granular solid.

The measurement axes in Figure 4 are V_p^2 and V_s^2. Adjacent to each axis are the Hooke's law coefficients, bulk modulus, shear modulus, and bulk density that are equivalent to those squared velocities. The point being emphasized is elastic-wave velocities for porous granular sedimentary rocks and slurries closely follow Hooke's law extended elastic equations over the entire (0 to 100%) porosity range.

The log data sets from which the claystone elastic coefficients were extracted had porosities ranging from 38 to 2%. All these lithologies were reasonably hard rocks, not slurries. The general slurrylike acoustic behavior for low-porosity claystones is reasonable considering the acoustic-wave travel path on the molecular scale. Each clay particle has an associated interlammelar electrostatically bound-water layer.

An elastic wave propagating through water-wet clay in any direction must pass through the much slower water phase in the interlammelar pore space. Even in very hard claystones, the individual clay la-

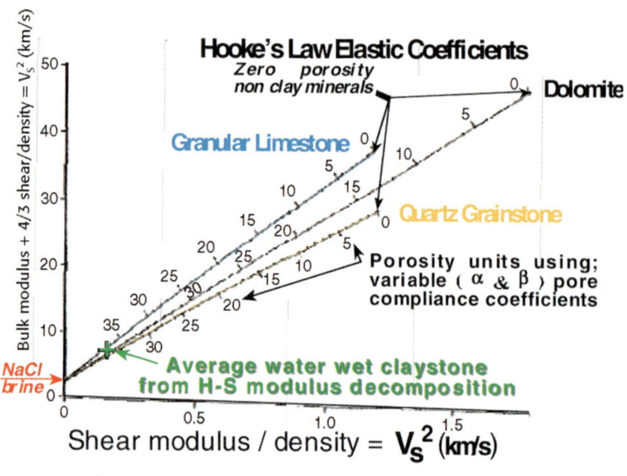

Figure 4. Extended elastic equation velocities ($V_p^2 - V_s^2$) crossplot of single-mineral grainstones and claystones with NaCl brine. Elastic coefficients–mineralogy–porosity–($V_p^2 - V_s^2$) relationships. This figure is modified from Kreif et al. (1990). The average zero porosity claystone coefficients were derived from $V_p - V_s$ sonic data through Hashin-Schtrikman modulus decomposition by Goldberg and Gurevich (1998). The claystone velocities and elastic coefficients fall in the same initial slurry to granular solid amalgamation zone as do the nonclay minerals.

mellae are generally not in direct solid-solid contact. Water-wet claystone elastic behavior is overall slurry-like, as each clay lamella is encased in its own electrostatically bound-water layer. The effective elastic modulus of water-wet claystones includes electrostatically bound water until almost all porosity is lost. Water-wet claystones are only slightly above the Hashin-Schtrikman lower grain contact limit in several large log data sets.

The extended elastic equations portrayed in Figure 4 can be used to invert porosity from V_p^2 and/or V_s^2 in-situ petrophysical data where used in concert with petrophysically measured bulk density (ρ_{bulk}) and mineralogy (V_{clay}) from natural gamma-ray petrophysical data (Holbrook et al., 1999).

PETROPHYSICAL SENSOR-DERIVED POROSITY (ϕ) LINKAGE TO NEWTONIAN IN-SITU FORCE BALANCE AND STRAIN ($1.0 - \phi$)

Figure 5 shows a flowchart to calculate in-situ force balance from petrophysical measurements. The upper region contains three vertical sensor specific flow paths that calculate porosity with respect to mineralogy. Each vertical flow path uses the appropriate in-situ density, conductivity, and elastic coefficients of the minerals and fluid that compose a sedimentary rock.

The grain framework bulk modulus elastic limit is a parameter in the porosity from sonic Δt flow path. The $PV/T - x$ variable sodium chloride brine coefficients are shown across the three sensor-specific flow paths. The diagenetic clay mineral grain density is an adjustable parameter in the porosity from γ-γ density sensor flow path. Mineral and fluid coefficients in all the sensor flow paths are consistent across the upper half elastic stress-strain domain.

The resistivity and sonic sensor flow paths contain nonlinear pore volume and shape coefficients as mentioned in the sensor to porosity transforms sections. If the three vertical sensor to porosity transforms are executed properly and there is no significant borehole wall damage, all three sensor-specific flow paths should indicate the same in-situ true rock porosity of water-filled sedimentary rocks. True rock porosity is the most important petrophysically derived reservoir parameter and is also the central force balance vs. in-situ strain ($1.0 - \phi$) consideration in Figure 5.

Below the true rock porosity midpoint in Figure 5 is the Newtonian closed-form force-balance load vs. strain relationship for normal fault regime approximately biaxial basins. The load elements of the Newtonian closed formulation are on the left side of the individually force-balanced equations. On the left, confining loads are denoted with an "**S**." Force-balanced corresponding effective-stress loads are denoted with a "$\boldsymbol{\sigma}$." Both **S** and $\boldsymbol{\sigma}$ are subscripted vectors. The "v" subscripts denote vertical gravitational loads, and "h" subscripts denote the two corresponding orthogonal horizontal vectorial loads. Average confining load (S_{ave}) and average effective stress (σ_{ave}) are force-balanced scalars.

The effective stress theorem is the fifth equation in the Newtonian force-balanced closed formulation. The scalar pore pressure (P_p) is calculated as the difference between the two load scalars [$(S_{ave}) - (\sigma_{ave})$]. Fracture propagation pressure ($P_F = P_p + \boldsymbol{\sigma}_h = \mathbf{S}_h$) is thereafter calculated in the sixth equation using force-balanced pore pressure (P_p) calculated using the effective stress theorem. All the load terms (**S**, $\boldsymbol{\sigma}_h$, and P_p) to the left of the diagonal load vs. strain (=) signs are a Newtonian closed-form force balance.

All the earth strain terms are on the right side of the equal signs in Figure 5. Absolute volumetric in-situ strain ($1.0 - \phi$) is in each of the individually force-balanced stress-strain equations. The descending arrow in the strain region of Figure 5 indicates the algebraic linkage of these equations to petrophysically

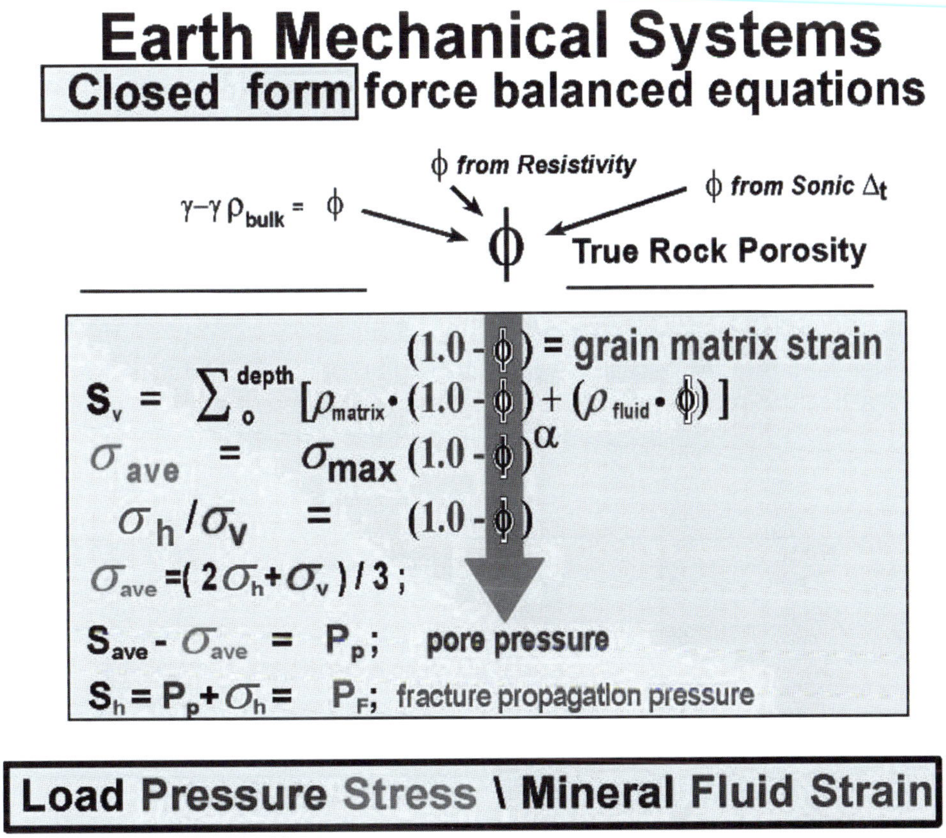

Figure 5. Linkage of petrophysical sensor readings through mineralogically sensitive porosity transforms to the Newtonian closed-form force-balanced stress-strain relationship in normal fault regime approximately biaxial basins. Taken from Holbrook, 2001.

measurable strain. The remaining strain terms (ρ, σ_{max}, and α) are mineral and fluid coefficients that are compositionally linked to each other.

The equal (=) signs denoted by background shading divide in Figure 5 mathematically relate force-balanced loads to absolute in-situ strain in the earth. Force balance–measurable strain linkage is unique to this Newtonian formulation and leads to simplicity and accuracy of calibration, prediction, and detection of pore pressure.

Regional Effective-Stress Calibration and Real-Time Pore-Pressure Prediction

Six wells in a Far East area were used to establish a regional unloading-limb stress-strain exponent (α_{offset}). Three porosity-sensitive sensors, resistivity, bulk density, and P-wave interval transit time were used to determine in-situ strain (solidity). A base-line–normalized gamma-ray sensor reading was used to estimate solid fraction shale volume. A regional average shale grain-density profile and formation-water conductivity profile were established. Mud-weight profiles and repeat formation testers (RFTs) were available to calibrate relative and absolute pore-fluid pressure. Leak-off tests were available to calibrate fracture propagation pressure. Pore-pressure and fracture-pressure data from all six wells were evaluated and weighted equally to determine the regional mechanical and petrophysical relationships. Holbrook (1996) described this calibration procedure up to the point of unloading-limb seal recognition and regional stress-strain exponent (α_{offset}) calibration.

Figure 6 shows one of the six regional calibration wells used. Mineralogy sensitive raw and normalized gamma-ray readings are shown in track 1. Porosity and mineralogy sensitive bulk density and transit time readings are also shown in track 1. Porosity from the m variable resistivity is shown in track 1 on the same 0–50 porosity units scale as porosity from bulk density. If the separate sensor–porosity transforms are correct, and there are no borehole-related sensor problems, the separate sensor porosity curves should be identical.

Porosity was calculated from resistivity using the second-order Archie relationship that accounts for the variable mineralogy dependent tortuosity-cementation m coefficient. Petrophysical sensor conflicts are resolved. The best calculated porosity enters both the power law effective-stress scalar (σ_{ave}) calculation and the integrated bulk density overburden (S_v) calculations. The raw Δ_t log is displayed as a blue trace in track 1. The raw Δ_t log tracks the red m variable Archie porosity log in shales. This correspondence tends to confirm that the input C_w profile used to calculate porosity from resistivity is correct.

Track 3 displays overburden, fracture-propagation pressure, mud weight, and pore-fluid pressure in ppg as fluid-pressure gradients. The calculated traces are closed-form force balanced. Leak-off tests and RFTs are annotated in track 3 to demonstrate the combined borehole pressure measurements match the closed-form force-balanced calibration. The RFTs closely match the force-balance calculated pore pressures. The mud-weight profile is generally greater than the calculated pore-fluid pressure. The green fracture pressure trace exactly matches the upper leak-off test, indicating that initial overburden is correct.

Evaluating all the traces, the comparable loading-limb, peak loading point, unloading-limb features from Figure 1 are readily interpreted. The porosity-sensitive sensors all indicate a decreasing porosity trend to the peak loading point followed by an increasing porosity trend. As in Figure 2, the peak loading-limb point is where the calculated effective stress reaches a maximum, and the calculated porosity reaches a minimum.

The five other calibration wells had these same recognizable features but at different depths. Considering

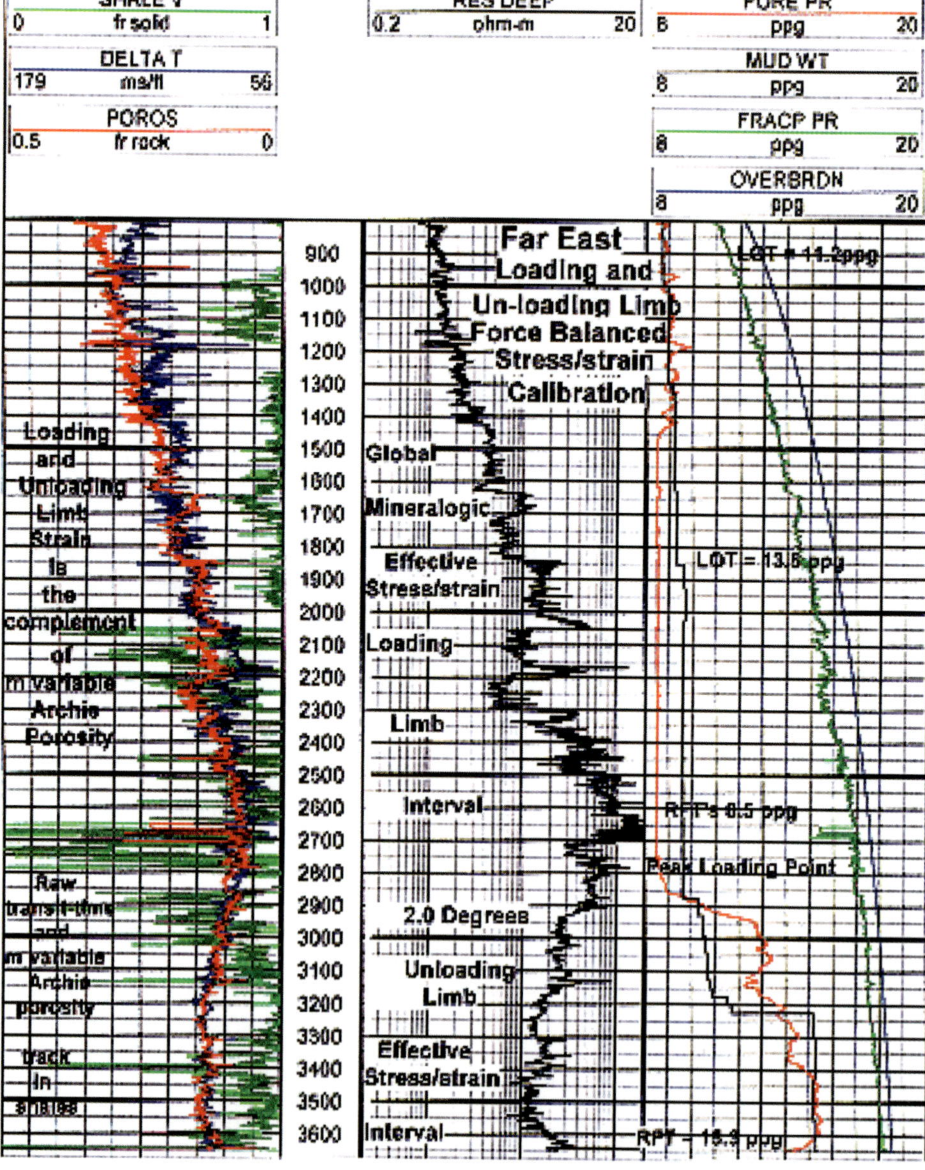

Figure 6. Loading- and unloading-limb force-balanced calibration well-log example. Shale volume and porosity, the complement of strain, are displayed in track 1. The deep resistivity data shown in track 2 is converted to porosity using a second-order *m* variable Archie relationship that accounts for the mineralogic tortuosity-cementation effect on electrical conductivity. The effective stress–strain loading- and unloading-limb intervals both use power-law linear stress-strain coefficient (α_t) or (α_{offset}) to calculate pore-fluid pressure. The peak loading point shown on the logs separates the loading and unloading effective-stress–fluid-pressure regimes. The peak loading point is on the regional high fracture propagation pressure-sealing surface.

the regional unloading pressure regime as a whole, a 2.0° (α_{offset}) relative unloading-limb stress-strain exponent angle was determined. Force-balanced pore-pressure, fracture-gradient, and overburden values in Figure 6 were calculated from the 2.0° unloading-limb stress-strain exponent (α_{offset}) joined at the peak loading point.

Figure 6 shows a long interval of increasingly underbalanced drilling from 3000 to 3500 m. This was also true in several other calibration wells. Drilling mud gas did not provide adequate warning of underbalanced drilling in these wells or the well that was logged measured while drilling (MWD). Some of the calibration wells were shut in to control unpredicted pore pressure. One of these wells required hole abandonment and side tracking to reach their drilling objective.

Pore-Pressure and Fracture-Gradient Prediction from MWD Data

Figure 7 shows the final log generated from combined wire-line and MWD petrophysical data. No adjustments were made to any of the regionally derived constants or control profiles for the entire duration of this well. This well also shows the same upper global mineralogic stress-strain loading-limb interval. An easily recognizable peak loading point is annotated in Figure 7. The same 2.0° relative unloading-limb stress-strain exponent (α_{offset}) was used below the peak loading-limb point.

Some initial skepticism in the force – balanced method was removed by three predicted gas cut mud incidents. The real-time pore-pressure log indicated underbalanced mud weight at 3380 m. The first gas-cut mud incidents occurred at 3400 m. The kill weight of 10.5 ppg was in agreement with that predicted by the real-time pore pressure log.

Mud weight was raised again to 10.7 ppg at 3480 m in preparation for setting casing. The primary objective of setting casing in the overpressured interval was achieved. Note that the onset of elevated pore pressure occurred 200 m below the peak loading limb point and 200 m below the resistivity reversal on the raw resistivity log. Based upon the previously mentioned mud cut and kill weights the regional 2.0° unloading-limb stress-strain calibration was correct.

Mud weight was raised to 11.6 ppg after casing was set. A sharp pore-pressure increase to 11.6 ppg occurred at 3725 m. Underbalanced mud weight was predicted at 3750 m. This preceded gas-cut mud, and mud weight was raised to 12.3 ppg that temporarily controlled the mud gas cut.

Heavy gas mud cut occurred again at 3800 m. The eventual kill weight for this gas cut mud was 14.8 ppg in agreement with the real-time pore pressure. The flow and kill mud weights bracketed the force-balance calculated pore pressure to 0.2 ppg accuracy.

CONCLUSIONS

Pore-pressure estimation accuracy resulted from the application of the closed-form force-balanced calculation method. The method corresponds to a regional loading-limb regime over a fluid-expansion pressure regime that was related to petrophysical data. The predictions were made through a regionally calibrated closed-form mechanical system. Both wire-line measured and MWD petrophysical data were used as measurement input to the mechanical system.

Drilling efficiency benefits were as follows: (1) the gas-cut mud incidents were circulated out without well shut-in; (2) the operator was able to safely reach his drilling objective with one less casing string by setting casing in the upper part of the unloading-limb pressure transition zone. Safety and cost objectives were met by applying force-balanced physics in the calculation of pore pressure.

Using force-balanced stress-strain physical relationships to directly determine pore pressure is a sound drilling engineering choice. Pore pressure is mechanically related to porosity, mineralogy, and density that can be estimated from several petrophysical sensors in the closed mathematical form described. The calculation procedure applied incorporates known rock stress-strain relationships with Newtonian physics that is mathematically related to absolute in-situ strain. This is the only pore-pressure method that involves a defined mechanical stress-strain system.

Loading- and unloading-limb pore-pressure regimes have distinct separate stress-strain exponents (α) and (α_{offset}). The loading-limb/unloading-limb regimes join along a surface that corresponds to a continuous high-solidity cap rock that partially seals an expanding fluid phase pore pressure. Pore pressures in both regimes are treated with mechanically sensible mineralogic stress-strain relationships.

Closed-form force balance comprehensively accounts for both fluid- and solid-borne subsurface loads. Pore pressure below the peak loading-limb surface is limited by the minimum work fracture propagation pressure of the sealing cap rock. Pore pressure in both loading and unloading regimes is determined from regionally calibrated effective stress–strain mathematical functions.

Pore pressure in the subsurface results from a force-balanced mineral and fluid load-sharing relationship. The Newtonian stress-strain formulation is also a

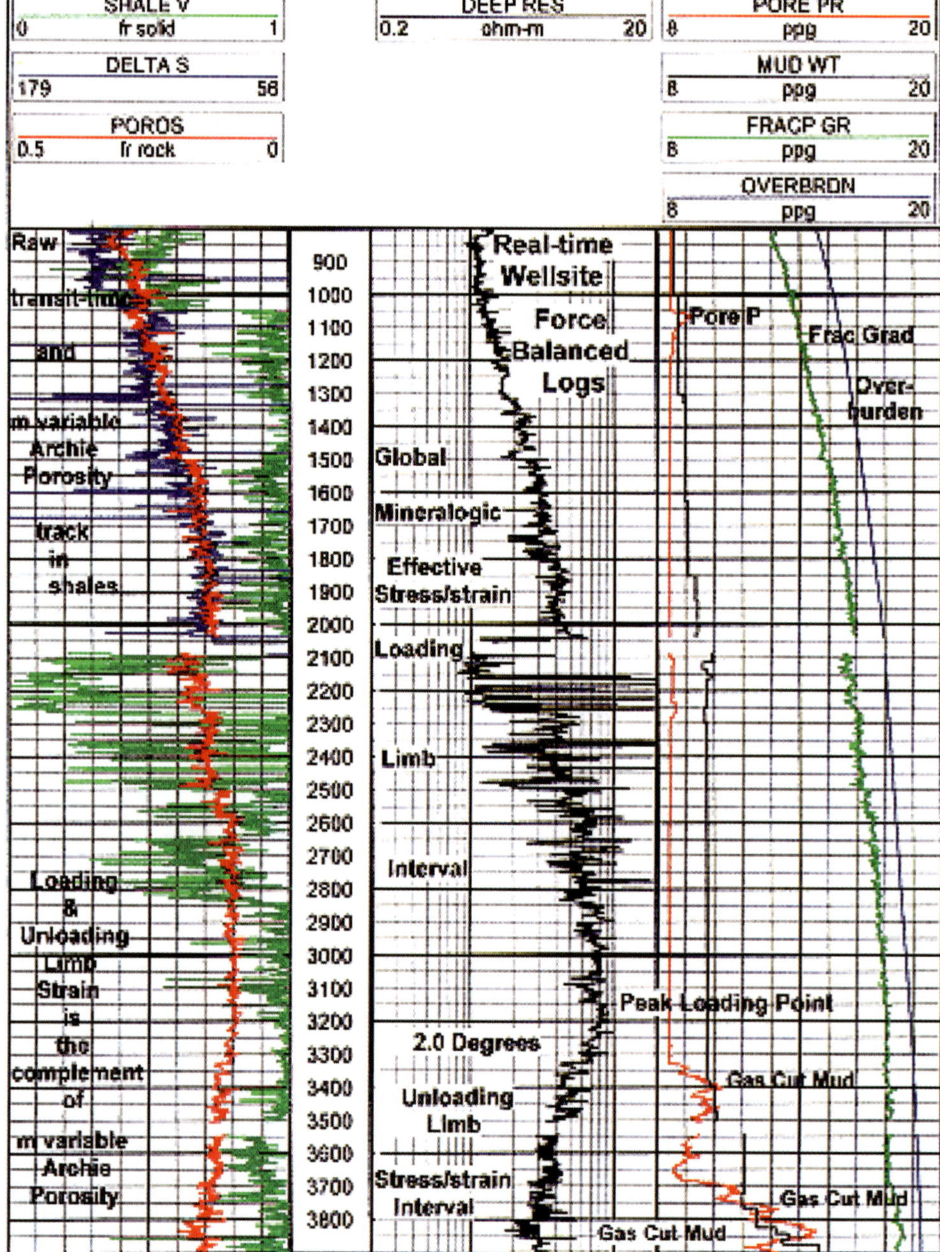

Figure 7. Real-time force-balanced logs in the same format as Figure 6. All petrophysical calibration constants are identical with the regional calibration used for Figure 6. The only operator intervention applied during the entire well was to switch to the regionally calibrated 2.0° (α_{offset}) unloading-limb stress-strain coefficient at the peak loading-limb point shown.

closed-form solution to many other oilfield mechanical problems that inherently depend on in-situ borehole vs. Earth formation force balance.

REFERENCES CITED

Aja, S. U., and P. E. Rosenberg, 1992, The thermodynamic status of compositionally-variable clay minerals: a discussion: Clays and Clay Minerals, v. 40, p. 292.

Archer, D. G., 1992, Thermodynamic properties of NaCl + H2O system II: thermodynamic properties of NaCl(aq), NaCl.2H2O(cr), and phase equilibria: Journal of Physical Chemistry, v. 21, no. 4, p. 793–829.

Archie, G. E., 1941, The electrical resistivity log as an aid in determining some reservoir characteristics: Transactions of the American Institute of Mining Metallurgical and Petroleum Engineers, v. 14.

Borai, A. M., 1987, A new correlation of cementation factor in low-porosity carbonates: Society of Petroleum Engineers 14401, p. 10–14.

Carmichael, R. S., 1982, Handbook of physical properties of rocks: Boca Raton, Florida, CRC Press, 404 p.

Gassmann, F., 1951, Elastic waves through a packing of spheres: Geophysics, v. 16, p. 673–685.

Goldberg, I., and B. Gurevich, 1998, Porosity estimation from P and S sonic log data using a semi-empirical velocity-porosity-clay model: Society of Professional Well Log Analysts 39th Annual Logging Symposium, paper QQ.

Grim, R. E., 1968, Clay Mineralogy: New York, McGraw-Hill, 596 p.

Hashin, Z., and S. Shtrikman, 1963, A variational approach to the theory of the elastic behavior of multiphase materials: Journal of Mechanics and Physics of Solids, v. 11, p. 127–140.

Holbrook, P. W., 1989, A new method for predicting fracture propagation pressure from MWD or wireline log data: Society of Professional Engineers Paper 19566.

Holbrook, P. W., 1995, The relationship between porosity, mineralogy and effective stress in granular sedimentary rocks: Society of Professional Well Log Analysts 36th Annual Logging Symposium, paper AA.

Holbrook, P. W., 1996, The use of petrophysical data for well planning, drilling safety and efficiency: Society of Professional Well Log Analysts 37th Annual Logging Symposium, paper X.

Holbrook, P. W., 1998, The universal fracture gradient/pore pressure force balance upper limit relationship which regulates pore pressure profiles in the subsurface: American Association of Drilling Engineers Industry Forum on Pressure Regimes in Sedimentary Basins and their Prediction.

Holbrook, P. W., 1999, Physical explanation of the closed form mineralogic force balanced stress/strain relationships in the Earth's sedimentary crust: Bulletin Centre Recherche Elf Exploration and Production, Memoir 22, p. 61–67.

Holbrook, P. W., 2001, Pore pressure through Earth mechanical systems: Houston, Texas, Force Balanced Press, 135 p.

Holbrook, P. W., 2002, The primary controls over sediment compaction, *in* A. R. Huffman and G. L. Bowers, eds., Pressure regimes in sedimentary basins and their prediction: AAPG Memoir 76, p. 21–32.

Holbrook, P. W., D. A. Maggiori, and R. Hensley, 1995, Real-time pore pressure and fracture pressure determination in all sedimentary lithologies: SPE Formation Evaluation, v. 10, no. 4, p. 215–222.

Holbrook, P. W., I. Goldberg, and B. Gurevich, 1999, Velocity—porosity—mineralogy Gassmann coefficient mixing relationships for water saturated sedimentary rocks: Society of Professional Well Log Analysts 40th Annual Logging Symposium, paper T.

Huang, W. L., J. M. Longo, and D. R. Pevear, 1991, An experimentally derived kinetic model for smectite to illite conversion and its use as geothermometer: Clay Minerals Society Annual Meeting.

Krief, M., J. Garat, J. Stellingwerf, and J. Ventre, 1990, A petrophysical interpretation using the velocities of P and S waves (full waveform sonic): The Log Analyst, v. 31, p. 355–369.

Mao, Z. Q., C. G. Zhang, C. Z. Lin, J. Ouyang, Q. Wang, and C. J. Yan, 1995, The effects of pore structure and electrical properties of core samples from various sandstone reservoirs in Tarim Basin: Society of Professional Well Log Analysts 36th Annual Logging Symposium.

Ward, C. D., K. Coghill, and M. D. Broussard, 1995, The application of petrophysical data to improve pore pressure and fracture pressure determination in North Sea Central Graben HPHT wells: Society of Petroleum Engineers Annual Technical Conference and Exhibition, SPE paper 28297.

15

Forward Modeling of Log Response in Geopressured Formations Reveals Valuable Insights to the Various Pore-Pressure Prediction Techniques

J. C. Rasmus
Schlumberger Product Center,
Sugar Land, Texas

ABSTRACT

A forward model for log response in a geopressured shale is developed. The state of compaction, lithology, water depths, salinity profiles, compaction coefficients, and clay characteristics are all user modifiable. Effective-stress relationships for geopressured shales that are currently widely practiced in the industry all assume constant log response is indicative of constant porosity and, therefore, constant effective stress. This modeling reveals that, although the porosity of a geopressured shale may remain constant for a constant effective stress, its log reponse does not. Therefore, this chapter illustrates that any effective-stress technique that uses the log response directly cannot be used to accurately compute the effective-stress state of the shale. Instead, the log response must first be characterized in terms of lithology, salinity profile, fluid moduli, water depth, and compaction coefficients. The modeling shows that this is much more critical for resistivity than velocity measurements.

INTRODUCTION

A self-consistent model of a sedimentary formation in various stages of undercompaction has been developed to forward model various logging-tool responses. The sedimentary formation model includes user-modifiable expressions for temperature, clay content, lithology, salinity, sediment water volume, water depth, and compacting stress as a function of sediment burial depth below the mud line. The formation model is input to various logging-tool response equations to predict log response as a function of any of the model characteristics. Valuable insight is gained by comparing these modeled logs to those expected from conventional pore-pressure interpretation techniques.

METHOD

The geopressure formation forward model consists of user-modifiable expressions for lithology, porosity, pore fluid temperature, salinity, and bulk modulus properties as a function of sediment vertical depth below the mud line. The formation is taken to be composed of quartz, wet clay, and effective (filled with moveable fluids) porosity. This is the porosity that contains the interstitial water that escapes from the formation while it is being compacted. The relationship, rock depth $= C \times 10^{(-b\phi)}$ is used to model the effective porosity of the formation with depth, where rock depth is the depth below mud line (Rasmus, 1991).

Rasmus, J. C., 2002, Forward Modeling of Log Response in Geopressured Formations Reveals Valuable Insights to the Various Pore-Pressure Prediction Techniques, in A. R. Huffman and G. L. Bowers, eds., Pressure regimes in sedimentary basins and their prediction: AAPG Memoir 76, p. 159–164.

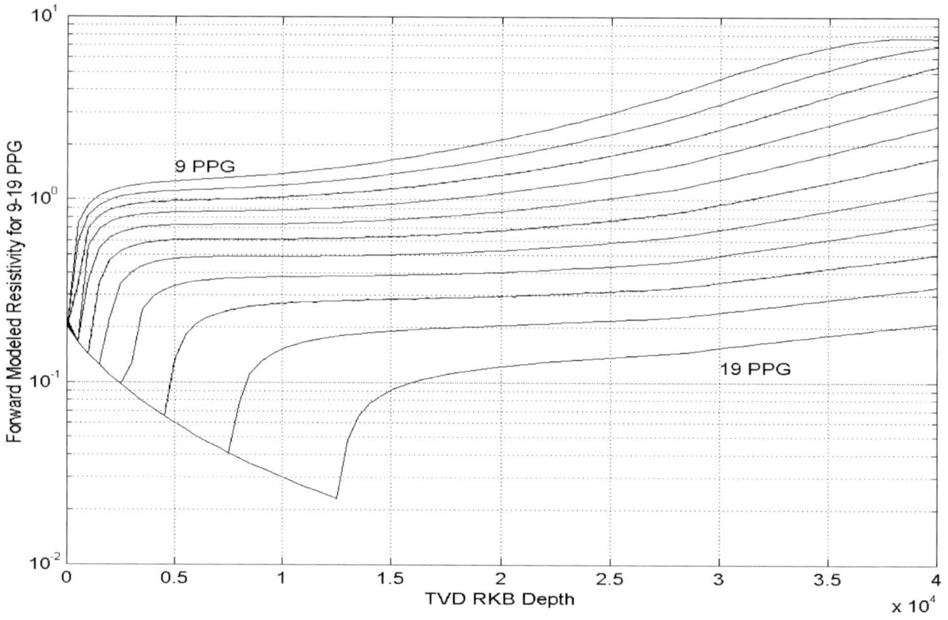

Figure 1. Forward-modeled resistivity vs. depth for various pore-pressure equivalent mud weights for a water depth of 0 ft and a quartz/wet clay ratio of 1.0. Notice how the resistivity trends are similar to other published overlays. True vertical depth (TVD) is in feet; resistivity is in ohm m.

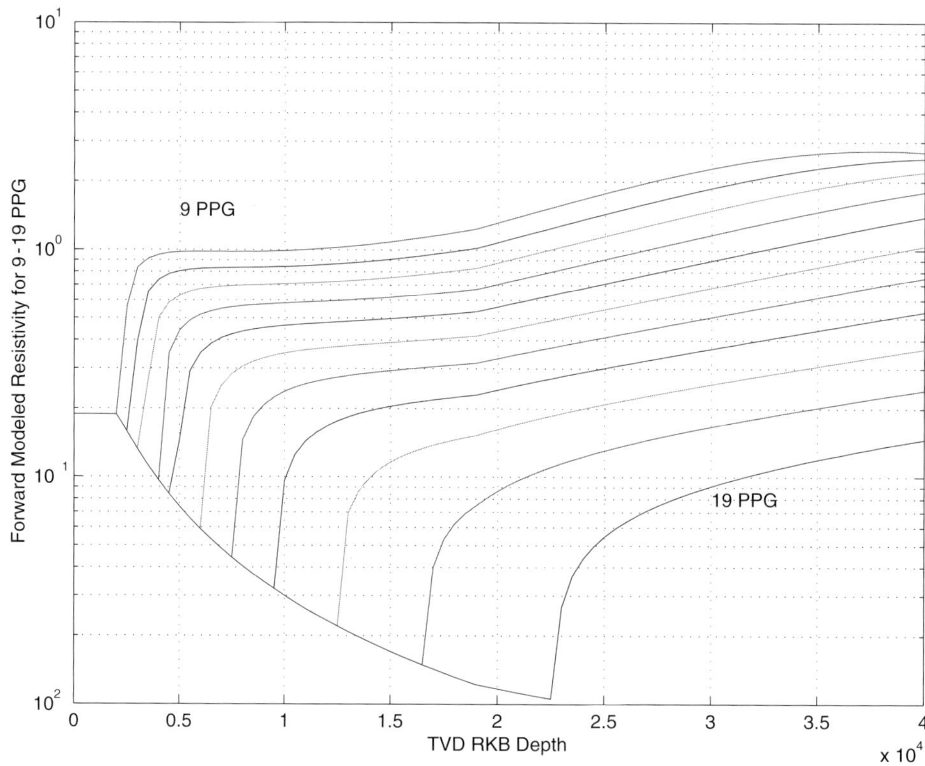

Figure 2. Resistivity vs. depth for various pore-pressure equivalent mud weights for a water depth of 2000 ft (610 m) and a quartz/wet clay ratio of 0.25. The deviation from Figure 1 is significant and illustrates the importance of proper characterization of the lithology and water depth for pore-pressure computations. True vertical depth (TVD) is in feet; resistivity is in ohm m.

C is the depth at which ϕ equals 0, and b is the rate of compaction. Water depth and air-gap parameters are input so that true vertical depth relative to kelly bushing (TVD RKB) can be plotted. The solids $(1 - \phi)$ are placed in the volume that is not taken up by porosity. They are divided into quartz and wet clay and represented by a parameter that is the ratio of quartz to clay, staying constant with depth to accurately reflect the fact that the solids mass is conserved. The temperature is modeled with an offset (surface temperature) and a gradient. An expression for the salinity fluid property allows it to range from seawater at the mud line to salt saturated at a given depth. The fluid bulk modulus is modeled as a function of pressure, temperature, and salinity. A model has also been developed for the compressional and shear dry frame modulus as a function of porosity and lithology.

The volumes and parameters from this forward model as a function of depth are input to the various logging-tool response equations containing these modeled parameters and volumes. The response of resistivity, density, gamma ray, and neutron log vs. depth

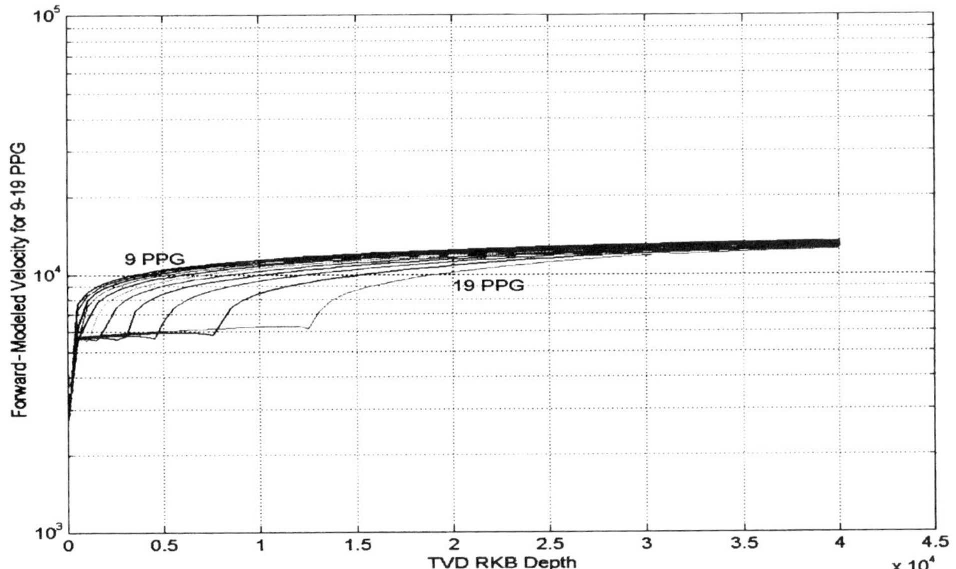

Figure 3. Velocity vs. depth for various pore-pressure equivalent mud weights for a water depth of 0 ft and a quartz/wet clay ratio of 1.0. Total vertical depth (TVD) is measured in feet, and velocity is in ft/sec.

can then be computed and displayed. The sonic compressional and shear response is modeled using Gassmann's equations (Gassmann, 1951). The modeled formation bulk density is integrated to compute the overburden pressure as a function of air gap, water depth, and sediment depth.

RESULTS

Forward-Modeled Resistivity

Figure 1 shows the forward model of resistivity for various pore-pressure equivalent mud weights vs. TVD RKB using a water depth of zero. The lines represent pore pressure equivalent mud weights ranging from 9 to 19 ppg, with 19 ppg being the right-most line. Notice how the constant equivalent mud weight lines have a similar appearance to the resistivity overlays developed by past authors (Matthews and Kelly, 1967) when shallow waters were being drilled. The resistivity rises quickly at shallow depths as the decreasing porosity effect outweighs the increasing salinity effect then rises more gradually as the porosity decrease becomes more linear. The difficulty in drawing a straight-line normal trend as required for some techniques can be seen at shallow depths. At depths greater than several thousand feet below the mud line and for intervals of less than 5000 ft (1524 m), however, the normal trend is fairly linear on this typical logarithmic resistivity scale. Figure 2 shows the effect of changing the quartz/wet clay ratio from 1.0 (0.5/0.5) in Figure 1 to 0.25 (0.2/0.8) and the water depth to 2000 ft (610 m). Notice how the slopes of the resistivity curves for each equivalent mud weight have changed, as well as the values at any particular TVD. It is these changes in lithology that cause shifts in the normal trend lines and can make these plots look noisy when real data are plotted at this scale. That is why previous authors have stressed that consistent-lithology shale points be plotted on these charts. Plots like Figures 1 and 2 using the forward-modeled resistivity allow a prediction of pore pressure as a function of lithology, water depth, and resistivity before drilling a well and help in defining the normal trend for other techniques.

Forward-Modeled Velocity

Figure 3 shows a plot of compressional velocity vs. depth for the various pore-pressure equivalent mud weights. Figure 4 is the same data plotted vs. effective stress, the difference between overburden stress and pore pressure. Normally, the velocity is assumed to be a function of effective stress independent of the value of pore pressure. In this case there would be only one line on this plot with all of the pore pressures lying on top of each other. Note that for the higher pore pressures, this is not the case. This is because the fluid bulk modulus, being a function of temperature, salinity, and pressure, can be different for the same effective stress. Consider two formations with equal effective stress, one shallow at normal pressure and one deeper with overpressure. The deeper one has a greater overburden, temperature, pore pressure, and possibly salinity. The fluid modulus is therefore larger at the deeper depth because of these environmental differences. A larger fluid modulus (less compressibility) gives rise to a greater velocity for the deeper formation although the effective stress is the same. This phenomenon has the same signature as unloading reported by others (Bowers, 1994). Figure 5

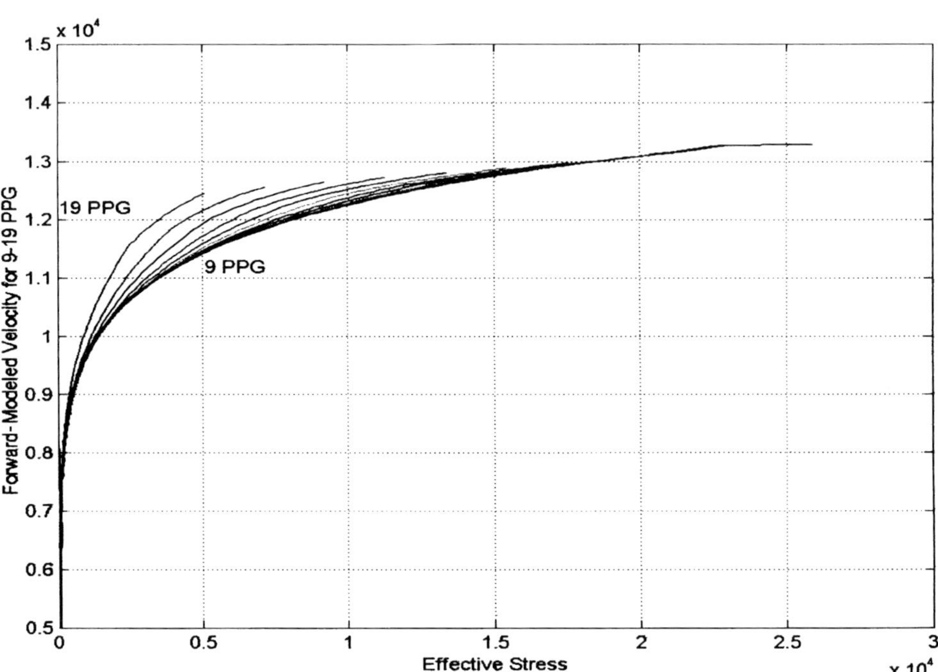

Figure 4. Velocity vs. effective stress for various pore-pressure equivalent mud weights for a water depth of 0 ft and a quartz/wet clay ratio of 1.0. The velocity depends not only on the effective stress but also on the fluid properties at any particular effective-stress state. This means that the velocity is a function of both effective stress and depth and has not been addressed in other pore-pressure techniques to date. Effective stress is measured in psi, and velocity is measured in ft/sec. Temperature in all cases was assumed to linearly increase at 1.0°F/100 ft from a surface temperature of 50°F.

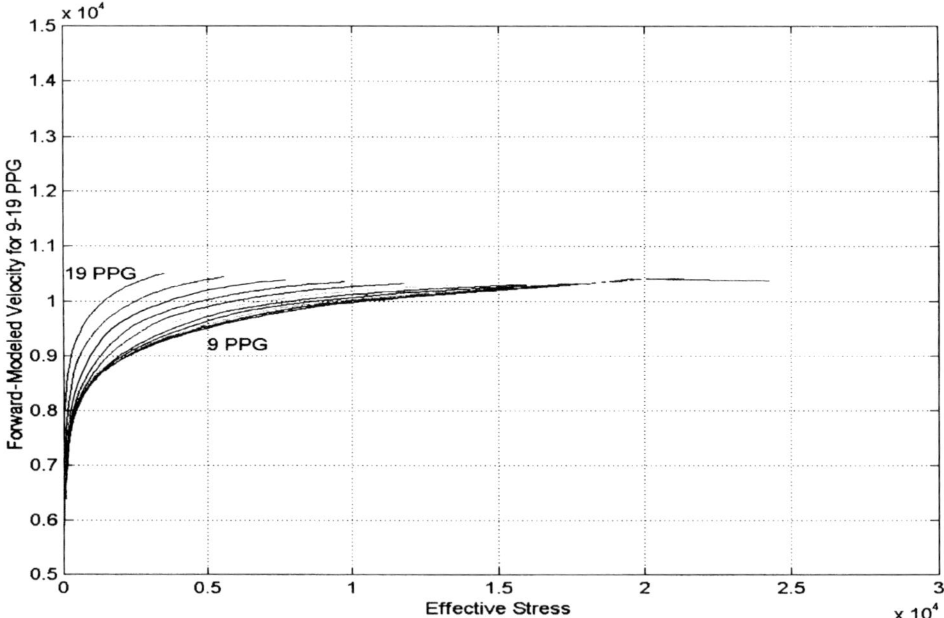

Figure 5. Velocity vs. effective stress for various pore-pressure equivalent mud weights for a water depth of 2000 ft (610 m) and a quartz/wet clay ratio of 0.25. Effective stress is measured in psi, and velocity is measured in ft/sec. Temperature in all cases was assumed to linearly increase at 1.0°F/100 ft from a surface temperature of 50°F.

shows the effect of changing the water depth to 2000 ft (610 m) and the quartz/wet clay ratio from 1.0 to 0.25 as was done for Figure 2. The large effect of relatively small changes in lithology can be seen.

INSIGHTS INTO OTHER TECHNIQUES

The dependence of velocity on both effective stress and depth affects other pore-pressure techniques as follows. In an undercompacted shale, the effective stress of the rock is the difference between the overburden and pore pressures. The equivalent-depth technique assumes that two formations with equal effective stresses have the same porosity and therefore the same log response. The first assumption is logical and valid, but the second is not as shown in Figure 4. The result is that these techniques underestimate pore pressures. Below about 15 ppg, however, and over limited depth ranges, these techniques should give good results. Other techniques (Eaton, 1975) use the ratio of an observed to normal log response to compute pore pressures. In the following

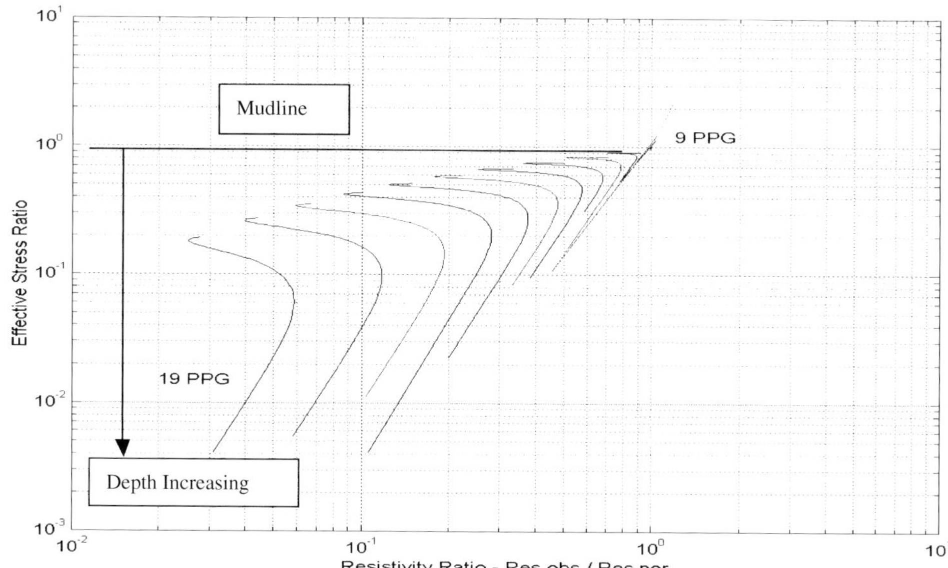

Figure 6. Resistivity ratio vs. effective stress ratio for various pore-pressure equivalent mud weights for a water depth of 0 ft and a quartz/wet clay ratio of 1.0. Effective stress ratio is dimensionless and is the right side of equation 5. Resistivity ratio is dimensionless and is the left side of equation 5. The fact that the curves are double valued and unique for each pore pressure illustrates the futility of using resistivity ratios for effective-stress and pore-pressure calculations.

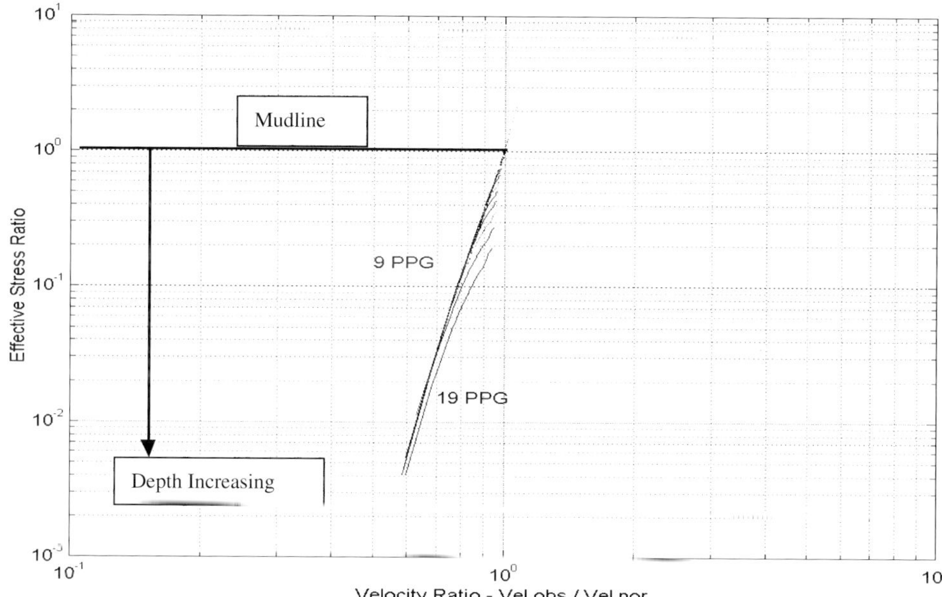

Figure 7. Velocity ratio vs. effective stress ratio for various pore-pressure equivalent mud weights for a water depth of 0 ft and a quartz/wet clay ratio of 1.0. Effective stress ratio is dimensionless and is the right side of equation 5. This plot illustrates that velocity ratios in moderate pore pressures can be used for effective stress and pore-pressure calculations.

paragraphs, I show that this is not an entirely independent technique from the equivalent-depth technique.

The definition of the equivalent depth method is that the effective stress of a geopressured rock at depth D_2 is assumed to be equal to the effective stress of the same type rock at a shallower, normally pressured depth D_1. This is the definition of the equivalent-depth method, sometimes referred to as a vertical method (Traugott, 1997). One can compute the pore pressure of the rock at depth D_2 knowing the overburden pressures and normal water pressure. An expression for the equivalent-depth method is

$$P_{\text{eff2}} = P_{\text{eff1}} \tag{1}$$

or

$$P_{\text{ovb2}} - P_{\text{wpore2}} = P_{\text{ovb1}} - P_{\text{wnor1}} \tag{2}$$

where P_{eff} is effective pressure, subscripts 1 and 2 refer to depths D_1 at normal pore pressure and a deeper depth D_2 at a higher pore pressure. P_{wpore} is the pore pressure at D_2, P_{wnor} is the normal water pressure, and P_{ovb} is the overburden pressure. Dividing by D_2, this can be rewritten as

$$P_{\text{wpore2}}/D_2 = P_{\text{ovb2}}/D_2 - (P_{\text{ovb1}} - P_{\text{wnor1}})/D_2 \tag{3}$$

which can be compared to Eaton's equation

$$P_{wpore2}/D_2 = P_{ovb2}/D_2 - ((P_{ovb2} - P_{wnor2})/D_2)(X_{obs}/X_{nor})^{(Y)} \quad (4)$$

where X_{obs} and X_{nor} are the observed and expected normal log responses at a given depth. The two expressions are equivalent if set as

$$(X_{obs}/X_{nor})^{(Y)} = (P_{ovb1} - P_{wnor1})/(P_{ovb2} - P_{wnor2}) \quad (5)$$

The right-hand side of this equation is the ratio of the normal effective stresses, which is also equivalent to D_1/D_2 when the slopes of the overburden stress and normal pore-pressure curves are constant with depth. Therefore

$$(X_{obs}/X_{nor})^{(Y)} = D_1/D_2 \quad (6)$$

Thus, Eaton's method is an attempt to map the normal and observed log response (sometimes called a horizontal technique [Traugott, 1997]) at one particular depth to the ratio of depths at equal effective stresses (sometimes called a vertical technique). This means that Eaton's method relies on the validity of the equivalent-depth method. Previously, it was shown that the equivalent-depth method is valid only over limited depth ranges due to the fact that fluid properties change with depth, causing the log response to change with depth even when the porosity (or effective stress) stays constant. Therefore Eaton's method is under the same limitations. The widespread success of Eaton's method lends credence to the equivalent-depth technique. As pointed out in Figure 4, however, the equivalent-depth method's assumption of constant log response for constant effective stress is not always valid and may explain why Eaton's exponent (Y) and/or normal trend lines may sometimes have to be altered to compute the correct pore pressure. The limitations of the equivalent-depth technique can be shown graphically by plotting data in the format given by equation 5. Plotting (X_{obs}/X_{nor}) vs. ($P_{ovb1} - P_{wnor1})/(P_{ovb2} - P_{wnor2}$) on a log-log plot should result in a straight line with a slope of Y. Figure 6 shows the results of plotting forward-modeled resistivity data representing a depth range from mud line to 40,000 ft (12,192 m) and pore-pressure equivalent mud weights from normal to 19 ppg. Where the pore pressure is normal, the effective-stress ratio and X_{obs}/X_{nor} are both unity, and all of the normally pressured data at each depth level plots at this point as seen on the plot in Figure 6. The various lines plotting away from this point represent pore-pressure equivalent mud weights from 9 to 19 ppg at various depth levels. The lines not only have changing slopes but go through a reversal and are doubled valued for a given resistivity ratio. This essentially shows that resistivity cannot be used in an equivalent-depth method. Figure 7, however, shows that velocity is better behaved and better suited than resistivity for use in an equivalent-depth technique. This is because the lines of constant pore pressure are close to each other and have similar slopes.

CONCLUSIONS

1. Forward modeling is used to produce overlays for a particular formation model and water depth. For shallow water depths, these overlays mimic the trends seen in earlier empirical overlays.
2. Forward modeling has shown that logging-tool response is not always constant with constant porosity and effective stress as heretofore assumed because of the temperature, pressure, and salinity changes that occur as a geopressured shale is buried deeper. Techniques using the equivalent-depth method are affected by this phenomena.
3. Horizontal techniques or those that use ratios of observed to normal log response are in reality an approximation to the equivalent-depth method and are under the same limitations.

REFERENCES CITED

Bowers, G. L., 1994, Pore pressure estimation from velocity data: accounting for overpressure mechanisms besides undercompaction: International Association of Drilling Contractors/Society of Petroleum Engineers 27488, p. 515–530.

Eaton, B. A., 1975, The equation for geopressure prediction from well logs: Society of Petroleum Engineers 50th Annual Fall Meeting Proceedings, SPE 5544, 11 p.

Gassmann, F., 1951, Uber die Elastizitat poroser Medien: Vier. der Natur. Gesellschaft, v. 96, p. 1–23.

Matthews, W. R., and J. Kelly, 1967, How to predict formation pressure and fracture gradient: Oil & Gas Journal, February 20, p. 92–106.

Rasmus, J. C., 1991, The use of real time pore pressure and drilling derived rock strength to optimize ROP: Society of Petroleum Engineers Drilling Engineer, v. 6, no. 4, p. 264–272.

Traugott, M. O., 1997, Pore/fracture pressure determination in deep water: Deepwater Technology 218 (8), supplement to World Oil, p. 68–70.

16

Pore Pressure ahead of the Bit: An Integrated Approach

Nader C. Dutta
WesternGeco, Houston, Texas

William H. Borland
Schlumberger Wireline and Testing Services, Gatwick, United Kingdom

W. Scott Leaney
Schlumberger Wireline and Testing Services, Gatwick, United Kingdom

Richard Meehan
Schlumberger Wireline and Testing Services, Sugar Land, Texas

W. Les Nutt
Schlumberger Wireline and Testing Services, Fuchinobe, Japan

ABSTRACT

Undercompacted shales generally have a lower acoustic impedance (product of density and velocity) than those that follow a normal compaction trend. Departure from the normal compaction trend may indicate potential drilling hazards due to overpressure. Techniques that can monitor acoustic impedance can be used to indicate the existence of such potential hazards, and thereby, help in designing the casing and mud program.

Prediction of pressure ahead of the bit starts with the best predrill model. In frontier wells, commonly seismic data are the only data available. Seismic velocities from analysis of stacking velocities and impedances from reflection sequence analyses, in conjunction with a predrill rock model, can be used to develop a predrill pressure vs. depth profile. This has been used with considerable success in deep-water wells. The limitations, however, are the lack of resolution in the reflection seismic data and uncalibrated velocity models. Thus, a strategy is developed that can update this so-called static model in real time using borehole data.

Conventional wire-line vertical seismic profile (VSP) measurements are commonly used to provide high quality reflection data within and below the bottom of the well. Inversion of VSP data for acoustic impedance has been demonstrated to be a reliable way to accurately predict acoustic impedance below the bit, with more resolution than the conventional velocity data from stacking-velocity analyses. This has been found to yield pressure vs. depth profiles, at the bit level, with more resolution. Downtime on the rig is required to acquire the wire-line data.

Vertical seismic profile inversion allows the location of the overpressured zone to be accurately determined in two-way traveltime. This time estimate can be converted to depth if the formation acoustic velocity ahead of the bit is known. The drill bit seismic technique, which uses a working drill bit as the seismic source, provides continuous time to depth information. These data can be used to estimate the formation acoustic velocity continuously, in real time, and to calibrate seismic velocity at the bit

Dutta, Nader C., William H. Borland, W. Scott Leaney, Richard Meehan, and W. Les Nutt, 2002, Pore Pressure ahead of the Bit: An Integrated Approach, in A. R. Huffman and G. L. Bowers, eds., Pressure regimes in sedimentary basins and their prediction: AAPG Memoir 76, p. 165–169.

level and hence allow an accurate, continuously updated, prediction of the depth to the overpressure hazard in real time at the well site.

In this chapter we present a methodology to quantify and predict overpressure hazards ahead of the bit using surface seismic, VSP, and drill bit seismic.

INTRODUCTION

Prediction of pore pressure is critical at several stages in the exploration and development process. During exploration it can be used to assess seal effectiveness and map hydrocarbon migration pathways. It can assist in analyzing trap configuration and basin geometry and provide calibration to basin modeling. During the exploration, appraisal, and development drilling phases, accurate pore-pressure prediction can be vital for safe and economic drilling. It is essential for optimized casing and mud-weight programs and helps avoid well-control problems. This chapter discusses techniques for predicting overpressure both before and during the drilling phase. The first part concentrates on the predrill problem, whereas the second looks at while-drilling methods.

PREDRILL

Conventional pressure-prediction techniques relate some measurable attribute, commonly a porosity indicator, to fluid pressure. A set of calibration curves for a particular region are established, relating deviation from the normal trend of that attribute to changes in pore pressure (Hottman and Johnson, 1965). In frontier areas, commonly the only data available are the surface seismic data. A new integrated geological and geophysical technique (Dutta, 1997) that uses surface seismic data is presented in this chapter.

The development of overpressure suggests that fluid movement is retarded, both vertically and horizontally. This can be due to the rapid burial of sediments, or lithology change, or both. Some of the important mechanisms that cause overpressure are as follows:

- Mechanical compaction disequilibrium (undercompaction) (Hubbert and Rubey, 1956)
- Clay dehydration and alteration due to burial diagenesis (Dutta, 1987)
- Dipping or lenticular permeable beds embedded in shales (Fertl, 1976)
- Buoyancy (Fertl, 1976)
- Tectonism (Dutta, 1987)
- Aquathermal pressuring (Dutta, 1987)

The method of pressure prediction discussed in this chapter uses a model that includes the first four of these mechanisms.

The velocity of a given lithology is related directly to effective stress and temperature. This relationship is based on BP's extensive database of wire-line logs, cores, and repeat formation tester (RFT) measurements in the Gulf of Mexico. The data have been carefully quality controlled and corrected for environmental effects. Two fundamental relations were developed: (1) bulk density vs. slowness for a given lithology and (2) velocity vs. effective stress and temperature for a given lithology. The first relationship enables the calculation of bulk density, and hence overburden pressure, from velocity data. The second gives effective stress directly from velocity. Pore pressure is then given as the difference between overburden pressure and effective stress.

Figure 1 shows the prediction procedure. The quality of the prediction is heavily dependent on the accuracy of the interval velocities calculated from the surface seismic data. Where well data are available, the process can be better constrained, and the calculated interval velocities can be validated. Examples of this approach can be found in Dutta (1997).

Although the results of this type of processing are encouraging, the lack of resolution in the seismic data and the difficulties involved in correctly identifying true interval velocities in areas of velocity anisotropy mean that the technique is in general limited to large-scale applications. To address the while-drilling problem, higher resolution techniques must be employed.

VERTICAL SEISMIC PROFILE (VSP) INVERSION

One of the most popular and important applications of traditional borehole seismic VSP data is the prediction of overpressure ahead of an intermediate total depth (TD). This is achieved by inverting the VSP data for acoustic impedance. The VSP data are more suitable for this task than surface seismic data because they commonly have higher bandwidths and better signal-to-noise ratios (SNR), and therefore, greater vertical resolution capabilities. A seismic trace is an indication of variations in acoustic impedance (velocity times density). These variations depend upon formation ve-

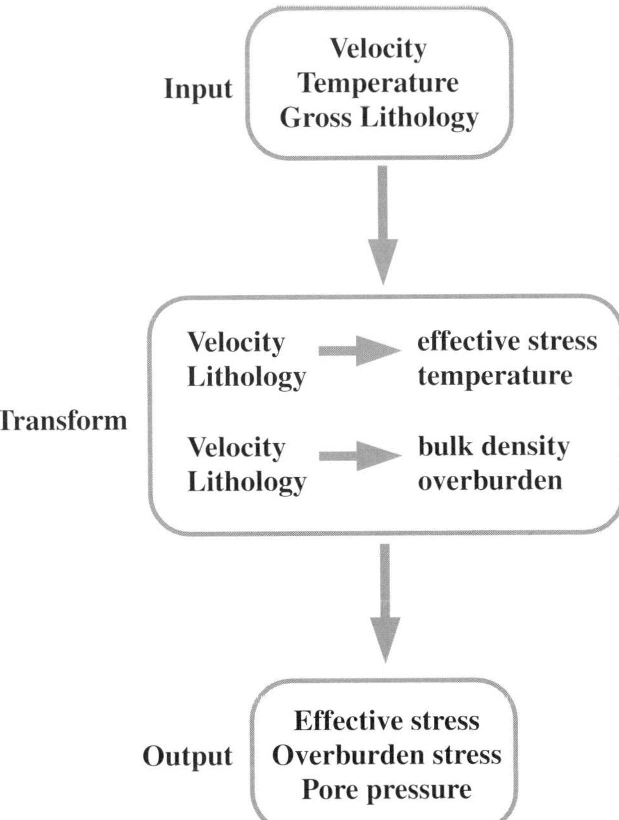

Figure 1. A schematic showing the procedure for the prediction of pore pressures from surface seismic data.

locity and density. In undercompaction environments, a decrease in acoustic impedance can indicate an increase in porosity, hence a potential overpressured zone.

Various techniques exist for inverting VSP data for acoustic impedance. In general they require that the data have a good SNR and contain low frequencies. The output is generally in the form of a prediction of acoustic impedance vs. two-way traveltime. Using an empirical velocity-density relationship, such as that proposed by Gardner (Gardner et al., 1985), calibrated by data from nearby wells if available, allows the acoustic-impedance profile to be converted to interval velocities. The formation velocity prediction can be transformed to a pore-pressure estimate, or minimum mud-weight recommendation, by using local information on the relationship between formation velocities (or slownesses) and pore pressure. If there are insufficient local data, various empirical methods have been suggested in the literature, for example, the well-known Hottmann-Johnson (Hottmann and Johnson, 1965) relationship.

The resultant pore-pressure vs. two-way traveltime profile can be converted to a pore-pressure vs. depth profile by using the calculated formation velocities. The accuracy of the predicted depth to the interpreted overpressure zone depends upon the validity of the assumptions in the velocity-density relationship and the efficacy of the VSP inversion technique. The accuracy of this depth prediction can be greatly enhanced by actually measuring the local formation velocity as drilling progresses. This can be achieved by using the drill bit seismic technique.

DRILL BIT SEISMIC

Drill bit seismic uses the acoustic energy radiated by a working rollercone drill bit to determine the seismic time to depth ratio as the well is being drilled. The energy required for drilling is supplied to the bit by rotation of the drillstring, causing the cones to roll over the bottom of the hole. As the cones roll over, the teeth penetrate and gouge the formation, destroying the rock. Each tooth impact applies an axial force to the bottom of the hole and an equal and opposite force to the drillstring. The succession of axial impacts as the bit drills radiate compressional or P waves into the formation and cause axial vibrations to travel up the drillstring. The bit acts as a dipole source for P waves (Hardage, 1992), radiating energy upward toward the surface and downward ahead of the bit. At the surface for land wells, and at the sea floor for offshore wells, geophones, hydrophones, or a combination of both are used to detect the P waves. Accelerometers, placed near the top of the drillstring, detect the axial vibrations traveling up the drill pipe. See Meehan et al. (1998) for a detailed description of this technique.

The bit-generated signal is continuous in nature, and timing information must be extracted. Referring to Figure 2, correlating the drillstring sensor signal with the seismic sensor signals gives the traveltime difference between the formation path and the drillstring path. Once ΔT_{rel} is known, if the time taken for the axial vibrations to travel along the drillstring, ΔT_{ds}, can be determined, the absolute traveltime from bit to surface, ΔT_f, can be calculated.

The time-to depth ratio is calculated using the direct radiation from the drill bit to the surface. The energy that propagates downward ahead of the bit is reflected back to the surface by impedance changes in the formation. This energy can also be detected and processed to produce a seismic image of the formation ahead of the bit. Where used in combination with the surface seismic, such look-ahead images allow the approach to critical horizons to be monitored as drilling progresses.

Although simple in concept, significant technical hurdles must be overcome to produce a robust and

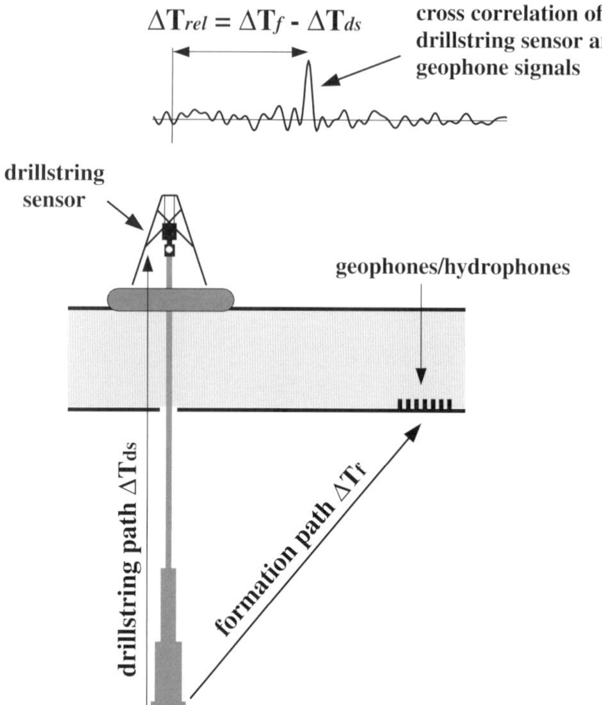

Figure 2. A schematic of the drill bit seismic system. Correlation of the rig sensor signal with the seismic sensors gives the relative traveltime difference between the formation path and the drillstring path. If the drillstring traveltime is known, the traveltime from the bit to the surface through the formation can be determined.

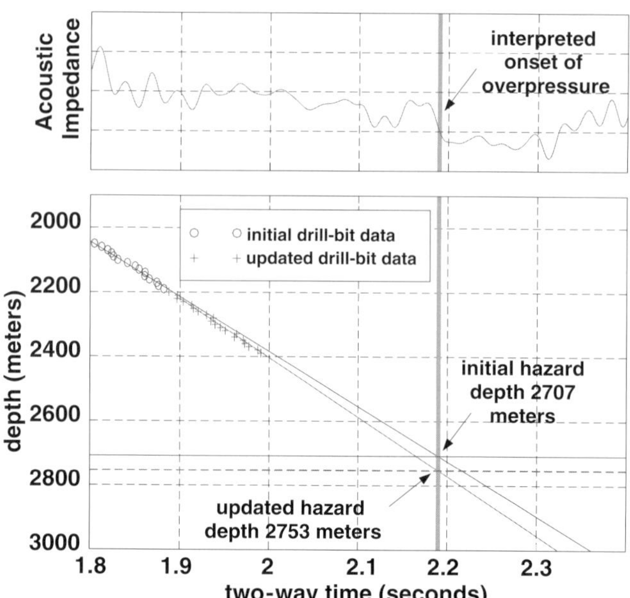

Figure 3. An example of the depth-to-hazard prediction technique.

reliable measurement technique. The most important information derived is the formation traveltime, and it is essential that all factors that affect the accuracy, both relative and absolute, of this measurement are understood. Quantifying the size of the possible timing errors gives confidence in the measurement and helps ensure the technique is correctly applied (Meehan et al., 1998). Particular attention must be paid to the drillstring traveltime measurement and to the effects of processing on the phase of the signals. Working rigs create a great deal of noise, and sophisticated signal-processing methods must be employed to extract the drill-bit signal (Meehan et al., 1998). Differentiation of the time to depth measurement with respect to depth gives an estimate of the local formation velocity.

The drill bit seismic technique can be used in conjunction with a conventional wire-line VSP to enable real-time prediction of the depth to overpressured zones. Figure 3 shows an example of this. The top part of the plot is an acoustic-impedance inversion of an intermediate wire-line VSP acquired at a TD of 2000 m. The sudden drop in acoustic impedance just before 2.2 s two-way traveltime is interpreted as the onset of overpressure. The bottom section of the plot shows how the time-to-depth information from the drill bit seismic is used to update the predicted depth to the top of the overpressured zone. Where the bit has reached a depth of 2200 m, the hazard depth is predicted to be at 2707 m (using a least-squares extrapolation). As drilling progresses, more time to depth information becomes available. Where the bit has reached 2400 m, the new depth-to-hazard prediction is 2753 m. The closer the bit approaches the hazard, the more accurate the prediction becomes.

This technique relies upon a successful inversion of the wire-line VSP data and the updating of the current time to depth ratio using the drill bit seismic data. If the drill bit seismic look-ahead image could be inverted for acoustic impedance, it would be more convenient, eliminating the need for the intermediate wire-line VSP. The poorer SNR of the drill bit seismic data and the lack of control over the source signature, however, mean that the data are not commonly suitable for inversion. As the methodologies for acquiring and processing drill bit seismic data improve and evolve, this situation will change.

CONCLUSIONS

In undercompacted areas, where pore pressure can be related to changes in porosity, seismic techniques can provide valuable pressure-prediction tools. For the predrill situation, this can be at the basin scale using surface seismic data or at the reservoir scale where surface seismic data are combined with borehole information. For real time, while-drilling prediction, a combination of the well-established wire-line borehole seismic techniques and the emerging seismic while-drilling technologies, can be used. Together these methods provide an accurate, continuously updated depth to overpressure hazard prediction in real time.

REFERENCES CITED

Dutta, N. C., ed., 1987, Geopressure: Society of Exploration Geophysicists Geophysical Reprint Series 7, 365 p.

Dutta, N. C., 1997, Pressure prediction from seismic data: implications for seal distribution and hydrocarbon exploration and exploitation in the deepwater Gulf of Mexico: Norwegian Petroleum Foundation Special Publication 7, p. 187–199.

Fertl, H. W., 1976, Abnormal formation pressures: New York, Elsevier, 382 p.

Gardner, G. H. F., L. W. Gardner, and A. R. Gregory, 1985, Formation velocity and density—the diagnostic basics for stratigraphic traps: Geophysics, v. 50, no. 11, p. 2085–2095.

Hardage, B. A., 1992, Crosswell seismology and reverse VSP: London, Geophysical Press, 41 p.

Hottmann, C. E., and R. K. Johnson, 1965, Estimation of formation pressures from log-derived shale properties: Journal of Petroleum Technology, v. 17, p. 717–722.

Hubbert, M. K., and W. W. Rubey, 1956, Role of fluid pressure in mechanics of overthrust faulting: Geological Society of America Bulletin, v. 70, p. 115–166.

Meehan, R., L. Nutt, and N. Dutta, 1998, Drill-bit seismic: a drilling optimisation tool: Proceedings of the International Association of Drilling Contractors/Society of Petroleum Engineers Drilling Conference, SPE 39312, p. 177–190.

17

SubSea MudLift Drilling: A New Technology for Ultradeep-Water Environments

Kenneth L. Smith
Conoco, Inc., Houston, Texas

Allen D. Gault
Conoco, Inc., Houston, Texas

ABSTRACT

Ultradeep-water drilling is characterized by several very significant challenges.

1. The available rig fleet for water depths greater than 914.4 m (3000 ft) is limited, resulting in very high day rates and very long lead times for those rigs.
2. Certain deep-water areas of the world require several casing strings in the shallower formations, resulting in unacceptably small holes at total depth (TD). In some cases, geological objectives simply cannot be reached because of the number of casing strings needed.
3. In these same areas, if the objectives can be reached, the small hole size at total depth (TD) results in a completion size that can be so restrictive that field development economics are severely impaired.
4. Some areas of the world have excessively strong currents, resulting in unacceptably high drilling riser loads and reduced operations periods, further driving up well costs.

The SubSea MudLift Drilling Joint Industry Project (SMDJIP) was formed to develop equipment and procedures that overcome these challenges. Two primary component systems are being developed:

- The near-seabed component, the SubSea MudLift system, offers the ability to simulate a dual-gradient mud system. This feature eliminates the need for several of the casing strings in deep water imposed by the low fracture gradients.
- The return riser system replaces the large marine drilling riser with smaller ones, significantly reducing riser loads in deep water and high-current areas.

Additionally, the full complement of operational and well-control procedures and the necessary training programs for these procedures are being developed.

This chapter explains the limiting geological aspects of deep-water exploration and production, why successful implementation of the SubSea MudLift system part of the SubSea MudLift drilling concept overcomes these limitations, and then, briefly, what the ongoing joint industry project is attempting to achieve.

Smith, Kenneth L., and Allen D. Gault, 2002, SubSea MudLift Drilling: A New Technology for Ultradeep-Water Environments, in A. R. Huffman and G. L. Bowers, eds., Pressure regimes in sedimentary basins and their prediction: AAPG Memoir 76, p. 171–175.

THE GEOLOGICAL CHALLENGE IN ULTRADEEP WATER

In low-horizontal stress, high–pore-pressure environments such as the deep-water Gulf of Mexico, several casing strings are needed in the shallower hole intervals to maintain control over the inherently weak fracture gradients and abruptly increasing pore pressures. This is because conventional drilling practices use a marine riser that conducts a single mud density from the rig to the bottom of the well to maintain the required bottom-hole pressure. This single-gradient approach has the disadvantage of imposing too much pressure on the shallower parts of the well, increasing the likelihood of fracturing the weaker formations. To prevent this, casing is set to protect the weaker formations from fracturing.

Figure 1 shows a typical Gulf of Mexico casing program for 2743 m (9000 ft) of water depth. In this example, a 152.4 mm (6 in.) hole interval has reached its total depth (TD) only 2438 m (8000 ft) below the mud line. Any geological objectives below this point will not be reached and fully evaluated. Further, even if a high-productivity reservoir had been recently penetrated, any completion placed in the subsequent 127 mm (5 in.) production casing or liner would likely restrict potential production rates, which are a key driver for economic projects in ultradeep water. Therefore, economic field development will be much more difficult to achieve.

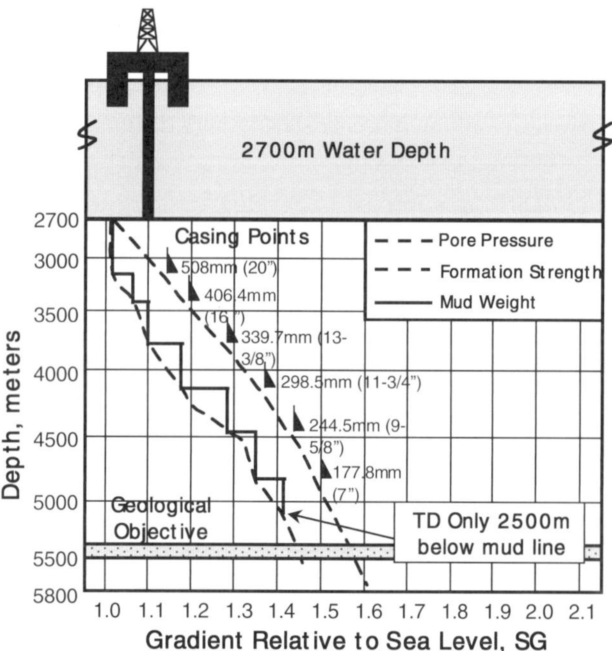

Figure 1. Typical ultradeep-water casing program. Ultradeep-water geological objectives cannot always be met.

In other words, increasing water depth is a barrier to successful exploitation of the ultradeep-water environment using today's drilling technology.

DUAL-GRADIENT DRILLING SYSTEMS

Recognition of this led Conoco and Hydril to revisit the development of an old concept that simulates a dual-gradient mud system: the SubSea MudLift concept.

To gain an initial understanding of the pressure profiles of a dual-gradient system, it is helpful to express them as pressure vs. depth (as opposed to pressure gradient vs. depth). As shown in Figure 2, a dual-gradient mud system's hydrostatic pressure profile is fundamentally different from a conventional single-gradient system. Using the dual-gradient system, a mud-density gradient is seen by the well from the bottom of the well to the seabed. In the riserless drilling concept, the MudLift system will be used to pump the mud from the seabed to the rig, simulating a seawater gradient. From the perspective of the well, the rig has been lowered to the mud line.

In dual-gradient systems, pressure gradients will be specified relative to the seabed. So, in comparison to conventional wells drilled with a single gradient system, wells drilled with a system simulating a dual gradient will use higher mud weights. The higher densities are needed to compensate for the apparent 1.03 specific gravity (SG) (8.6 ppg) seawater above the mud line. Figure 3 shows the impact of referencing gradients to the seabed. Note that the margin between fracture gradient and pore-pressure gradient is greatly increased, and this reduces the need for several casing strings.

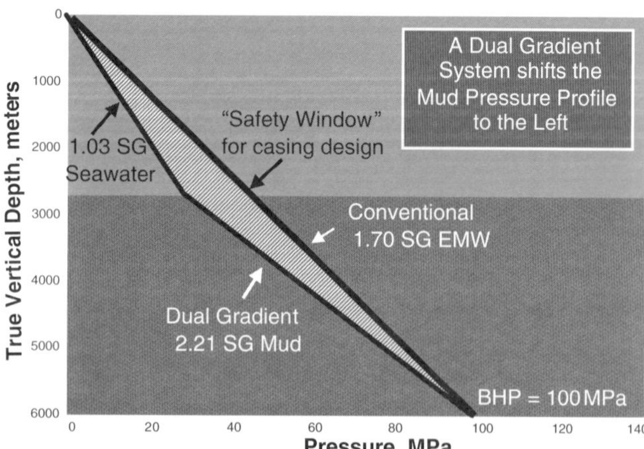

Figure 2. Hydrostatic profiles of single- and dual-gradient mud systems.

Consequently, water depth is less of a factor in reaching TD. Further, TD is reached with a hole size that allows almost any completion configuration, for example, horizontal and multilateral wells or 177.8 mm (7 in.) production tubing, that is needed to generate maximum economic returns in development. (In fact, the design basis for the SubSea MudLift Drilling Joint Industry Project [SMDJIP] is 244.5 mm (9.625 in.) casing at TD.)

THE ESTABLISHMENT OF THE SUBSEA MUDLIFT DRILLING JOINT INDUSTRY PROJECT

The geological limitations are just one of the reasons that dual-gradient systems are of interest to industry. With either increasing water depth, or with very high currents, the loads imposed on the riser and the drilling rig become tremendous. Consequently, most current rig capacities are exceeded or the operations are exposed to high, very costly, down times because of riser disconnects.

The initial configuration with SubSea MudLift Drilling will include a mud-lift system that has been attached to or integrated with the subsea BOP and conventional 533.4 mm (21 in.) outside diameter (OD) drilling riser. This riser will be filled with seawater rather than with mud. An evolution of the system may be to replace the standard 533.4 mm (21 in.) OD drilling riser with smaller return lines and removing them from the proximity of the drillstring, as shown in Figure 4. In the case of replacing the riser with smaller return lines, current loads on a drilling riser are no longer an issue. In the case of filling the 533.4 mm (21 in. riser with seawater, the tension loads required for the drilling riser are significantly reduced. In either case, the rig's capabilities are greatly extended, and the available rig fleet for ultradeep water is effectively increased.

This universal appeal to both operators and drilling contractors led to the establishment of the SMDJIP with a strong representation from most major deepwater operators and drilling contractors. The first phase of the JIP had 22 members and began in October 1996. The second phase of the JIP commenced in April 1998 and is comprised of four operators, four drilling contractors, Minerals Management Service (MMS), Ocean Drilling Program (ODP), and Hydril, the project designer.

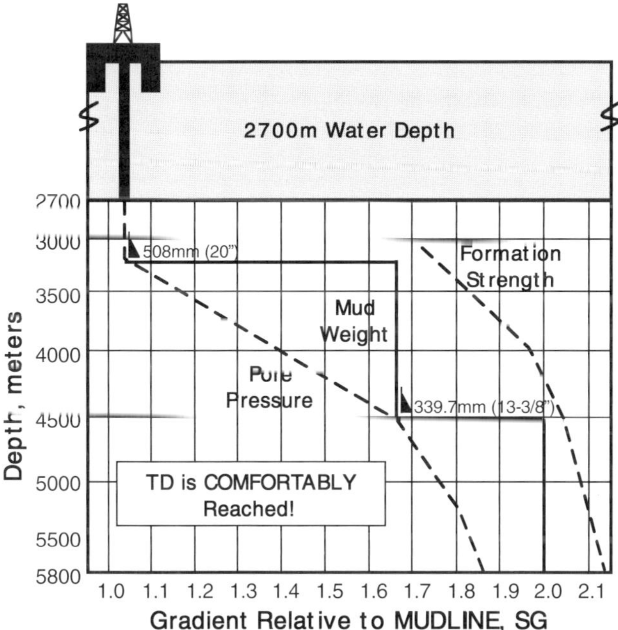

Figure 3. Benefit of referencing pressure gradients to the mud line.

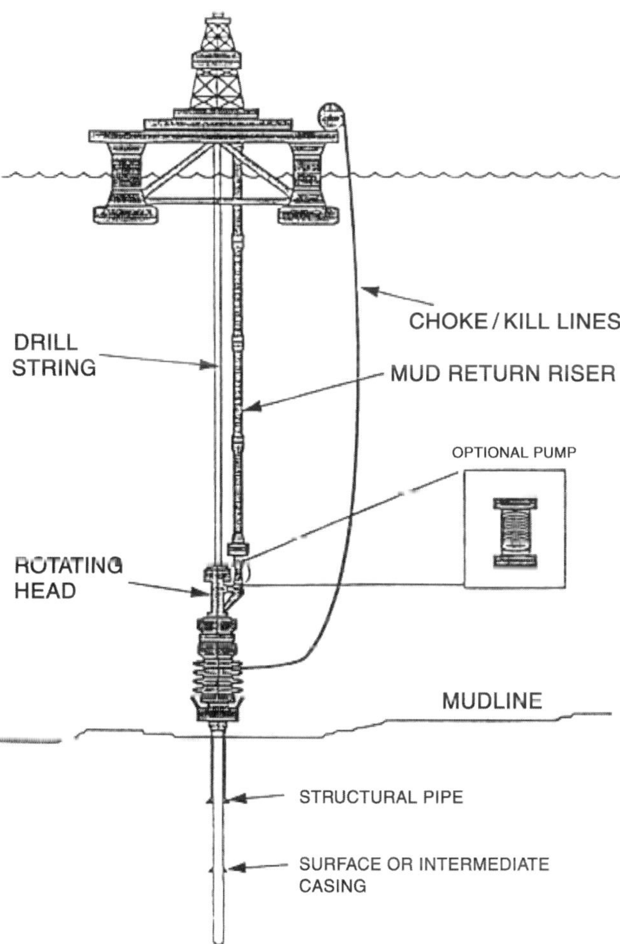

Figure 4. SubSea MudLift Drilling system may evolve into a marine-riserless configuration.

RESULTS OF PHASE I OF THE JIP

The first phase of the JIP lasted approximately one year. During that phase, the membership studied various equipment configurations and pursued various alternatives on lifting devices, instrumentation packages, and other components. All drilling and well-control procedures were detailed to a very high level. A well-control model was built to simulate the complicated hydraulics in the well, and well-control sensitivities were run.

In short, every effort was made to find the weaknesses in the concept and determine whether any insurmountable obstacles existed in developing the technology into a commercially viable and available product.

In conclusion, there appears to be nothing about SubSea MudLift Drilling that cannot be overcome. All equipment is feasible, and much of it is an extension of equipment that already exists. All routine and nonroutine drilling, completion, and well-control procedures can be performed. In fact, many operations are actually much easier or offer a wider margin of safety when compared to conventional deep-water drilling practices.

FUTURE PHASES OF THE SMDJIP

The JIP has three remaining phases. The overall project schedule for each is shown in Figure 5.

- Phase II was ongoing and continued through early 2000. The overlying goal for this phase was to advance all equipment components and procedures such that the JIP membership has about an 80% confidence in the delivery of technology. This was successfully achieved.
- Phase III will continue for about a year beyond that, and will finalize all system designs and procedures, finalize the training schools and secure the regulatory approvals necessary to take the equipment out to the field for testing.
- Phase IV will be the actual field testing of the MudLift system and procedures and will conclude in 2002.

Once Phase IV has been completed, the equipment and procedures will be modified as necessary to support the overall project goal of commercial availability for all of its participants.

PHASE II OF THE PROJECT

From an engineering standpoint, one of the most exciting aspects of dual-gradient drilling technology is the mind-set changes that people experience. In many aspects of the technology, conventional practices simply do not work. These aspects were identified in Phase I of the project, and the techniques, equipment, or other approaches were also identified that enabled us to achieve the specific operation. Phase II began the work of detailing these techniques and equipment designs. For example,

1. The system simulates a dual-gradient mud system, but there is actually only one mud density being used, which is far beyond what the well could withstand without fracturing. How is the well isolated from the pressures?

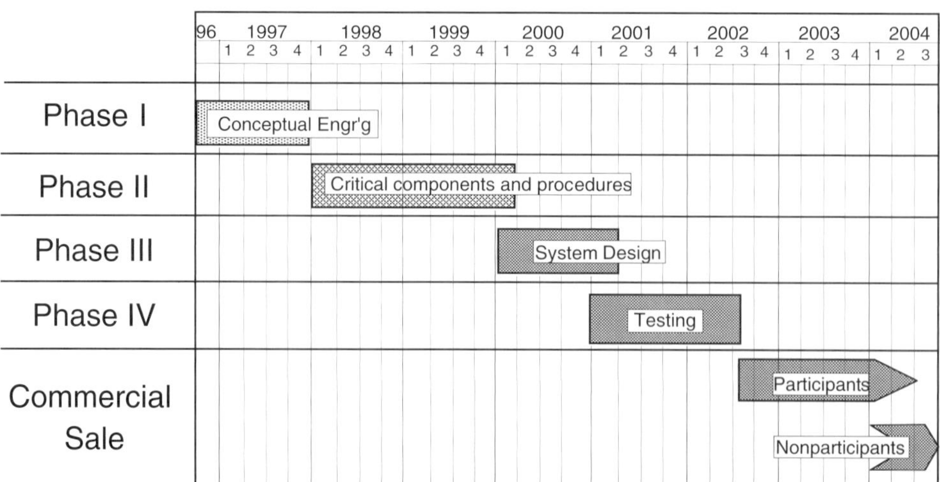

Figure 5. SubSea MudLift Drilling JIP schedule.

2. A tremendous pressure imbalance exists from the drill pipe to the annulus side of the well. When the mud pumps are shut down, the well may u-tube for several minutes. Is this acceptable? If not, how is it controlled?
3. If a u-tube situation is allowed after shutting down the pumps, how is a suspected well flow checked?
4. How is the difference between a well-control situation (kick or lost returns) and a MudLift pump that has lost synchronization with the surface mud pumps determined?
5. How is a kick killed?
6. How are cuttings pumped?

Among the deliverables of this phase are the following:

1. Every critical MudLift component will be fully designed, built, and tested for reliability.
2. All operational procedures, both for routine and nonroutine drilling operations, such as logging, fishing, and squeeze cementing, will be detailed, and operational hazards analyses will be performed. This will carry over into Phase III.
3. The control system logic for MudLift will be finalized.
4. Well-control procedures will be further developed and integrated with the control system logic.
5. The necessary training programs for well control and general MudLift system usage will be initiated. This will ultimately include the development of the training simulator that will be fully integrated into the MudLift control system.

PHASE III OF THE PROJECT

Phase III will finish the design and procedures to the point that no further work will need to be done prior to manufacturing the final prototype test equipment and taking it offshore for the field test. This will include the following:

1. System design of the equipment will be completed. The fully tested components will be integrated into a system on the shop floor, controlled by software and sensors in the shop environment.
2. All procedures development and the training school for the operational procedures will be completed.
3. The well-control school and simulator will be completed, and the required regulatory approvals will be secured.
4. A test rig will be selected and the prototype SubSea MudLift system for test on that rig will be customized.

PHASE IV OF THE PROJECT

Phase IV will be the field test itself. It will include the following:

1. Purchase of all remaining equipment necessary to build the MudLift system.
2. Integration of the MudLift system into the selected rig's systems.
3. Training of the necessary personnel in well control and operational procedures.
4. Installation of the MudLift system onto the selected rig.
5. Testing of the system on a deep-water well.
6. Revision of the procedures to reflect the learning from the test.
7. Designing the equipment upgrades that reflect the learning from the test and incorporating them into the final commercial product.

Once the commercialized product is finalized, it will be available to industry: first to participants in the JIP, and then to nonparticipants.

CONCLUSIONS

Development of a dual-gradient technology is needed to successfully exploit ultra–deep-water leases. The SubSea MudLift Drilling Joint Industry Project has been formed to develop and deliver the technology needed to simulate the dual-gradient technology. This technology is not new, but it is revolutionary. As such, the challenges are numerous and will require extensive effort for the next three years. Though numerous, all of these challenges can be overcome. (At the time of this publishing, this project has now successfully drilled the world's first dual gradient well. The SubSea MudLift Drilling system is now available commercially.)

ACKNOWLEDGMENTS

We gratefully acknowledge the support of the membership of the SMDJIP in allowing this chapter to be written and presented at the American Association of Drilling Engineers Industry Forum on Pressure Regimes in Sedimentary Basins and Their Prediction.

18

Velocity Estimation for Pore-Pressure Prediction

David W. Bell
Conoco Inc., Ponca City, Oklahoma

ABSTRACT

The speed of propagation of compressional-wave energy in the subsurface, known simply as "formation velocity," is strongly influenced by compaction, particularly in young clastic basins. Because pore pressures affect compaction, changes in velocity can be calibrated to changes in pore pressure. Velocities derived from surface seismic data provide indirect pressure measurements at undrilled locations. The accuracy depends on the validity of the relationship between pressure and velocity, the quality of the velocity measurements at enough points to perform the calibration and prediction, and the reliability of average velocities to correctly convert from seismic time to depth.

A key step is construction of a velocity profile with depth that simultaneously defines both the compaction characteristics and a valid time-depth curve. A linear fit to the logarithm of the sonic transit time with depth is commonly assumed to represent the normal compaction trend. Such a velocity-depth trend, however, does not produce a time-depth relationship that accurately converts seismic measurements in time to depth. A linear fit of velocity with time provides a consistent fit to both time-depth and velocity-depth data and is a better empirical representation of the normal compaction trend. The linear velocity-time model can be used to smooth through inaccuracies in seismic stacking-velocity picks where applied to geologically consistent units.

This chapter illustrates relationships between velocity and the geologic setting and establishes an empirical model for the normal compaction trend. It then reviews various assumptions and techniques for converting seismic stacking velocities into representative formation velocities. It concludes with a step-by-step recommendation for estimation and calibration of velocity from seismic data.

VELOCITY-PRESSURE RELATIONSHIPS

A starting point for pressure prediction from seismic velocity is to invoke an empirical relationship between velocity and effective pressure. Effective pressure, P_e, is the difference between the overburden, or confining, pressure, P_c, and the pore pressure, P_p. In zones of normal compaction, the overburden pressure is obtained by integrating a density function, whereas the pore pressure is obtained simply from the weight of a unit volume of the fluid times the total depth. Subtracting these two qualities gives values of effective pressure that can be crossplotted against velocity measurements over the same region. A curve fitted to these data provides a calibration between effective pressure and velocity. Pore pressure in anomalous zones can then be deduced from additional velocity measurements at that location or anywhere in the basin where the initial calibration curve is valid. Figure 1 illustrates these concepts with synthetic data.

In Figure 1A, velocity increases uniformly with depth in a zone of normal compaction and then decreases with depth where overpressure is encountered. Figure 1B plots effective pressure, pore pressure, and

Bell, David W., 2002, Velocity Estimation for Pore-Pressure Prediction, in A. R. Huffman and G. L. Bowers, eds., Pressure regimes in sedimentary basins and their prediction: AAPG Memoir 76, p. 177–215.

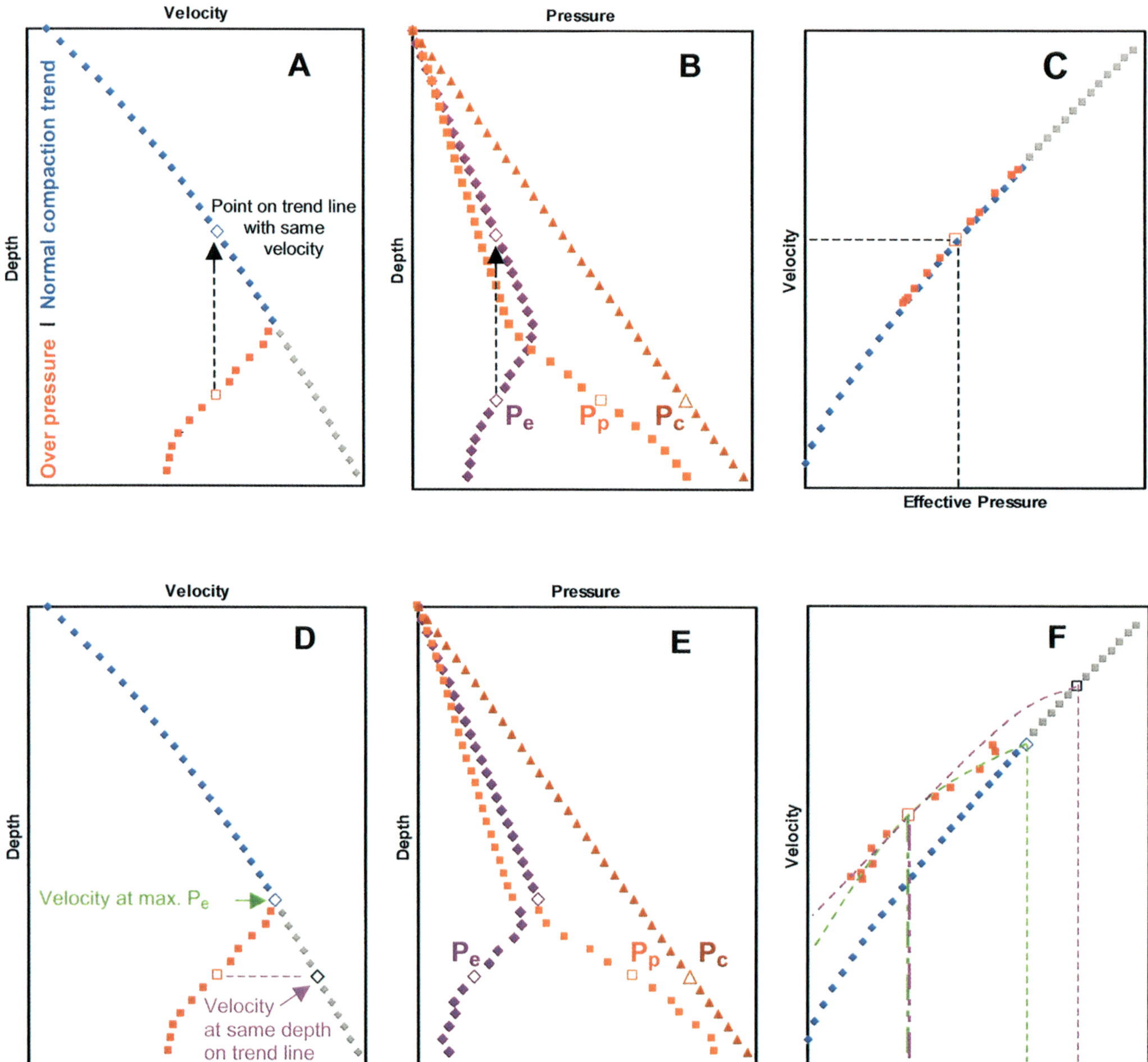

Figure 1. Velocity-pressure relationships. Parts A, B, and C illustrate the relationships between depth and velocity, depth and pressure, and velocity and effective pressure under the assumptions of compaction disequilibrium and the equivalent depth method. Parts D, E, and F show the behavior attributed to unloading.

overburden pressure over the same depth interval. Note that points at different depths that have the same velocity also have the same effective pressure. Velocity plotted vs. effective pressure defines a single, depth-independent curve (Figure 1C). No similar one-to-one correspondence to either the pore pressure or the overburden pressure exists.

The graphical scheme for pore-pressure prediction implied by Figure 1A and B is known as the equivalent-depth method. The velocity at a point with unknown pore pressure is projected vertically upward until it intersects the normal trend line at a shallower depth. The effective pressure at the new depth is equated to the effective pressure at the point of interest. Similar results can be obtained from Figure 1C, or analytically from the curve fit; however, no universally accepted functional form for the relationship between velocity and effective pressure exists. Some functions in use are concave, as implied in Figure 1C. Others are convex. To a large extent, the functional form assigned to the velocity-pressure relationship is dependent on the assumed form of the normal trend line for velocity vs. depth. This chapter reviews those relationships and suggests that some are more appropriate than others. In particular, the common assumption that the logarithm of velocity is linear in depth is not consistent with check shot data.

A single, definitive relationship between velocity and pressure is difficult to establish for various reasons. One is that the true variation of velocity with depth is not as simple as shown in Figure 1A. Variations caused by lithology exhibit large positive and negative excursions from the trend. Plots as simple as the one in Figure 1C are obtained only if the velocity data are significantly smoothed. A second reason is that several functions yield similar residuals where fit to a few noisy data points over a limited range of the independent variable. Any empirical technique that works is difficult to criticize. A third reason is that such a function may not physically exist.

The assumption of a one-to-one relationship between velocity and effective pressure is not always valid. The approximation appears to hold whenever the cause of abnormal pressure is simple compaction disequilibrium. Processes such as clay diagenesis, organic maturation, or aquathermal pressuring can modify the velocity-pressure relationship, as can unloading caused by uplift and erosion. Figure 1 also illustrates a case where the velocity-pressure relationship in the overpressure region differs from that of the normal compaction curve.

Figure 1D postulates the same velocity-depth values as Figure 1A. The effective pressure in Figure 1E, however, is no longer the same as that at a shallower depth with the same velocity (the equivalent-depth point). Velocities in the overpressure zone no longer retrace the velocity-effective stress curve of the normal compaction trend (Figure 1F). Application of the equivalent-depth method predicts too large an effective pressure and thereby underestimates the pore pressure.

Two points related to the normal trend are identified in Figure 1D that can be used to recalibrate the data. Bowers (1995) assumes a functional form for the unloading curve that is based on the equivalent-depth solution and the effective pressure at the start of the reversal. His technique reproduces the behavior of Figure 1C whenever an exponent is set to 1. For larger values of the exponent, it produces a curve similar to that shown in green in Figure 1F. The Eaton (1975) technique is based on extrapolated values of velocity and pressure along the normal trend line at the same depth as the point in question. Allowing the exponent in the Eaton equation to vary produces a curve similar to that shown in pink in Figure 1F.

Figure 2 uses real data to demonstrate some of the concepts introduced in Figure 1. Figure 2A shows the sonic velocities from two wells in the same basin, albeit more than 100 km apart. One is in shallow water, and the other is in deep water. A second set of curves in Figure 2A represents sonic values smoothed to extract the low-frequency compaction trend from high-frequency variations caused by lithology. Triangles and squares superimposed on the curves indicate depths at which pore-pressure measurements are available. The pore pressure in the shallow-water well follows a hydrostatic gradient, whereas the pore pressure in the deep-water well rises above hydrostatic pressure as the depth increases (Figure 2B). Figure 2C shows the density curves from the two wells. Fitting a curve to the measured densities allows extrapolation of the sediment density to the sea floor. Integration of the density function from sea level provides an estimate of the total overburden pressure, P_c. Subtracting the measured pore pressures from the calculated overburden pressure yields the effective pressure, P_e (Figure 2D). Now sufficient information is available to determine the relationship between velocity and pressure. Figure 2E plots smoothed velocity vs. pore pressure, whereas Figure 2F plots smoothed velocity vs. effective pressure. No simple relationship exists between velocity and pore pressure that connects one well with the other. A common trend in effective pressure exists that matches the assumed form in Figure 1C.

The previous discussion reveals that accurate prediction of pore pressure from velocity requires two critical sets of information. One is an accurate velocity trend. The other is sufficient data to calculate the effective pressure at enough points to establish and calibrate a relationship between velocity and effective pressure. That in turn requires density and pore-pressure measurements in a similar setting. The rest of this chapter concentrates on the characteristics of velocity-depth functions and how to accurately determine them from surface seismic data. Remember, however, that velocity is only one piece of the puzzle.

VELOCITY-DEPTH-TIME RELATIONSHIPS

Figure 1A implies that the presence of overpressure can be deduced from a reversal in the velocity depth curve. That is commonly the case in young clastic basins, but velocity reversals can also be caused by changes in lithology, particularly where salt, carbonates, or volcanics are present. Figure 2A demonstrates that significant overpressure can be encountered even if no velocity reversal occurs. Therefore, it is instructive to review several velocity-depth curves to determine clues to overpressure that are useful for quality control of seismic velocities.

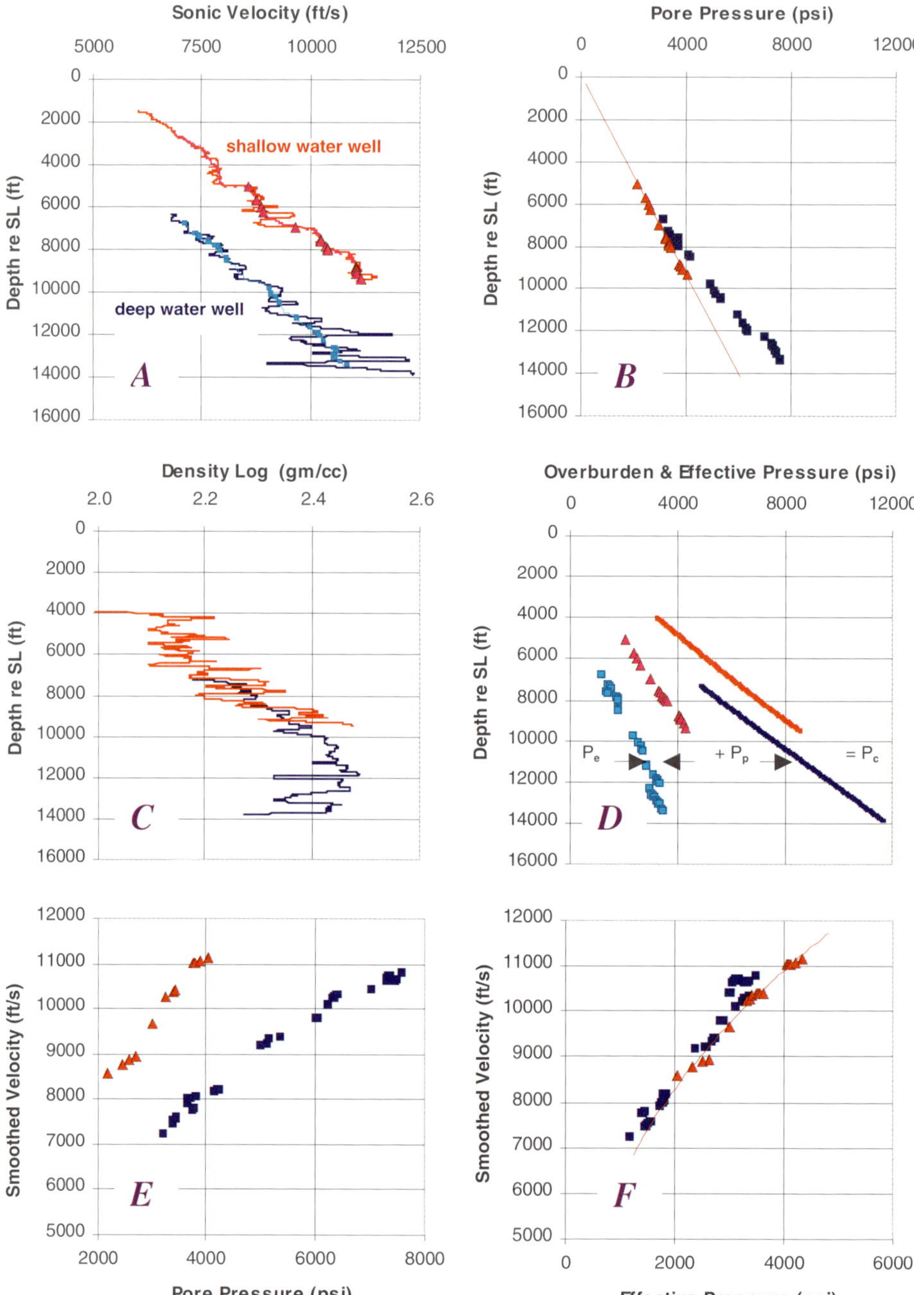

Figure 2. Examples of velocity-pressure relationships. (A) Sonic velocity vs. depth; (B) pore pressure vs. depth; (C) density log vs. depth; (D) overburden and effective pressure vs. depth; (E) pore pressure vs. smoothed velocity; (F) effective pressure vs. smoothed velocity. SL = sea level.

Sonic logs in Figure 3 illustrate velocity trends in four wells. Diamonds indicates overpressure zones. Two of the wells are from Africa, and two are from Indonesia. Figure 3A demonstrates a typical profile in a young clastic basin. The velocity increases uniformly with depth according to a smooth empirical compaction trend. Near a depth of 9000 ft (2743 m) there is an abrupt decrease in velocity away from the trend line where the wellbore intersects a fault. The region of low velocity below the fault corresponds to a region of high pressure.

Figure 3B shows a similar velocity decrease around 11,000 ft (3353 m) that again correlates with overpressure in a clastic sequence. Also, there is a positive break in the trend line at a major unconformity near 5000 ft (1524 m). Sediments below the unconformity have experienced tectonic uplift and therefore exhibit a higher velocity for a given depth than expected from the compaction trend evident above the unconformity. The analysis in Figure 1A–C, neglected the influence of a change in the history of the effective stress. The possibility of multiple compaction curves in the same well

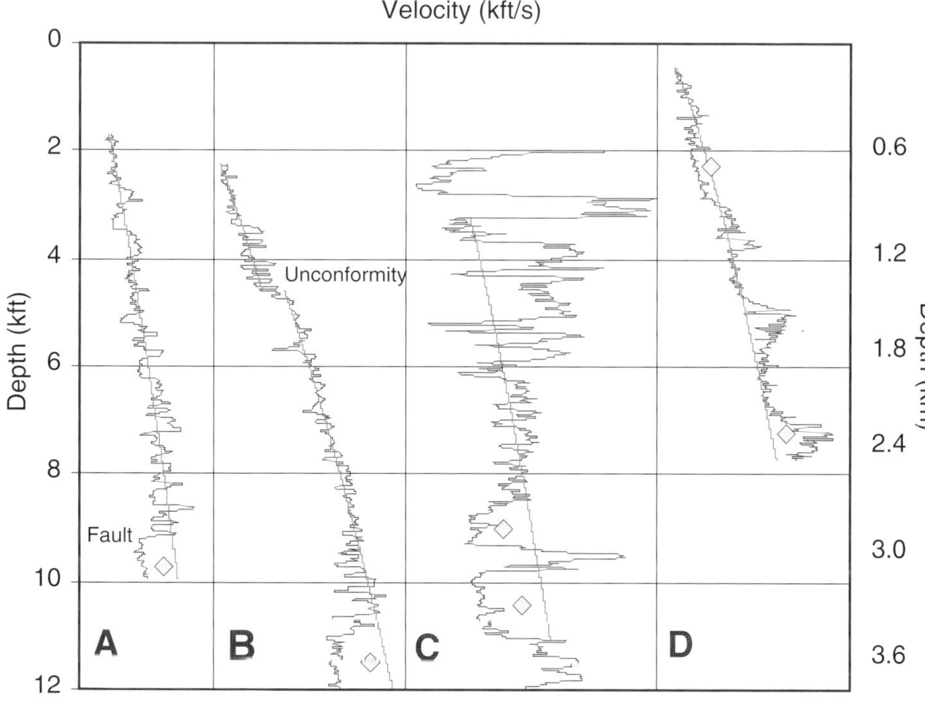

Figure 3. Sonic velocity vs. depth for four wells, A, B, C, and D. Approximate normal compaction trends shown as smooth lines. Overpressure zones indicated by diamonds. kft = feet in thousands.

suggests that accurate pore-pressure prediction requires an understanding of the depositional and tectonic history of an area.

The velocity profile in the third well (Figure 3C) is complicated by variations in lithology. The velocity decrease near 6000 ft (1829 m) reflects a change from carbonates to clastics, rather than a change in pressure. A second velocity break near 9000 ft (2743 m) does result from excess pore pressure. Changes in lithology complicate pressure prediction from seismic velocities. In this example, the presence of carbonates is easily inferred from large seismic stacking velocities at shallow depths. The shallow, high-velocity zone, however, also interferes with accurate seismic velocity measurement at greater depths. Accurate pressure prediction from surface seismic data is difficult under such circumstances.

Not all velocity changes due to overpressure are as abrupt as those shown in the first three examples. The fourth well (Figure 3D) shows a modest deviation from the normal velocity trend near 2000 ft (610 m) that represents a change in mud weight from 9 to 11 lb (4–5 kg).

The fourth well was also chosen as a counterpoint to the previous examples. Near 7000 ft (2133 m), an increase in pressure correlates with a thick sand unit whose velocity is greater than the overlying shale. Sonic, density, and resistivity data in this well all indicate continued compaction within the deeper overpressure zone. Mechanisms that induce overpressure after compaction, and possible cementation, do not necessarily lower the velocity sufficiently to produce an obvious imprint.

Average Velocity

Normally, pressure predictions are based on point measurements, that is, formation velocities at particular depths. Various relationships dealing with average velocity also provide useful insights. Average velocity represents the total depth from a datum divided by the total one-way transit time of a seismic signal. (See Table 1 for a comparison of velocity terms.)

Average velocity is closely akin to seismic stacking velocity. Stacking velocity does not always need to be converted to interval velocity to infer pressure anomalies. Knowing the influence of overpressure on average velocity aids in picking and quality control of stacking velocities for pressure prediction, particularly in the presence of noise.

Figure 4 shows the average-velocity curves derived from the wells in Figure 3. All of the important trends observed in interval velocity are manifest as slope changes in the average velocity. An increase in slope (toward the vertical) indicates a reduction in interval velocity. A decrease in slope (toward the horizontal) indicates an increase in interval velocity. A sharp slope break indicates an abrupt interval-velocity change. A smooth transition in the slope of the average velocity implies the same in interval velocity.

Table 1. Comparison of Velocity Terms*

Term	Symbol / formula	Definition	Comments
Instantaneous velocity, formation velocity	$V_{inst,}\ V_i = dz/dt$	The speed of wave propagation at a point. Interval velocity in limit of small interval. May depend on frequency and direction.	Approximated from laboratory measurements on core and by sonic velocity.
Sonic velocity	$V_{son,}\ V_i = 1/d\tau$	$d\tau$ = sonic transit time obtained from downhole logging tools. Provides interval velocity over a few feet.	Best velocity for detailed work involving variations with lithology, porosity, pore fluid, and so on.
Interval velocity	$V_i = \Delta z/\Delta t$	Average velocity over a depth interval.	Derivation from VSP data suffers from uncertainty in time over a small time increment.
Average velocity	$V_a = z/t = \Sigma V_{in}\Delta t_n/\Sigma\Delta t_n$	Distance divided by time. Useful for seismic time to depth conversion when measured from the surface.	Best obtained from check shot or VSP data. Integrated sonic suffers from missing shallow data and poor data zones.
Stacking velocity	V_{stack}	Velocity used to correct offset seismic traces to zero offset prior to CMP stacking and poststack migration. Normally based on some measure of image quality. May include higher order moveout terms.	Examines event travel times within a CMP separately. Need not represent any true velocity of the medium. Basis for pressure prediction work in good data quality areas. Results improved by prior application of DMO.
Normal moveout velocity	$V_{nmo} = [x^2/(T^2{}_x - T_0{}^2)]^{1/2}$	Stacking velocity assuming hyperbolic moveout in the limit of small offsets, x.	Equals V_a and V_i for a single, flat, constant velocity layer. Otherwise, depends on offset.
Rms velocity	$V_{rms} = [\Sigma V_{in}{}^2\Delta T_n/\Sigma\Delta T_n]^{1/2}$	Approximately equals V_{nmo} for horizontal layers without lateral velocity variations. Based on least traveltime principle, accounts for ray bending consistent with increased travel distance within faster layers.	Equals V_a and V_i for a single, flat, constant velocity layer. Exceeds V_a for a series of flat layers. Cannot be calculated for an interval without assuming $V(z)$ within the layer.
Dix velocity	$V_{Dix,}$ $V_i = [(V^2_{stack2}T_2 - V^2_{stack1}T_1)/(T_2 - T_1)]^{1/2}$	Common technique for deriving interval (actually interval V_{rms}) velocity from stacking velocity.	Based on assumption of flat, constant velocity layers. Estimates generally too high.
Migration velocity	V_{mig}	For poststack time migration, generally a smoothed version of the stacking velocity field that has been adjusted a few percent to improve image quality. Not generally a good starting point for pressure prediction.	For prestack time and depth migration, a velocity field, often based on modeling, that accounts for nonhyperbolic moveout from structure. Reliable pressure prediction may require additional layers within the model. It is possible to do conventional velocity analysis for high-quality pressure prediction if offset traces without moveout correction are generated from correctly positioned migrated data.
Tomographic velocity		Conventional moveout analysis examines one event in a single CMP at a time. Tomography calculates a velocity field that simultaneously matches travel times from several events over a range of CMPs.	Can reveal subtle velocity variations in regions with numerous good events, but will also yield a smooth velocity field in regions without picks.

*z = depth; t = time; x = offset; T = vertical two-way traveltime; T_x = two-way traveltime for a source/receiver offset of "x." See Appendix for definitions.

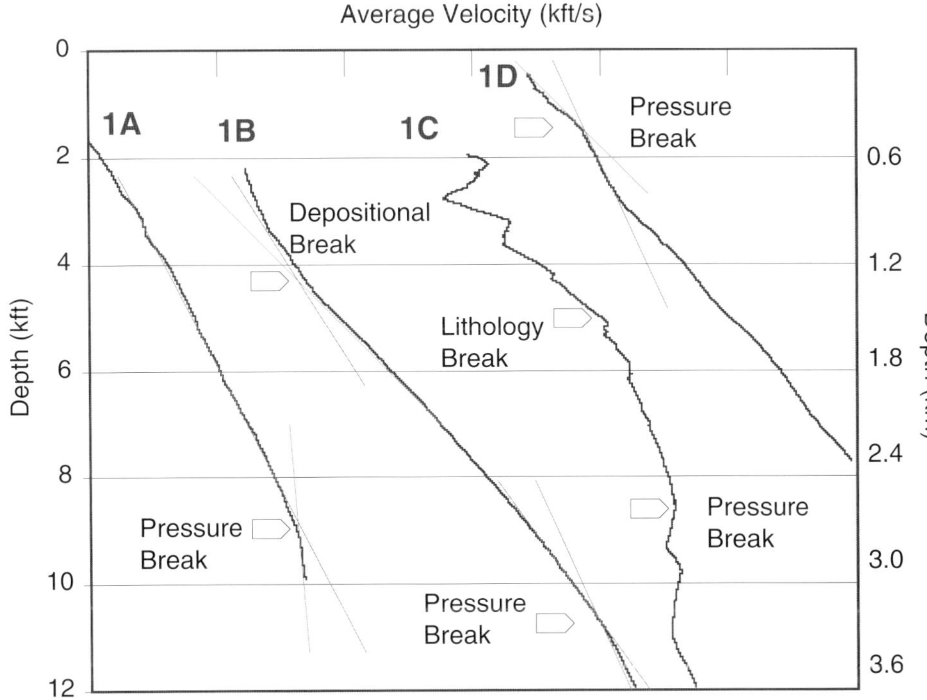

Figure 4. Average-velocity curves for wells from Figure 3. Slope breaks indicate changes in interval-velocity trends associated with changes in deposition, lithology, and pore pressure. kft = feet in thousands.

Interval velocities used in pressure prediction are commonly smoothed. A normal trend line is one form of smoothing, but deviations from the trend are also commonly smoothed to prevent the predicted pressure profile from exhibiting the large, rapid fluctuations seen in a sonic log. Average-velocity curves are a form of smoothing and provide guidelines for other filtering processes. Smoothing interval-velocity excursions is appropriate within regions where the slope of the average-velocity curve is reasonably constant. Velocity fidelity is compromised if averaging is performed across major slope breaks in the average-velocity curve.

Time-Depth-Velocity Relationships from Check Shots

Accurate prediction of the depth to a pressure anomaly from seismic velocities requires a valid time-depth relationship. Check shots provide the most accurate time-to-depth information. Check shots are direct transmission measurements at seismic frequencies of traveltime from a surface source to a receiver at a known depth. Uniformly spaced check shots at small depth intervals are characteristic of vertical seismic profiles (VSPs). (Processed VSPs also provide a seismic trace that ties waveform changes in time to formation boundaries in depth).

Figure 5 plots the time-depth relationship (Figure 5A), the average velocity (Figure 5B) , and the interval velocity (Figure 5C) derived from a VSP in a shallow-water Gulf of Mexico well. The onset of overpressure is manifest on a time-depth curve as a deviation from the trend in the shallower data. The effect, however, is more pronounced if the check shots are converted to average velocity, $V_a = z/t$. The break in slope of the average velocity is correlated with overpressure similar to the curves in Figure 4. Note that check shots provide more accurate average-velocity information than sonic logs alone. Missing sonic values at shallow depths, in washout zones, or near casing points lead to uncertainty in time-depth relationships because a depth integration is required over all intervals. Each check shot is an independent measurement. Errors in shallow measurements do not affect deeper values.

Average velocities calculated from good-quality, vertical-incident check shots are accurate within 0.5% or less. Interval velocities between closely spaced check shots are subject to much greater uncertainty. The time difference between two levels in a VSP approaches the sample rate. A picking error that is small relative to the total time can be large compared with the differential time. Extreme interval-velocity excursions implied by check shot data should be verified with the sonic log data. Check shot velocity information should be smoothed before being used to calibrate the sonic log. Nevertheless, overall velocity trends should still be evident in an interval-velocity curve derived from check shots (Figure 5C). Check shots also

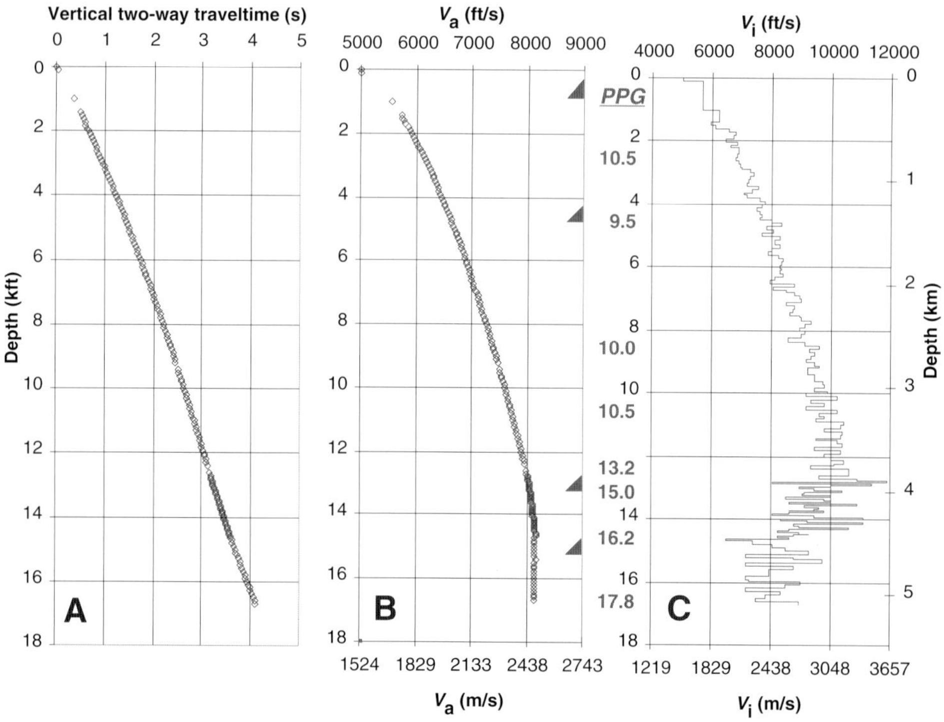

Figure 5. (A) Time-depth, (B) average velocity–depth, and (C) interval velocity–depth relationships from check shots in a Gulf of Mexico well. Seismic prediction of depth to overpressure requires accurate time-depth conversion. Check shots provide the best time-depth calibration. Casing points and mud weights are shown in decimal numbers. Major pressure transitions occur across faults. kft = feet in thousands.

provide data very well suited to test various equations used to represent the normal compaction trend for velocity vs. depth.

Various properties are desired of an empirical model used to characterize noisy data. First, the model parameters should be fairly insensitive to the range of the data and the number and placement of individual data points. A corollary is that the model can be used to confidently extrapolate data into regions where the model applies but data are lacking. Second, the differences between the values predicted by the model and the actual data points (the residuals) should be small and randomly distributed about zero. Third, consistent values of the parameters should be obtained where fitting the data with mathematically equivalent statements, for example, using sonic transit time ($d\tau$) rather than V_i. A fourth criterion is a bias toward a small number of parameters. Any given data series can be fit perfectly with enough degrees of freedom, but extra parameters commonly fit noise rather than signal and tend to violate the previous criteria.

This chapter examines four two-parameter models, all of which yield a reasonable starting value, V_0, at zero depth: an exponential curve, linear on a semi-log plot

$$V_i = V_0 e^{bz} \tag{1}$$

a power law, linear on a log-log plot

$$V_i = V_0 + z^n \tag{2}$$

a linear model, exponential in time (see equation 10)

$$V_i = V_0 + kz \tag{3}$$

and a square-root model, linear in time (see equation 7)

$$V_i = (V_0^2 + 4Az)^{1/2} \tag{4}$$

Because velocity is the rate of change of distance traversed for a given unit of time (dz/dt), the previous expressions can be used to obtain formulas for equivalent time-depth and time–average-velocity curves, provided that the zero datum is consistent with the model (see following paragraphs).

Figure 6 displays how well the various models match the data from Figure 5C. The data are displayed both as velocity on a linear scale (Figure 6B) and as $d\tau$ on a semi-log plot (Figure 6A). Coefficients for the curves were determined from least-squares fits of the original check shot time-depth data from 3000 to 12,000 ft (914–3658 m). Near 9000 ft (2743 m), all the curves predict approximately the same velocity value. Below that point they begin to diverge significantly. That has implications for pore-pressure prediction schemes based on the ratio of the normal trend and actual value at a given depth (Eaton, 1975), or the difference between the two (Hottmann and Johnson, 1965).

In Figure 6A, the exponential model (blue curve) yields a straight line. Close examination of the figure, particularly at a larger scale, suggests that a different slope better fits the velocity data over the range used

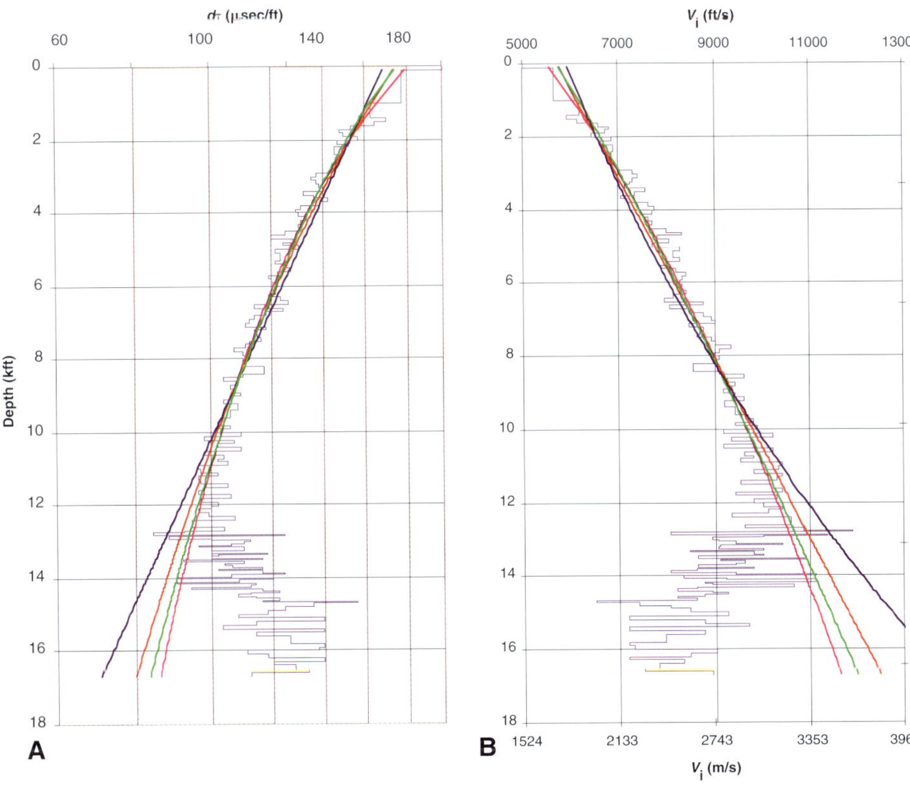

Figure 6. Comparison of check shot velocities with various trend lines fit to time-depth data (Figure 5A) from 3000 to 12,000 ft (914–3658 m). Blue = exponential; red = linear depth; green = power law; pink = linear time. Note significant variations in predicted values in overpressured region below 13,000 ft (3962 m). (A) sonic transit time with a log scale; (B) velocity with a linear scale. kft = feet in thousands; μsec = microseconds.

to calculate the coefficients. Figure 7 shows that such is indeed the case. The difference in implied velocity in the overpressure zone is significant. Of the four models under consideration, only the exponential expression shows such a large variation where fitting mathematically equivalent expressions for vertical two-way traveltime vs. depth rather than velocity vs. depth.

Figure 8 shows how well the various curves predict average velocity as a function of time. The input data range was approximately 1.0–3.0 s two-way traveltime. Once again there is significant variation in values extrapolated into the region of overpressure. The exponential model using parameters fit to interval velocity vs. depth from Figure 7 is clearly a poor representation of the data in this form. Restated, the velocity profile obtained from fitting the exponential function to velocity-depth data does not yield a suitable equation for converting from seismic time to depth.

The best visual fit appears to be the square-root parameterization, which yields a straight line as a function of time. Also, there is an indication for this well that none of the curves adequately represent the data at both shallow and intermediate depths. That could be due either to a change in the depositional history, that is, the rate of compaction, or a shortcoming of the empirical models. Additional data imply the former and support the linear time relationship as the best overall empirical fit for both time-depth conversion and prediction of V_i trends within zones of normal pressure.

Figure 9 presents data from 12 wells in a shallow-water area in the Gulf of Mexico. Coefficients for the models were determined for each well based on least-squares fits to check shot time-depth data from 3000 to 7000 ft (914–2134 m). The plots are of residual error, that is, the difference between predicted depths and actual depths. The prediction was extended by another 2000 ft (610 m) in each direction to test the ability of the models to accurately forecast trends into regions of known data in normally pressured formations. Over a short enough range, the data and all the models approach linearity. The coefficients of a robust model should not be unduly sensitive to either the range or the sample interval of the data available for curve fitting.

A perfect model including random noise would have the residuals randomly distributed about zero over the entire region fit by the model, not just over the range of data used to determine the coefficients. In such a case, the magnitude of the residual indicates the uncertainty in the input measurements. Figure 9A illustrates such behavior. A single function for each well based on a square-root variation of velocity with depth

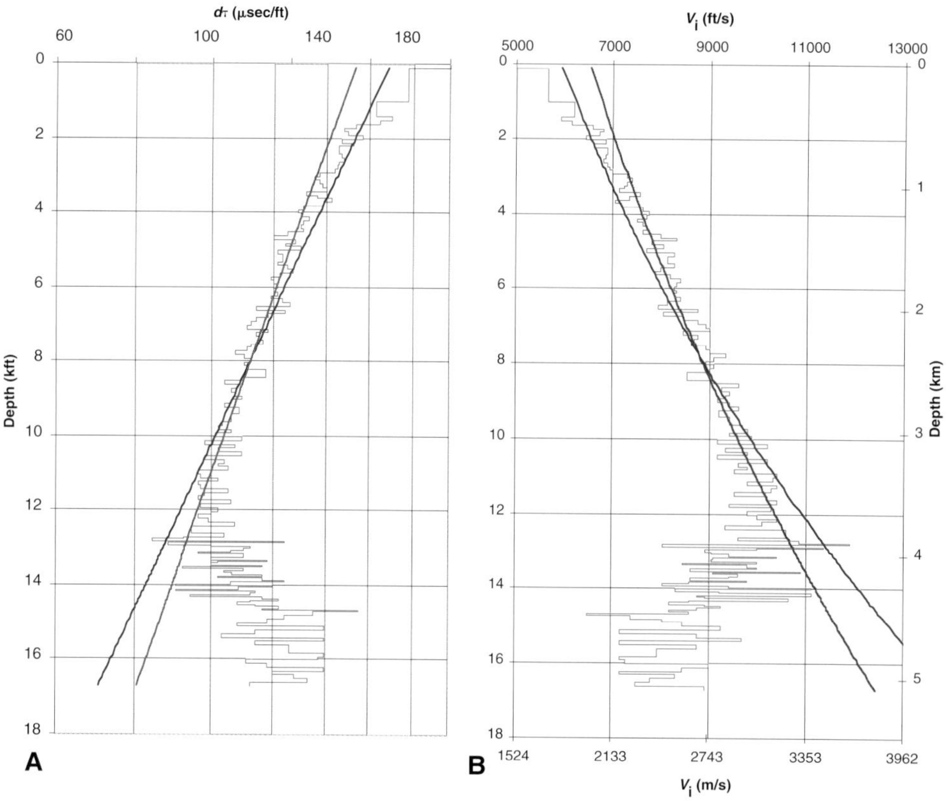

Figure 7. Comparison of exponential trend lines: (A) transit time with a log scale; (B) velocity with a linear scale. Coefficients for blue line determined from best statistical fit to time-depth data from 3000 to 12,000 ft (914–3658 m). Coefficients for brown line determined from best fit to interval-velocity data over same depth range. Lack of consistency implies exponential approximation is not a good empirical fit to the data. Note large velocity difference in overpressured zone. kft = feet in thousands; μsec = microseconds.

can reproduce the check shot data from depths of 1000–9000 ft (2743 m) to within 50 ft (15.3 m). Given the expected error in the check shot data itself, the accuracy of the model is probably within 20 ft (6.1 m). At 5000 ft (1524 m), that is less than one-half of 1% of the true value. Because the residuals fluctuate around zero, addition of more variables to the model would serve only to fit the noise. That is, three-parameter equations (e.g. Marsden et al., 1995) do not appear necessary to characterize either the time-depth or velocity-depth relationships representing normal compaction.

The depth residuals from the linear depth-velocity model (Figure 9B) are still relatively small. Note, however, that the residuals vary from positive to negative and back to positive. That implies structure in the data that is not explained by the model and the possibility that additional degrees of freedom would improve the fit. The same behavior is displayed in the residuals from the exponential model (Figure 9C). The residuals are still relatively small, particularly over the range used to determine the coefficients, but grow more than three times as large as those from the first model. The difference in predicted interval velocity is similar to that shown in Figure 6. At a depth of 9000 ft (2743 m), the difference is around 800 ft/s (244 m/s), from 9670 ft/s (2947 m/s) for the square-root model to 10,470 ft/s (3191 m/s) for the exponential model using a well in the middle of the distribution.

Figures 6–9 demonstrate that the square-root model is a better mathematical representation of the data than the exponential model. Both the square-root and linear models are commonly used by geophysicists for time-depth conversion, with about the same degree of success, particularly over limited depth ranges. The power law is a generalization of the linear fit. Empirically, the coefficient n is close to 1, and the power-law curve falls between the linear and square-root models. The power law is not as useful for time-to-depth conversion as the others because there is not a simple, closed-form evaluation of the integral of velocity for a general value of n (Kaufman, 1953).

The exponential model is somewhat a standard for pore-pressure analysis, although various authors, (e.g. Kumar, 1979; Heasler and Kharitonova, 1996) have noted problems with the formulation. An ever-increasing velocity gradient with depth is counterintuitive. The popularity of the exponential model apparently arose from the acceptance of Athy's formula relating porosity to depth (Korvin, 1984) and the simple assumption that sonic transit time for shale is proportional to porosity. The wide scatter in sonic data and the contraction effect of a log plot commonly provides a satisfactory visual fit where the traveltime implications are omitted. In the end, calibration of velocity with pressure is needed to account for several vagaries in the underlying assumptions. The accuracy of those assumptions is a distinctively different subject from the accuracy of the input velocities that this chapter addresses. For the purposes of this discussion, empirical trends are used to constrain the velocity in poor data zones, as indicators of changes in lithology as well as pressure, and to produce accurate depth predictions from seismic data. Eaton (1975) in his conclusion noted that the methodology "used to establish normal trends varies as much as the number of people who do it." The square-root model is one such variant that deserves closer examination.

That the best mathematical fit relating velocity to depth also gives the simplest expressions for working with seismic velocities in time is fortuitous. If it is assumed that

$$V_i = (V_0^2 + 4Az)^{1/2}$$

then given that $V_i = dz/dt = dz/d(T/2)$,

$$z = V_0T/2 + A(T/2)^2 \quad (5)$$

$$V_a = V_0 + AT/2 \quad (6)$$

and

$$V_i = V_0 + AT \quad (7)$$

where T = vertical two-way traveltime. Alternatively, if it is assumed that

$$V_i = V_0 + kz$$

then

$$z = (V_0/k)\exp(kT/2 - 1) \quad (8)$$

$$V_a = (2V_0/kT)\exp(kT/2 - 1) \quad (9)$$

and

$$V_i = V_0\exp(kT/2) \quad (10)$$

The first set of equations gives depth, average velocity, and interval velocity as functions of vertical two-way traveltime for the square-root model. Extra factors of 2 and 4 arise from using two-way traveltime in anticipation of comparisons with seismic data. The time-depth curve is a simple quadratic. Both the interval velocity and average velocity are linear functions of time with V_a having one-half the slope of the V_i curve. This behavior simplifies interpretation of seismic velocities. The time expressions in the second set of equations are derived from the linear velocity-depth model and are not as easy to visualize or manipulate. (The linear depth model is preferred in some geophysical applications because traveltimes and travel paths are easier to calculate as a function of offset.)

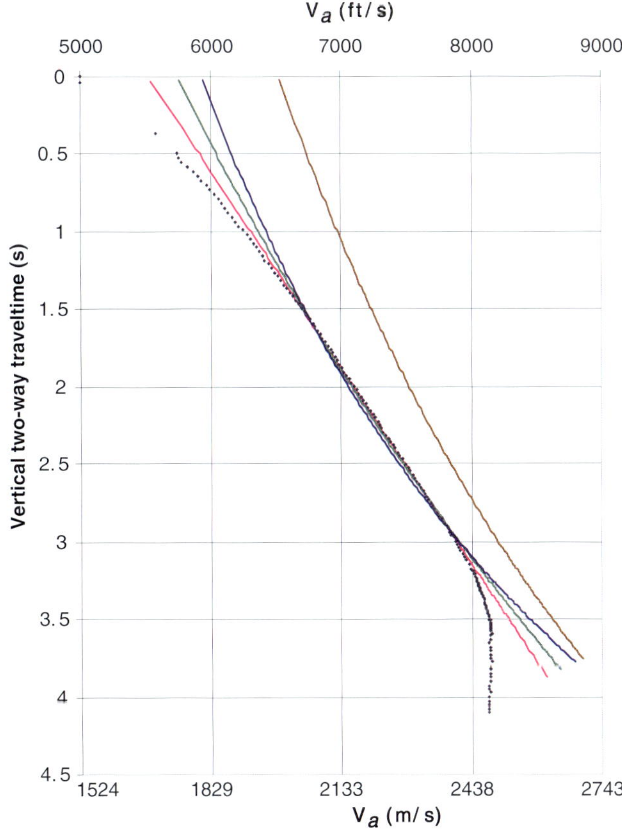

Figure 8. Comparison of trend lines with average-velocity data (Figure 5B) displayed as functions of vertical two-way traveltime. Brown curve is exponential model fit to interval-velocity data. It should not be used for time-to-depth conversion. Coefficients for other curves determined by fit to time-depth data. Blue: exponential, green: linear depth, pink: linear time. Note that linear time model predicts that both V_i and V_a are linear with vertical two-way traveltime.

Effect of the Water Column

Variations from basin to basin in the normal compaction coefficients should be considered. Before that is done, however, it is advantageous to remove the effect of the water column. In normally pressured areas, the height of the water column does not directly influence the

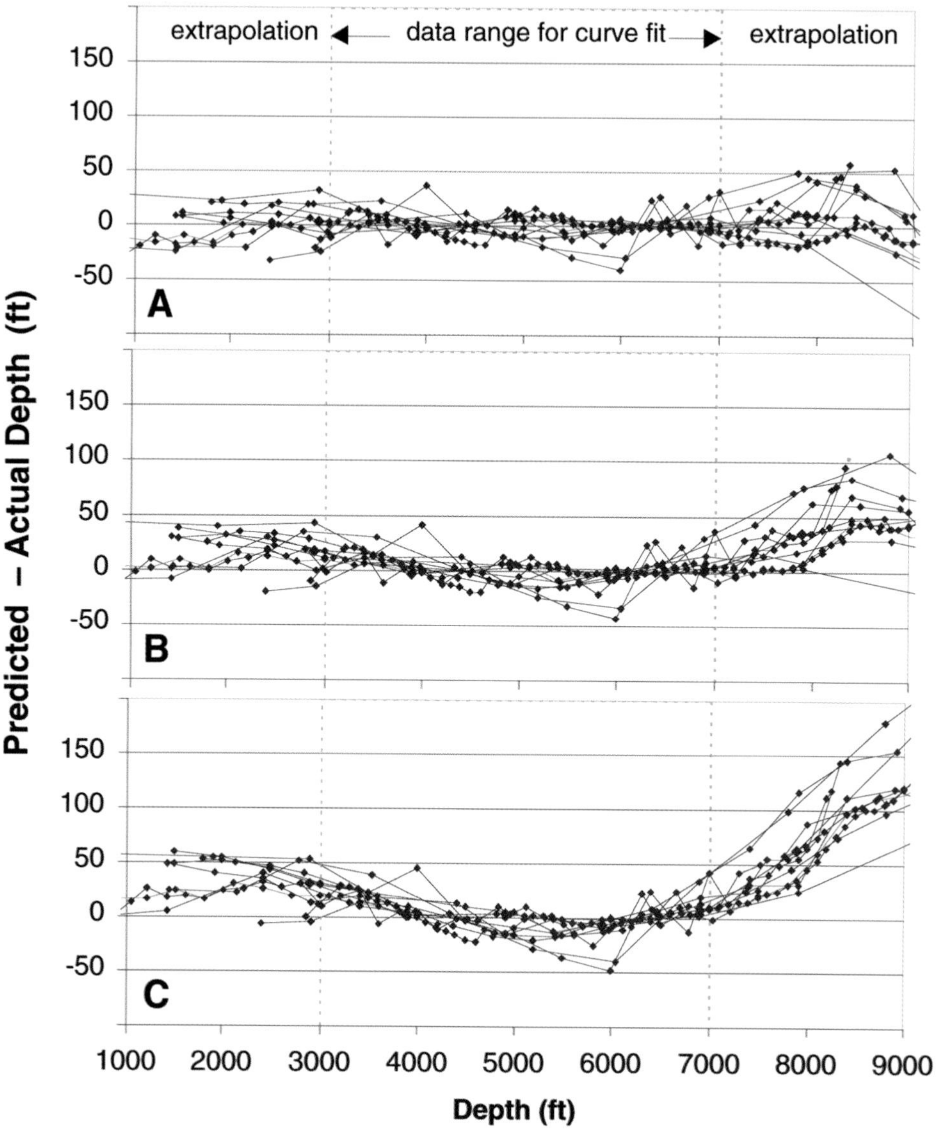

Figure 9. Comparison of depth errors obtained by individually fitting various velocity models to check shots in 12 GOM wells. Statistical fit of data from 3000 to 7000 ft. Values extrapolated outside of those bounds. (A) linear time model; (B) linear depth model; (C) exponential model.

trend in the formation velocity, other than to shift the datum to the mud line. (The effective stress is zero at the sea floor regardless of the water depth.) The water depth does impact the time-depth curve, the average velocity, and the seismic stacking velocity. Changing the velocity datum to the mud line makes it easier to infer geographic variations in formation velocity directly from average or stacking velocities. Figure 10 demonstrates the effect using check shot data from a region of the Gulf of Mexico where there are significant changes in water depth.

Geographical Variations in Compaction Trends

Figure 11 displays average velocities from check shots from five areas in two different basins. Data from several of the wells have already been shown in previous displays. All of the data are referenced to the sea floor. The orange, blue, and purple data are from shallow-water locations that exhibit a well-defined normal compaction trend before encountering overpressure. The brown and green data represent deep-water wells. Most of the deep-water wells begin to build abnormal pressure at relatively shallow depths. Some of these have a second, abrupt, pressure build across faults.

Within the normal compaction zone of each of the shallow-water areas, both the intercept, V_0, and the gradient, A, vary only slightly. A single curve fit to all of the wells in a given area is not as accurate as individual curves for each well, but in general the depth residuals are still less than 100 ft (30.5 m) over the entire normally compacted depth range. That is within the accuracy expected from seismic velocities. Because the wells are separated by up to 20 mi (32 km), the

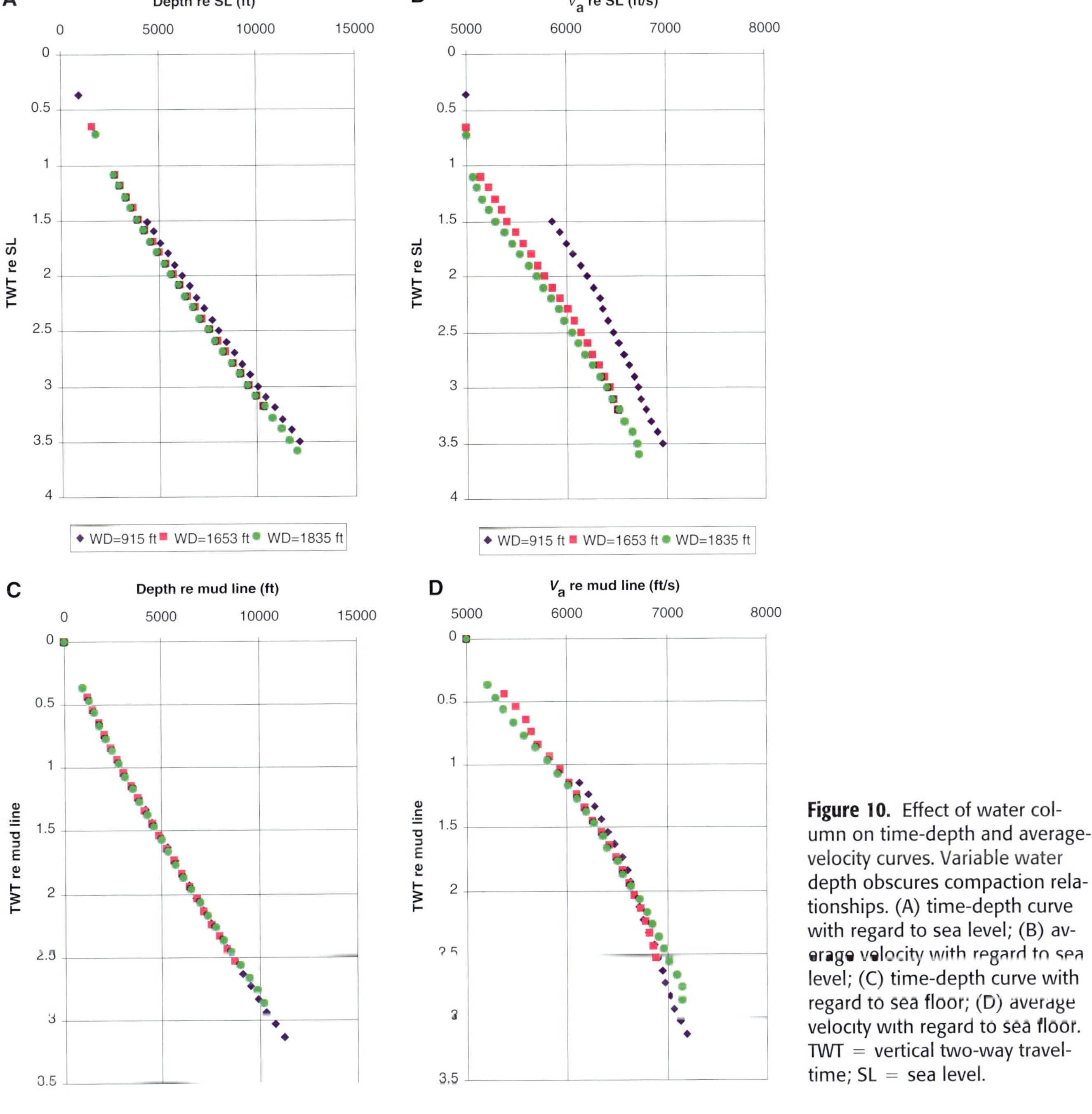

Figure 10. Effect of water column on time-depth and average-velocity curves. Variable water depth obscures compaction relationships. (A) time-depth curve with regard to sea level; (B) average velocity with regard to sea level; (C) time-depth curve with regard to sea floor; (D) average velocity with regard to sea floor. TWT = vertical two-way traveltime; SL = sea level.

implication is that velocity tracks depth rather than local lithologic, stratigraphic, or structural trends. That is, the formation velocity of a given sand or shale within a depositional sequence is not constant from well to well but varies with depth. The velocity can be reasonably constant across a fault although rocks of different ages and lithologies are juxtaposed. A single time-depth curve referenced to the sea floor can commonly be used to the top of overpressure. These observations are used in a following section to justify both vertical and lateral smoothing of seismic velocities.

Effects of Sand/Shale Ratio, Age, and Burial Rate

Several techniques are available, ranging from drilling rates to resistivity logs, for predicting pore pressure on the basis of a deviation from a normal compaction trend. Although direct pressure measurements are normally made in sands, most prediction

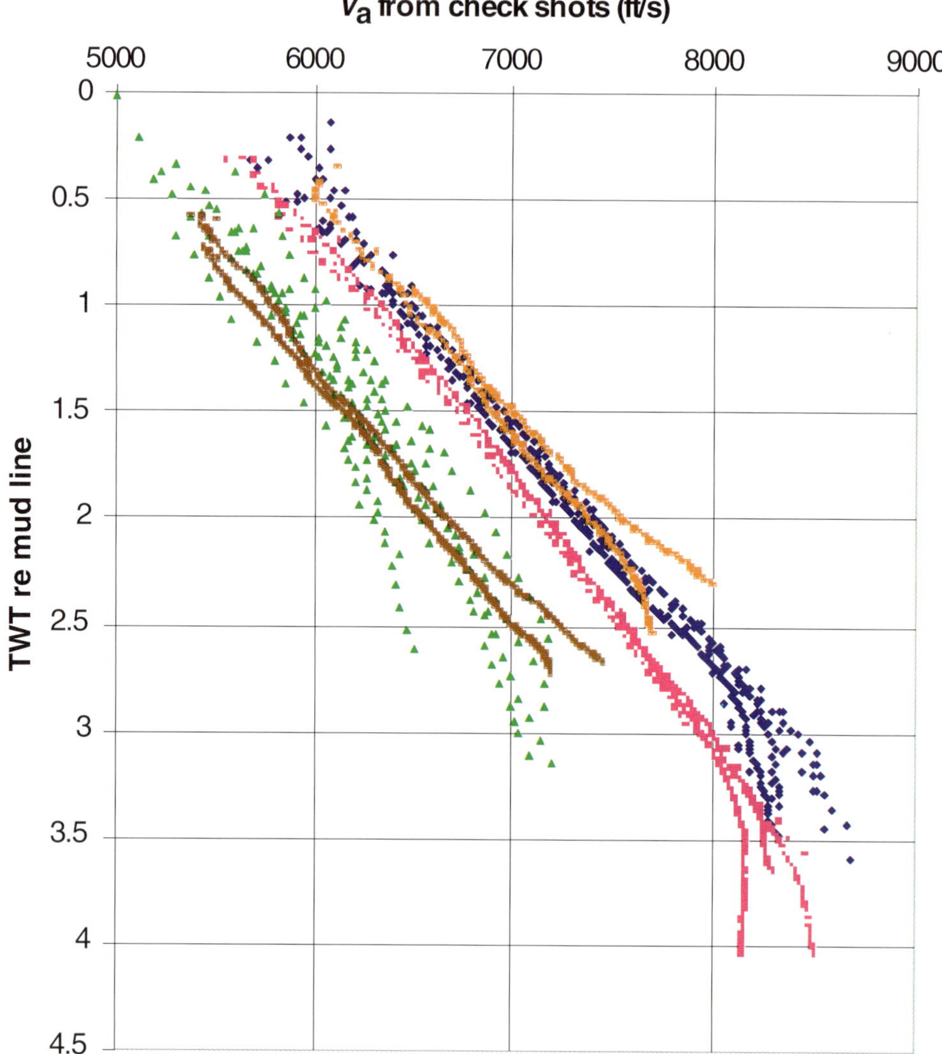

Figure 11. Geographic variation in average-velocity trends. Basin 1-A (blue diamond), basin 1-B (pink square), basin 1-C (green triangle), basin 2-A (orange square), and basin 2-B (brown square). TWT = vertical two-way traveltime.

techniques, including sonic methods, define empirical trends based on measurements in shale. Presumably, porosity and density variations with depth are better defined for the shales. Seismic velocity measurements using differences in moveout, however, seldom have the resolution necessary to separate lithology on a fine scale. Fortunately, velocity-depth trends for sand and shale are similar enough that lithology affects tend to average out unless thick, seismically resolvable units are present.

Shale velocities tend to fall between those of high-porosity sands containing compressible hydrocarbons and well-cemented, low-porosity, water-saturated sands. Statistically, for deposits of a given age, there is a depth at which the velocity of water-saturated sands transition from slower than shale to faster (Neidell and Berry, 1989). Given the wide range in both sand and shale velocities possible at a given depth, however, such a sand/shale crossover is hard to observe on a single sonic log. Figure 12 shows the variability of velocity with sand/shale ratio using the sonic log from the shallow-water well in Figure 2A.

DETERMINING VELOCITY FROM MOVEOUT ON SEISMIC GATHERS

Now that expectations have been established with well data, it is appropriate to determine how accurately the results can be reproduced using velocities determined from surface seismic data. Figure 13 compares well velocities with seismic velocities for the well shown in

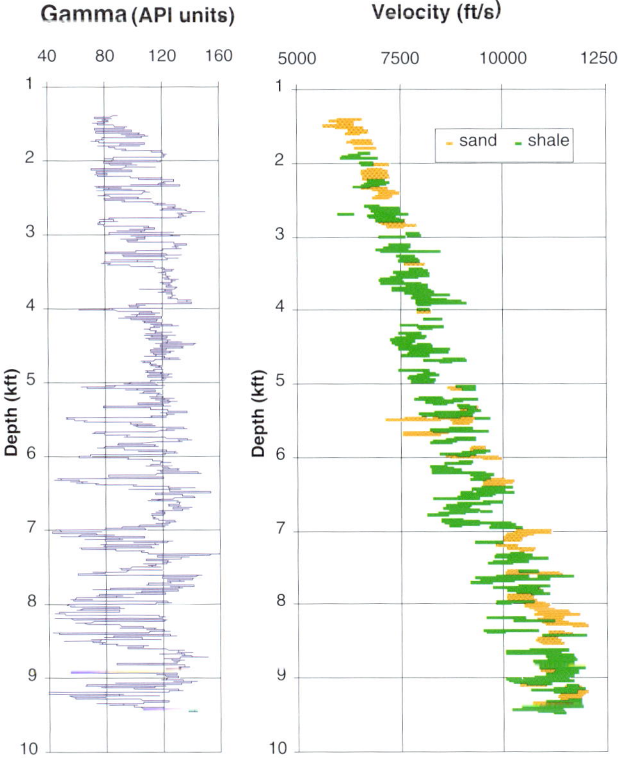

Figure 12. Sonic velocity as function of depth and lithology. Sands are in yellow. Shales are in green. kft = feet in thousands.

Figure 5. The seismic velocity data are taken from the three processing contractor velocity analysis locations closest to the well. The average-velocity trends are similar although the seismic values tend to be slightly faster than the sonic values. The interval velocities correctly locate the first pressure break in depth (around 12,500 ft [3810 m]). The difference in velocity between the three curves provides an estimate of the uncertainty in the technique. Also, note that the second pressure break near 14,500 ft (4420 m) has not been resolved. Seismic processors commonly err on the high side where there is uncertainty in the stacking-velocity trend. Otherwise, there is a danger of stacking multiply reflected energy in poor data zones. In this case, deeper picks were extrapolated such that the resulting interval velocities increased rather than decreased. Those values were excluded from the plot. Generally, velocities used to process seismic data reveal major pressure breaks. They may be, however, inadequate for detailed pressure prediction work. That is particularly true for speculative processing or fast track volumes. Ideally, seismic velocities should be repicked for the express purpose of pressure prediction, which can be costly and time consuming unless the appropriate data are saved during the original processing. At the very least, contractor picking of velocities should be quality controlled by someone familiar with the concepts of pressure prediction.

Several seismic-processing procedures derive velocity estimates to account for the moveout in reflection time observed with offset. The most common sorts the data into a common midpoint (CMP) gather and determines the best-fit normal moveout velocity, V_{nmo}, as a function of vertical two-way traveltime (T) for each major reflector. V_{nmo} assumes that the measured traveltime is a hyperbolic function of the offset distance between seismic sources and receivers. The moveout velocity is used to remove the traveltime variations, that is, to flatten the event at the zero-offset time, prior to stacking the data to obtain a seismic image. Generally, the stacking velocity, V_{stack}, is an estimate of V_{nmo} obtained by systematic repetition of the stacking procedure involving examination of variations in gather flatness, semblance across the gather, and stacked image continuity. (Refer to Table 1 for a comparison of velocity terms.)

Figure 14 shows common seismic velocity analysis displays from a good-quality, deep-water location. Figure 14B and C show a CMP gather both before and after moveout correction. Events should be flat after the moveout correction. All of the traces after moveout are then summed to form a single trace for that CMP location. The semblance contours in Figure 14A are a measure of stacked amplitude in a sliding T window for a large number of trial stacking velocities. The semblance is high where events are correctly flattened and low where they are not. The windowing process means that semblance maximums do not always directly correspond to event times seen on the CMP gather. Figure 14D shows the result of stacking several CMPs with constant velocities around the analysis location. Continuity of events at a given time is another indication that the correct stacking velocity has been obtained. In poor data areas, which unfortunately are commonly associated with overpressure, distinct events may not be seen on the CMP gathers. In those cases, constant-velocity stacks are an important tool.

Note the slope break at 5.5 s seen in the stacking-velocity trend in Figure 14A. Although no wells have been drilled in this area, it probably indicates overpressure. The trend above 5.5 s is not linear as expected because of the large depth of water. It is linear if the data are datumed to the sea floor. The resulting interval-velocity curve does reveal a linear trend down to the velocity reversal indicative of pressure. The linear trend in interval velocity results somewhat from a bias in the picking. Minor changes in stacking velocity within the contours of high semblance have little influence on the final stack but can cause large changes in the interval velocities, particularly if the picks are close

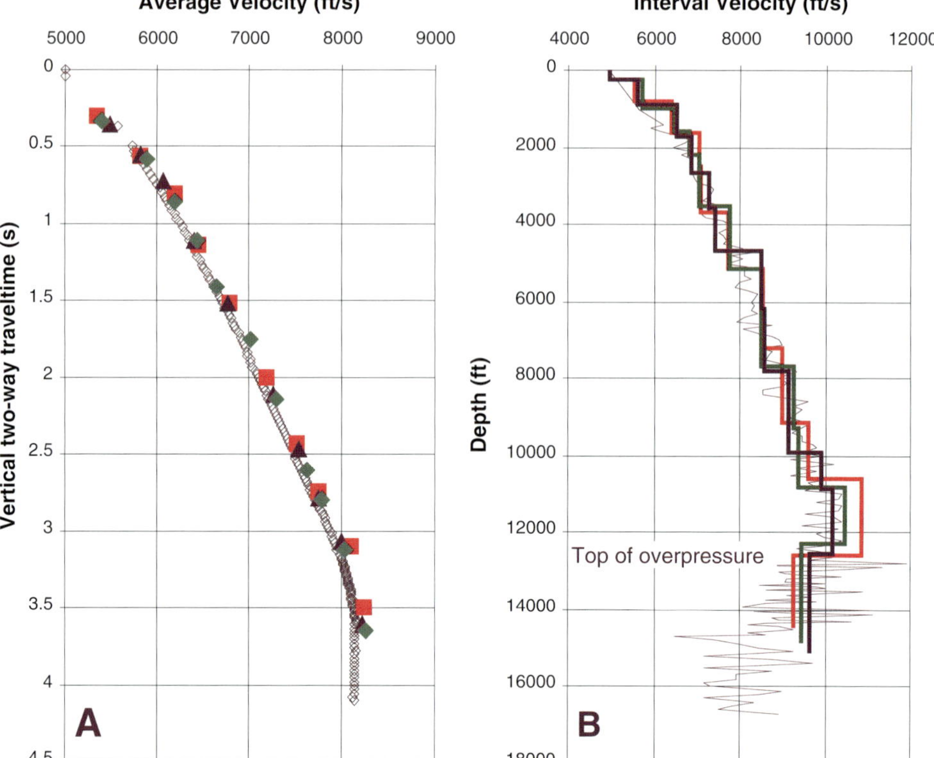

Figure 13. Comparison of check shot and seismic velocities using well data from Figure 5. Seismic stacking velocities from three locations near the well converted to V_a and V_i. Two-way traveltime converted to depth. (A) Average velocity vs. two-way traveltime; (B) interval velocity vs. depth.

together. Commonly, an unreasonably high excursion is immediately followed by a very slow interval velocity, or vice versa. The stacking velocity picks in this case have been adjusted to remove interval-velocity pairs that fluctuate about the trend. The thick, low-velocity water layer also impacts the sensitivity of the interval-velocity determination near the water bottom. In this example, a 1% change in the value of the first stacking velocity pick below the water bottom leads to a 20% change in the corresponding interval velocity.

Figure 15 shows additional semblance panels and CMP gathers to demonstrate how the quality of stacking velocity information can vary over relatively short distances. The semblance panel at the first location (CMP1) shows numerous distinct events that can be tracked both on the seismic section and the gather. The quality of the semblance picks begins to deteriorate at the second location. At the third location, little confidence should be placed on the deeper semblance picks. Events on the seismic section are chaotic and difficult to interpret. Coherent multiples interfere with primaries on the CMP gather. In areas such as these, it is important that the pressure analyst understand the variable uncertainty in the seismic velocities.

Calculating Interval Velocity from Stacking Velocity

For a single horizontal layer with constant velocity, the formation velocity at each depth, the interval velocity across the unit, the average velocity to any depth below the surface, and the normal moveout velocity for the base of the unit are all the same. All other situations require adjustments to the stacking velocity to obtain an appropriate representation of the formation velocity.

The conventional method to convert from V_{stack} in measured two-way traveltime to V_{i} vs. depth is to use the Dix equation to determine V_{i} between adjacent stacking velocities and then to multiply the interval one-way time by V_{i} to obtain a layer thickness. The running sum of the thickness for each of the preceding layers gives the depth. The Dix equation (Sheriff, 1991) for the interval velocity of the nth layer is as follows:

$$V_{in} = [(V_{\text{stack}n}^2 T_n - V_{\text{stack}n-1}^2 T_{n-1}) / (T_n - T_{n-1})]^{1/2} \quad (11)$$

where $V_{\text{stack}n}$ and T_n are the stacking velocity and vertical two-way traveltime, respectively, for the top of the nth layer.

The Dix equation is founded on two assumptions. First is that the earth is composed of multiple horizontal layers of constant velocity. Ray bending due to a continuous change of velocity with depth, as implied by compaction, causes the calculated interval velocities to be slightly too high. Anisotropy (a difference

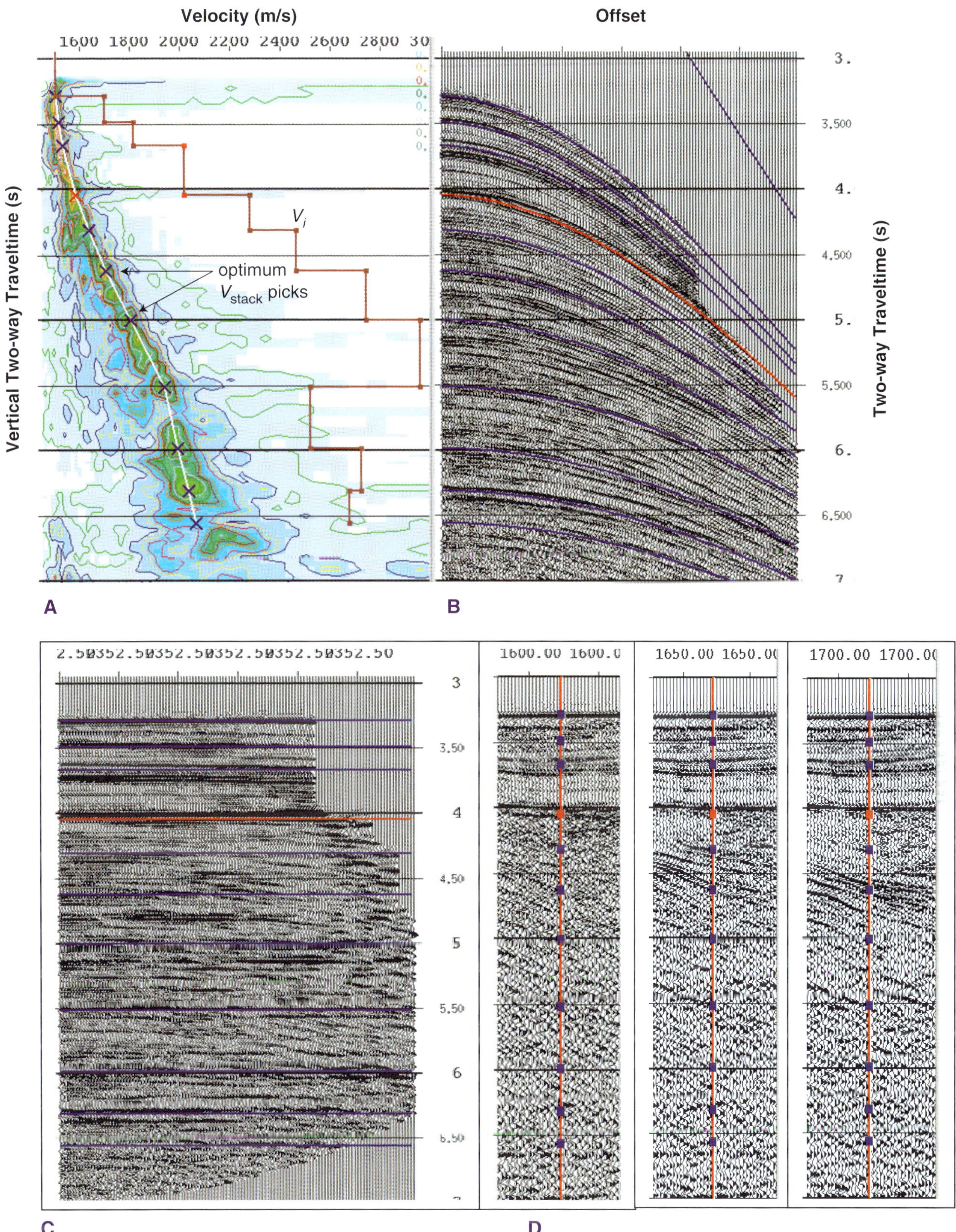

Figure 14. Determination of optimum stacking velocity from seismic data. (A) Semblance panel with stacking-velocity picks and resulting interval velocities; contours plot semblance as a function of stacking velocity and vertical two-way traveltime; (B) CMP gather showing hyperbolic moveout with offset; (C) CMP gather after normal moveout correction using stacking-velocity picks; (D) constant stacking velocity stacks at 1600 (left), 1650 (middle), and 1700 (right) m/s. Moveout curves properly fit by those stacking velocities appear flat.

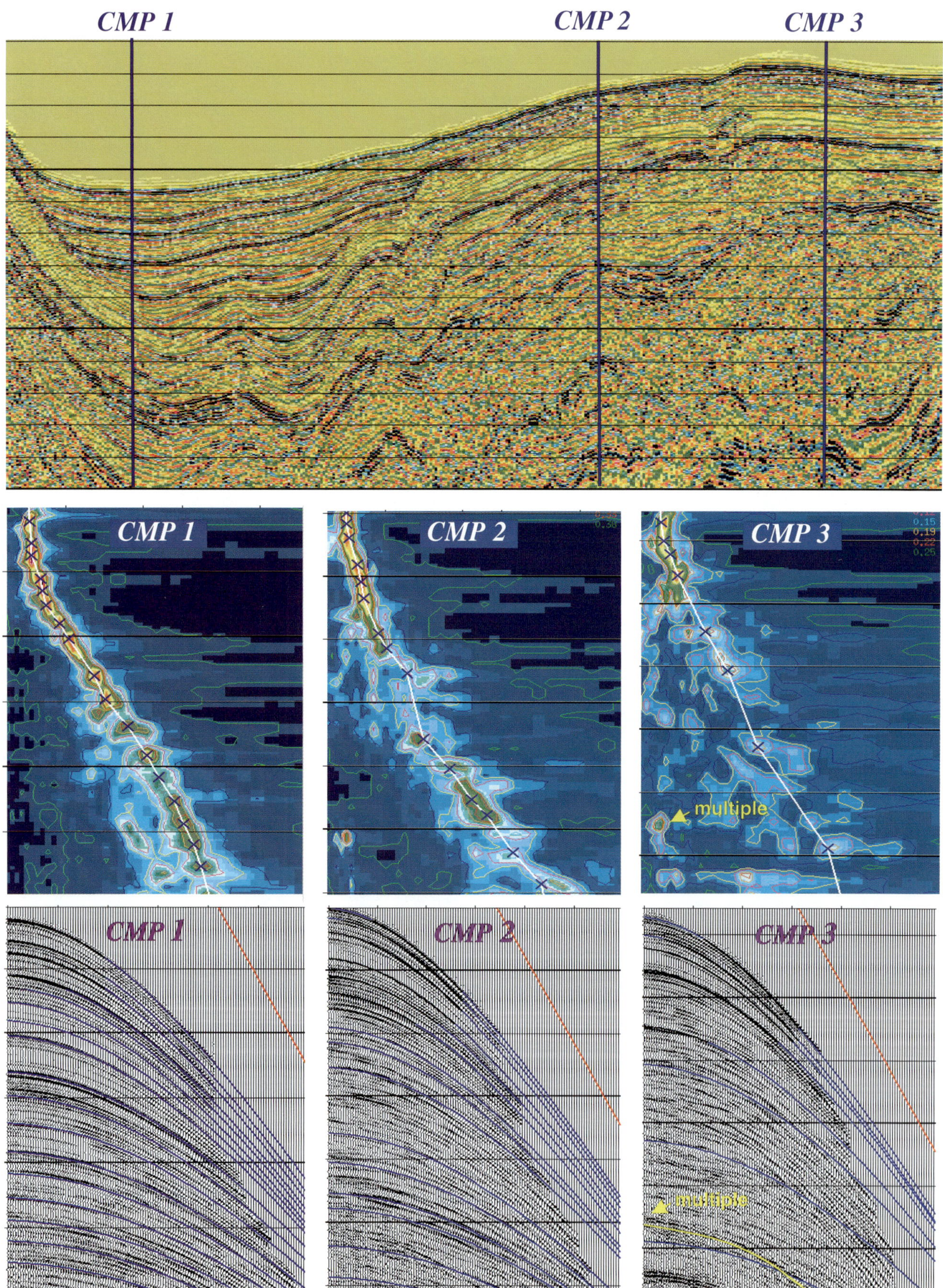

Figure 15. Seismic line (top) with velocity scan locations. Semblance panels (middle) and gathers (bottom) at selected locations. Confidence in velocity information decreases from left to right. Deeper coherent events in CMP 3 are multiples.

between the horizontal and vertical velocity at the same point) also introduces errors, as do dipping horizons and other lateral velocity variations.

The second, related, assumption made in deriving the Dix equation is that stacking velocity is the same as the root-mean-square (rms) velocity, V_{rms}. The rms velocity variations mimic the average-velocity behavior noted previously but is always slightly higher because rms averaging is biased toward higher velocities. The Dix equation accounts for that part of the difference between stacking velocity and average velocity, but only if the velocity is truly constant within the interval. Additionally, the moveout curves for multiple layers, even if flat and of constant velocity, are no longer strictly hyperbolic. This introduces a positive bias that depends on offset and the relative velocity changes between layers. In places, higher order terms are added to the moveout equation to better flatten events at long offsets. The Dix equation is more appropriate in the limit of short offsets.

Figure 16 illustrates the application of the Dix equation to a single 4000 ft (1219 m) flat layer with four simple variations of velocity within the layer. Figure 16A plots the four different velocity functions, all of which have the same vertical traveltime and average interval velocity of 6427 ft/s (1959 m/s). The first is simply a constant velocity. The second is a series of four constant-velocity steps. The velocity in the third example varies as the square root of the depth as per the linear time model in Figure 9. The fourth case is a single anisotropic layer with a vertical velocity of 6427 ft/s (1959 m/s) and a horizontal velocity of 7712 ft/s (2351 m/s).

Anisotropy with a vertical axis of symmetry, also known as transverse isotropy, is easy to think of in terms of vertical and horizontal velocities. Four parameters are necessary, however, to characterize such a medium (Alkhalifah and Tsvankin, 1995). A common parameterization consists of the vertical compression-wave velocity, the vertical shear-wave velocity, and two additional constants, δ and ε. In simple terms, the difference between δ and ε controls the stacking velocity. If δ equals ε, the moveout is exactly hyperbolic, and the stacking velocity for a single flat horizon gives the horizontal velocity. If δ does not equal ε, the moveout curve diverges from a hyperbola. Two cases are illustrated in Figure 16, both with the same vertical and horizontal velocities. Open triangles represent $\delta = 0.219$ and $\varepsilon = 0.22$. Filled triangles represent $\delta = 0.1$ and $\varepsilon = 0.22$.

Figure 16B shows the moveout curves. (These were calculated analytically except for the four-step function, which was ray traced.) Hyperbolas were fit to the moveout curves over offset ranges of 0–4000 ft (0–1219 m) and 0–10,000 ft (0–3048 m). As expected, the Dix equation yields the true velocity, independent of offset for the constant-velocity case. The Dix estimated velocity is too high for the two cases of a vertical velocity increase within the layer. The error in the estimate also grows as a function of the offset range over which the moveout curves are fit. The Dix equation applied to the anisotropic layer where δ nearly equals ε yields the horizontal velocity rather than the vertical velocity. Where δ does not equal ε, the velocity from the Dix equation falls between the vertical and horizontal velocities and depends noticeably on offset.

The raypaths for both the constant-velocity and anisotropic cases are straight (Figure 16C), which was the assumption used to derive the Dix equation. Raypath segments for the four-step function are straight lines that bend, or refract, at the velocity discontinuities such that the overall traveltime is minimized. A continuous variation in velocity with depth results in continuous ray bending. In the presence of ray bending, Dix velocities in the zero offset limit more closely approximate rms interval velocities rather than average interval velocities.

The cause of the variation of stacking velocity, and hence the Dix velocity, with offset is shown in Figure 16D. The difference between the best fit hyperbola and the actual moveout is plotted. The residuals are zero for the constant-velocity case and almost so for the brown anisotropic curve. The residuals are similar for the two cases with vertical velocity variations and are large enough to influence the stacking-velocity estimate without unduly impacting the image quality (4 ms is a normal wavelet trace sampling). If the offset range is large compared with the reflector depth, the deviation represented by the orange anisotropic case can impact the imaging. Unfortunately, the correct vertical velocity cannot be derived solely from the moveout curve.

Although the velocity differences in this simple example appear minor, the effect of velocity layering on estimation of true interval velocities from seismic stacking velocity are almost always noticeable (5–10%), more so if significant anisotropy is involved. An accurate prediction of the depth of an overpressure zone requires calibration of the Dix velocity function with check shots. Even so, the effects of vertical variations in velocity tend to be small compared to those introduced by lateral structural variations.

Effect of Structure on Seismic Velocity

Significant errors in the Dix estimation of interval velocity from stacking velocity arise where the earth departs from the simple flat layered model. Figure 17 compares a flat layer geometry with that of a dipping

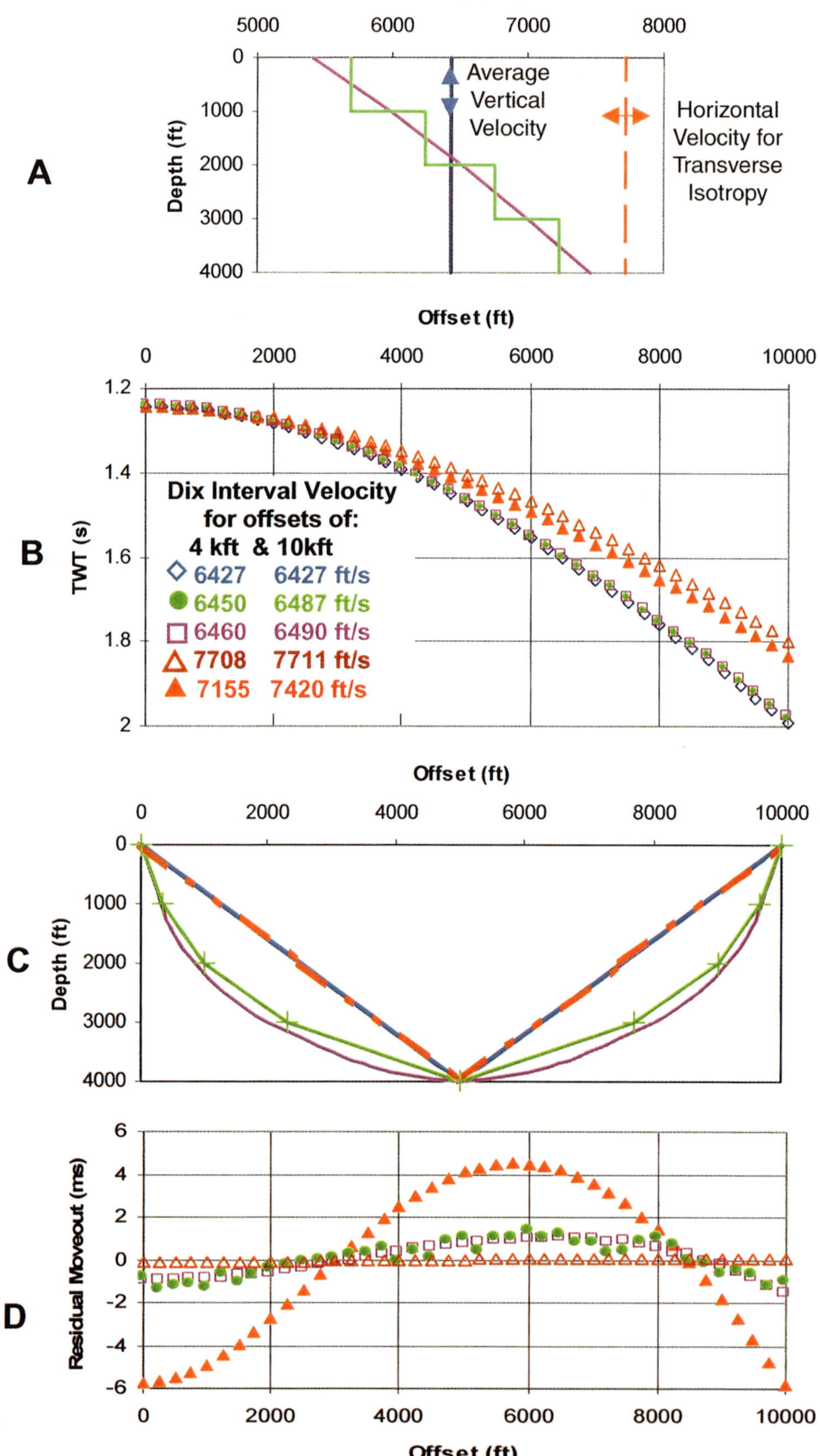

Figure 16. Effect of velocity model on interval velocities derived from seismic moveout. (A) Color-coded models all with same vertical average interval velocity. Blue = constant velocity; green = four velocity layers; pink = velocity gradient; orange = transversely isotropic velocity. (B) Corresponding moveout curves and derived interval velocities for each model with two offset ranges. kft = feet in thousands. (C) Corresponding raypath plots. (D) Residual moveout curves (deviation of moveout from assumed hyperbola) for three of the models. TWT = two-way traveltime.

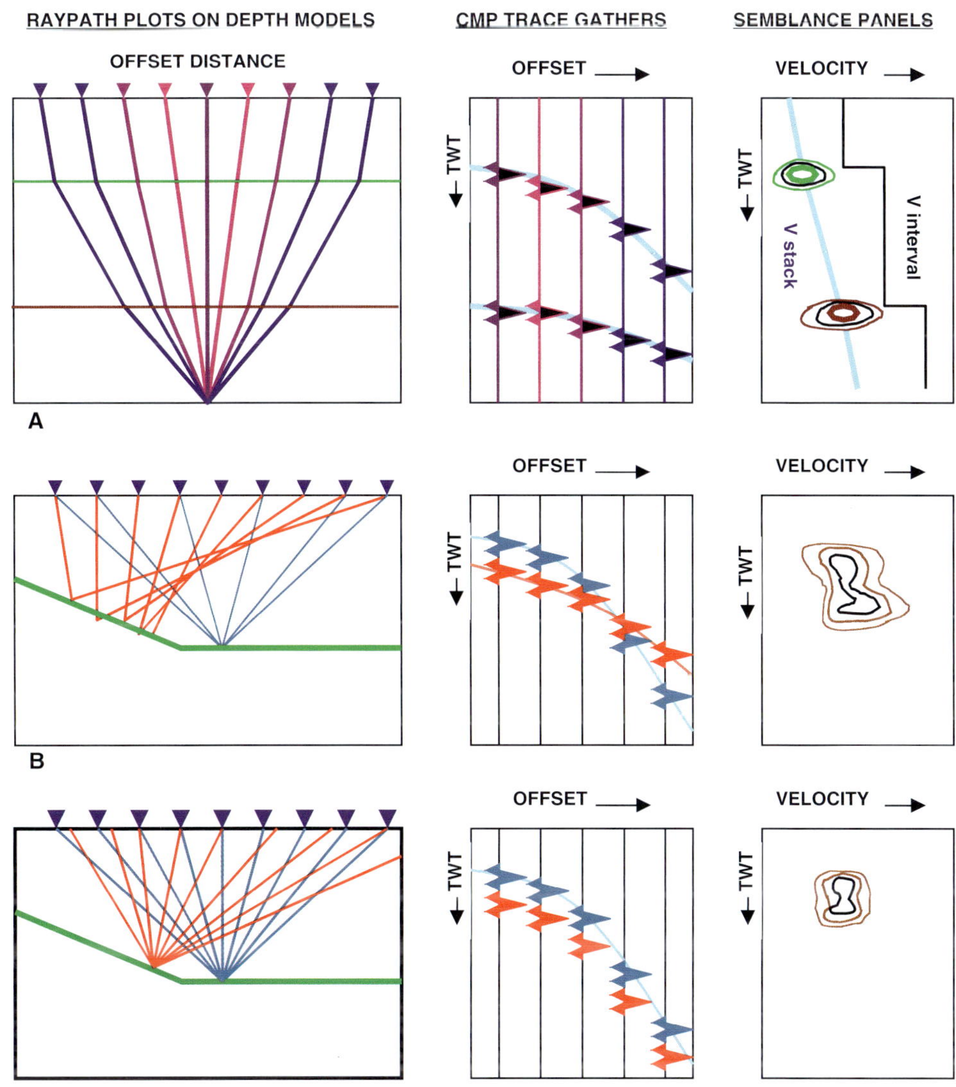

Figure 17. Influence of dip on normal moveout and effect of DMO. (A) Flat-layer model; (B) dip effects; (C) effect of DMO. TWT = two-way traveltime.

event. For a single interface, the stacking velocity is increased proportional to the cosine of the dip angle. For stacked layers with variable dips, simple corrections based on analytical expressions are not possible. Interpretation is further complicated where a change in dip occurs within a cable length. In Figure 17B, the dipping interface has a higher moveout velocity than the horizontal interface. If a single velocity pick is made for the pair of events, one or the other, or neither, stack correctly. If a velocity pick is made for each event, the Dix equation imposes an apparent velocity change where none actually exists.

Also note in Figure 17B that the rays representative of seismic energy for different offsets no longer converge at a single point directly beneath the midpoint of the gather. A seismic velocity profile does not always represent vertical sampling of the earth. Migration is a process applied during seismic imaging to correct for various items, including lateral positioning of events. Several types of migration algorithms exist that vary in their cost and ability to handle lateral velocity changes. All migration codes require an input velocity field. Commonly, migration velocities available from the seismic processor have been smoothed too much for detailed pressure work. In general, pressure analysis should be performed using stacking velocities picked after dip moveout (DMO) correction (Figure 15C).

The DMO is a prestack process that does a reasonable job of removing dip effects so that flat and dipping events can be stacked with the same velocity (see Deregowski [1986] and Liner [1999]). Application of DMO requires an initial estimate of the velocity field, but the operator is relatively insensitive to the velocity. A constant-velocity DMO operator is normally adequate for stacking purposes. If transverse isotropy is present, however, constant-velocity DMO does not yield the best image. The difference in moveout between a dipping event and a flat event can be used to estimate the transverse isotropy (TI) parameters (Alkhalifah and Tsvankin, 1995).

If neither DMO nor more sophisticated prestack migration processes have been applied to the data prior to NMO analysis (uncommon in modern data processing), then the pressure analyst should concentrate on areas in the seismic section that are flat over the length of a seismic cable. Velocities from those locations may then be extrapolated to prospect locations using geologic insight.

Although DMO does a reasonable job of removing dip effects, it does not correct for lateral changes in velocity. An important example for pressure work is how stacking velocities vary across a fault that bounds a velocity decrease due to overpressure.

Figure 18 shows how a simple fault model distorts the stacking velocity. The effect of a lateral velocity change is first seen on the far offsets of a CMP (Figure 18A). Successive CMPs are influenced to a greater extent until the middle of the spread moves past the discontinuity. Figure 18B shows how the moveout of the event at CMP 7 differs from the constant-velocity case at CMP 5. Similarly, the moveout at CMP 11 is noticeably different from the moveout for the constant-velocity case at CMP 15 (Figure 18C). Figure 18D tracks the lateral change in stacking velocity. For imaging purposes, it is important to honor such changes. For pressure work, it is commonly appropriate to extend lateral velocity trends up to the fault rather than to honor stacking-velocity picks within a cable length of an implied velocity discontinuity. This can be difficult where the target is a fault block that is smaller than typical cable lengths of 6 km. In such a case it is sometimes useful to carefully pick the location of the velocity analysis and to discard long offsets.

Figure 18D also illustrates another display useful in velocity work. If semblance calculations are made at close CMP spacing, it is possible to sort the data based on a digitized horizon and produce a plot of semblance vs. velocity and CMP location for a single event.

The lateral smearing of velocity information over a cable length has other implications with regard to velocity analysis. Velocity analysis for pressure work is typically done with a CMP interval of 1 km or so. Because that is already well below the length of a seismic cable, the velocity information from one point to the next is not independent. Velocity analysis at closer CMP spacing can provide useful statistical information but seldom provides increased lateral resolution using the Dix equation. Abrupt changes in stacking velocity from one profile to the next are suspect. Finer sampling may be warranted using tomographic techniques, but that is another topic.

Shooting geometries commonly deploy more receivers per shot than are sorted into a CMP bin. Adjacent CMP gathers may contain only every other or even every fourth offset. Because velocity information is already smeared laterally because of the cable length and reflector dip, it is commonly appropriate to combine several adjacent CMP profiles into one supergather prior to velocity analysis. See Figure 19.

The previous discussion of structural affects assumes that hyperbolic NMO velocities are still appropriate for stacking but illustrates situations where the best stacking velocity may not be appropriate for Dix derivation of interval velocities. If reflector moveout observed during velocity analysis cannot be reasonably flattened with a hyperbolic fit, then the concepts of NMO and the Dix equation are no longer appropriate. Depth migration is then required. Depth migration is appropriate whenever there are large abrupt changes in lateral velocity, for example, under the flanks of salt domes (see Lines et al. [1993] and Whitmore and Garing [1993]).

Figure 20 shows examples of displays used to determine velocities for depth migration. An initial model consisting of two layers with vertical velocity gradients is shown as a blue line in Figure 20A. The model is then used to migrate the data. If the velocity field is correct, then all events in a common image point (CIP) gather should be at the same depth. (A CIP gather represents a single reflection point in the earth directly below the surface location. Reflections on a CMP may arise from several locations away from the center, as in the dipping-layer example of Figure 17B.) Figure 20A displays the depth error in both vertical and horizontal semblance. After a tomographic update to the velocity model, the maximum semblance values are near zero residual depth, and the events in the migrated gather appear flat. Tomography (Sheriff, 1991) is a procedure in which a velocity model is automatically adjusted to simultaneously minimize the error between multiple horizon picks on multiple offsets and the values predicted by the model.

CASE STUDY: SEISMIC VELOCITY ANALYSIS

Figure 21 shows a seismic line with two wells. Well A was drilled first and did not encounter significant overpressure. There were problems completing well B after it encountered abnormal pressures across a fault. The

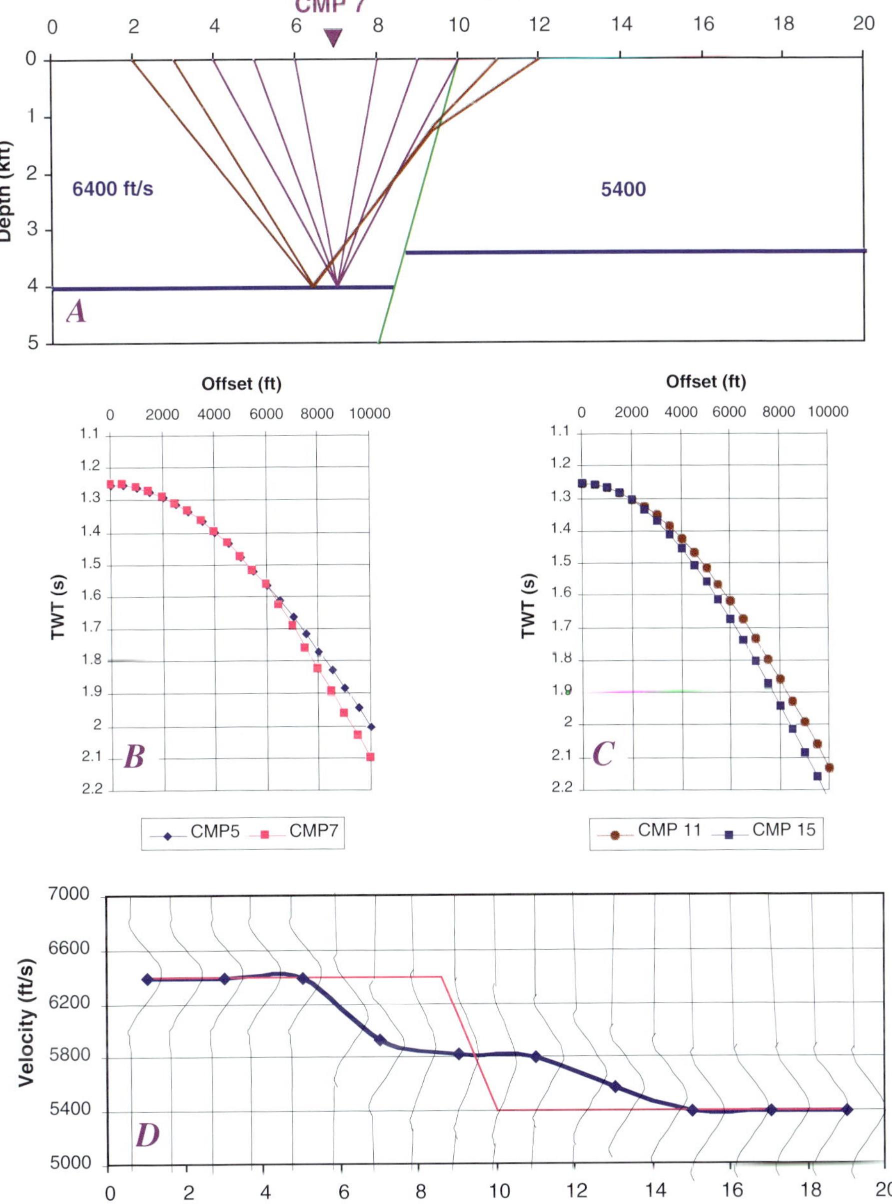

Figure 18. Effect of lateral velocity change across fault on event moveout and corresponding stacking velocities. (A) Cross section of depth model showing ray path distortion at CMP 7 due to fault; kft = feet in thousands; (B) moveout curves at CMP locations 5 and 7; (C) moveout curves at CMP locations 11 and 15; (D) cross section of velocity semblance picks vs. CMP locations. TWT = two-way traveltime.

sonic log for this well is shown in Figure 3A. Note the change in reflection continuity in the center of the section. As a rule of thumb, such changes in image quality are commonly associated with overpressure. Unfortunately, this also implies that reliable velocity estimates are more difficult to extract from those regions.

Figure 21B shows the CMP locations where velocity analysis was performed. Figure 22 shows the velocity analysis panel near the well B location. Only a limited number of picks are shown to prevent obscuring the event moveout seen on the CMP gather. Note that the slope break in the stacking-velocity trend occurs at the known location of the pressure transition.

Figure 23 compares the Dix derived interval velocities (blue stair step) with the sonic log (gray curve) from well B, both in time and in depth. The original interval velocities displayed in time (Figure 23A) correctly predict the decrease in velocity seen on the sonic log just above 2.5 s. Where those velocities are used for time-to-depth conversion (Figure 23B), however, the pressure transition is mislocated by several hundred ft. As indicated in the previous discussion, such a

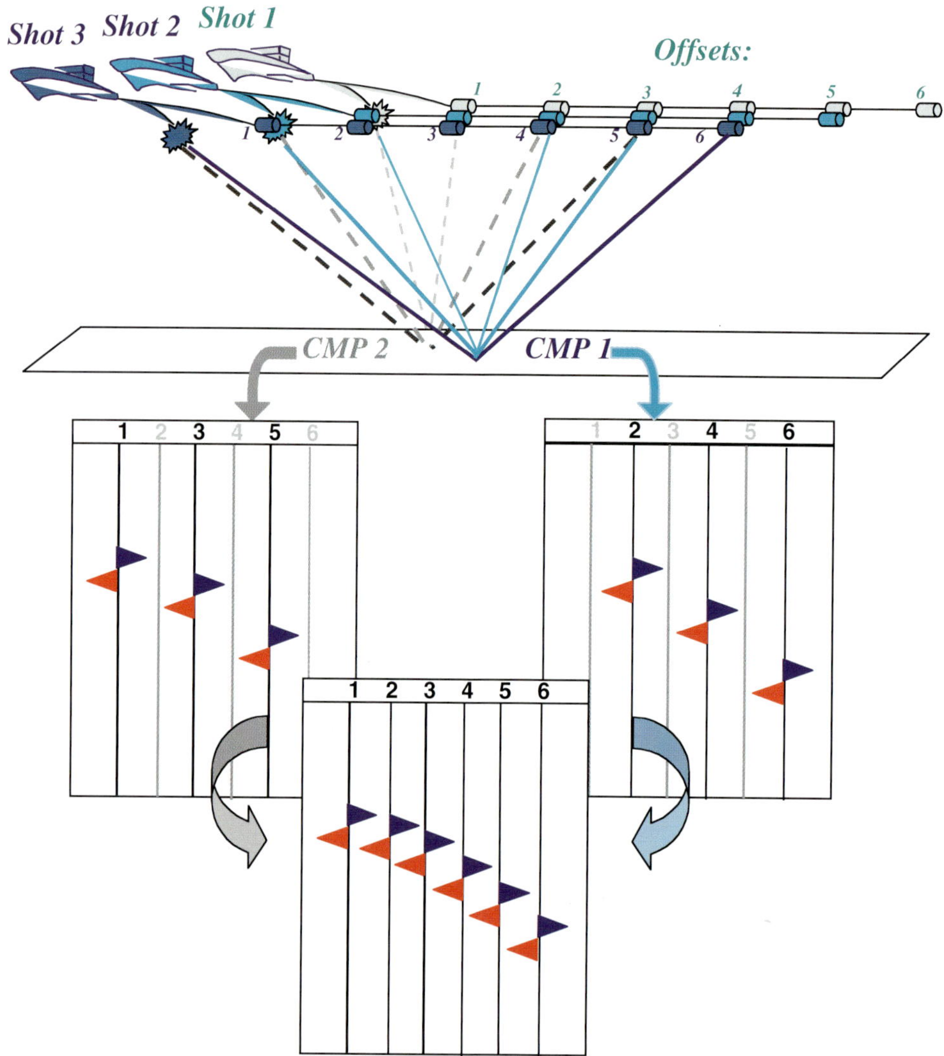

Figure 19. Common shooting geometrics result in adjacent CMP gathers having different offset distributions. Velocity analysis can be improved if supergathers are formed that contain all available offsets.

result is expected because of several factors that impart a positive bias to the Dix estimates.

Figure 23A also compares the original Dix calculated average velocity (pink curve) with check shot values in the same well (green points). To demonstrate analysis techniques ahead of the bit, check shots from well A were used to calibrate the average-velocity curve at the well B location. The calibrated curve (in red) is consistent with known well B values and corrects the error in the depth display of the interval velocity (Figure 23B). The final difference between the sonic and the calibrated Dix interval velocities is an indication of the uncertainty in the velocity analysis technique and the low resolution of seismic moveout velocities compared with sonic values.

To some extent, the Dix assumption that the earth is composed of a finite number of constant-velocity layers is at odds with the concept of a smoothly varying normal compaction trend. Pressure analysts sometimes use Dix velocities only at the midpoint of the layer to produce a smooth trend in the predicted pressures. Reality lies between the two simplistic models. A useful compromise is to regard the earth as a series of layers with velocity gradients. Constant-velocity layers simply have zero gradients, whereas all layers within a normally compacted region have the same gradient. Such a view fosters the concept of vertical and lateral smoothing of stacking velocities within geologic units to reduce the uncertainty in both time-to-depth conversion and pressure predictions. The gradient can be based on any analytic expression thought to represent the proper velocity-depth relationship. The key criterion is that the chosen expression is consistent with the stacking-velocity information.

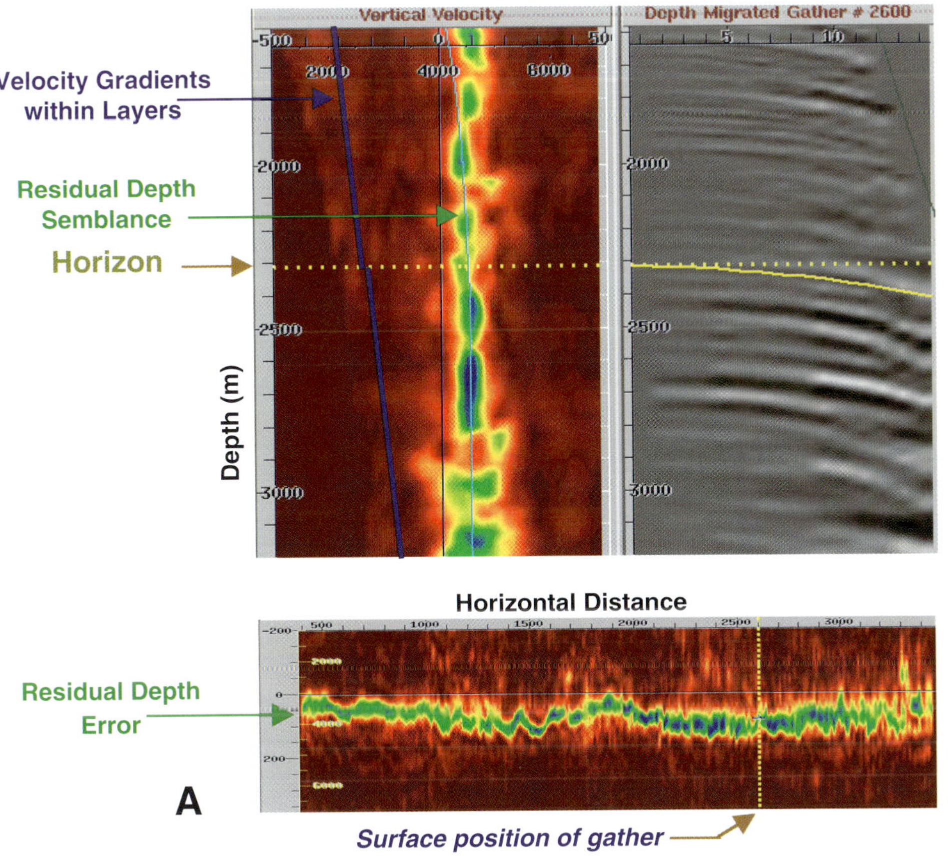

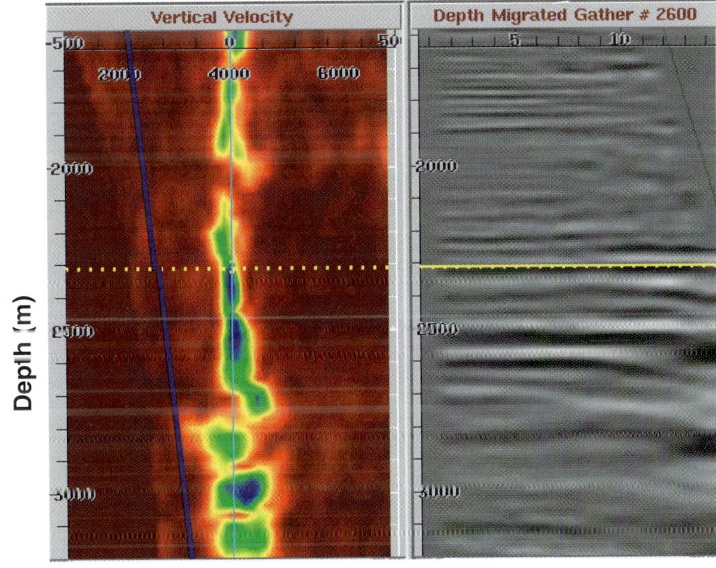

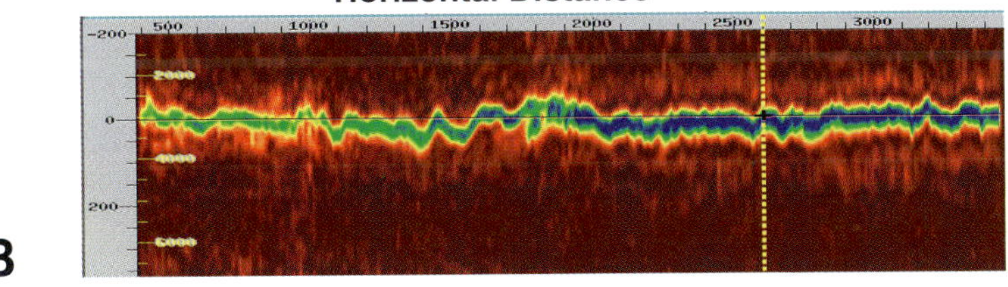

Figure 20. Horizon-keyed velocity analysis in the depth domain. (A) Diagnostic plots using initial layered velocity model; (B) diagnostic plots after tomographic update of velocity model.

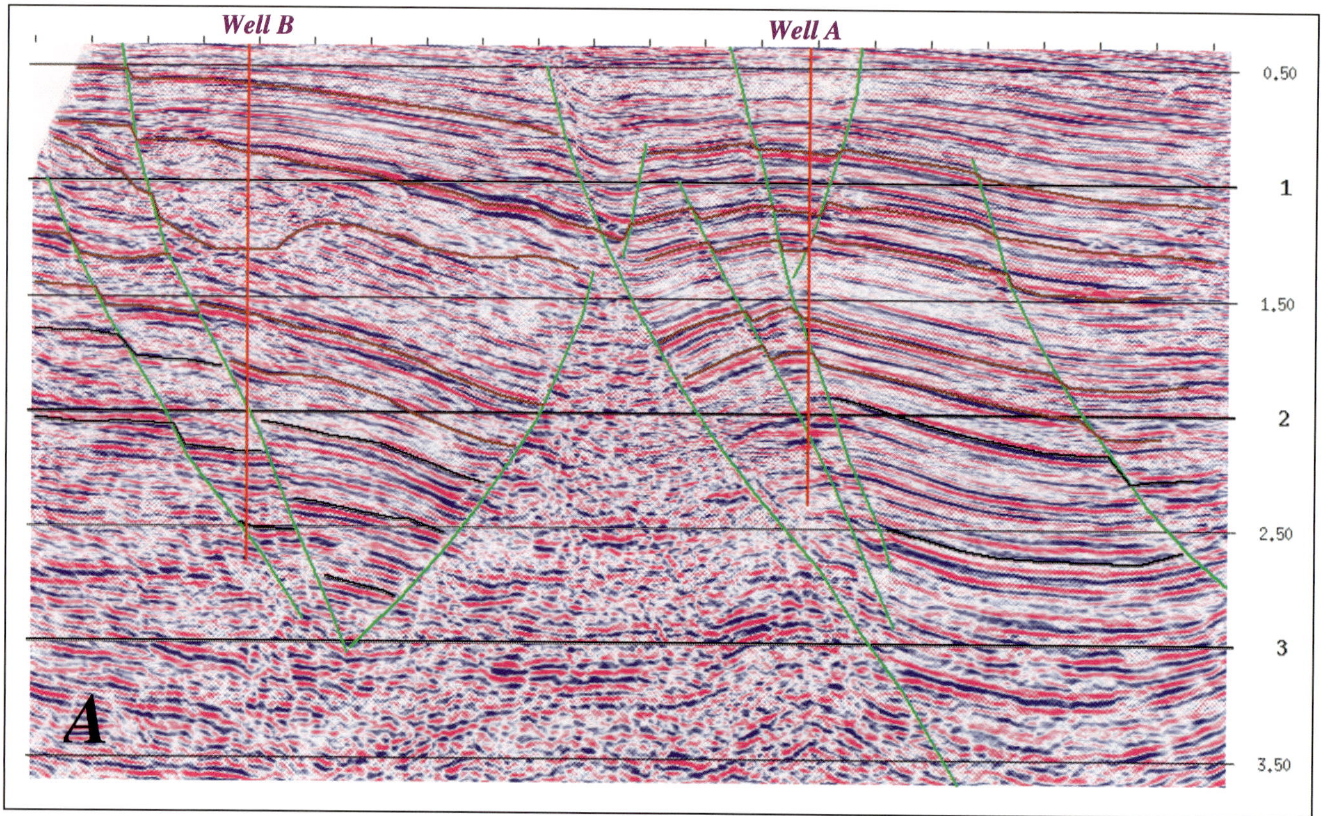

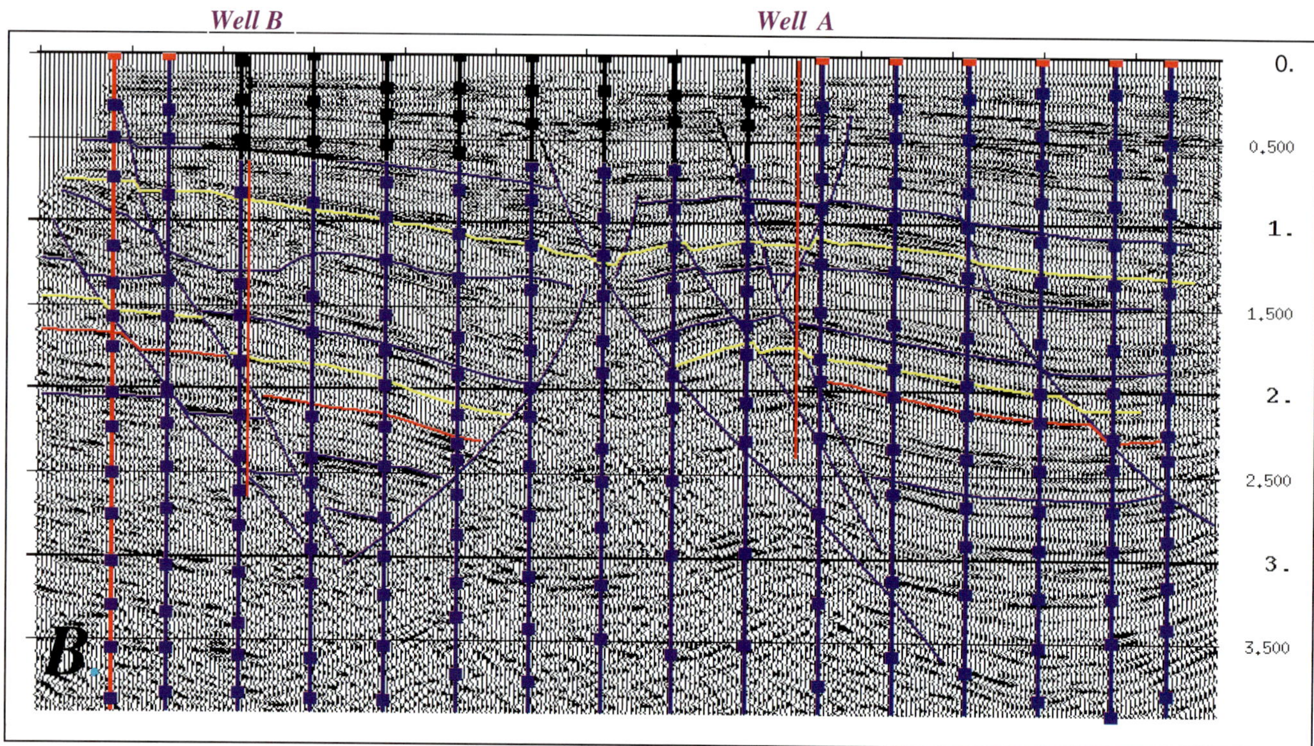

Figure 21. Seismic data from study area. (A) Interpreted section with well locations in red. (B) Section with location of velocity picks shown as blue squares. Sonic log for well B is first curve in Figure 3. Overpressure encountered after cutting second fault. Vertical axes are vertical two-way traveltime in seconds.

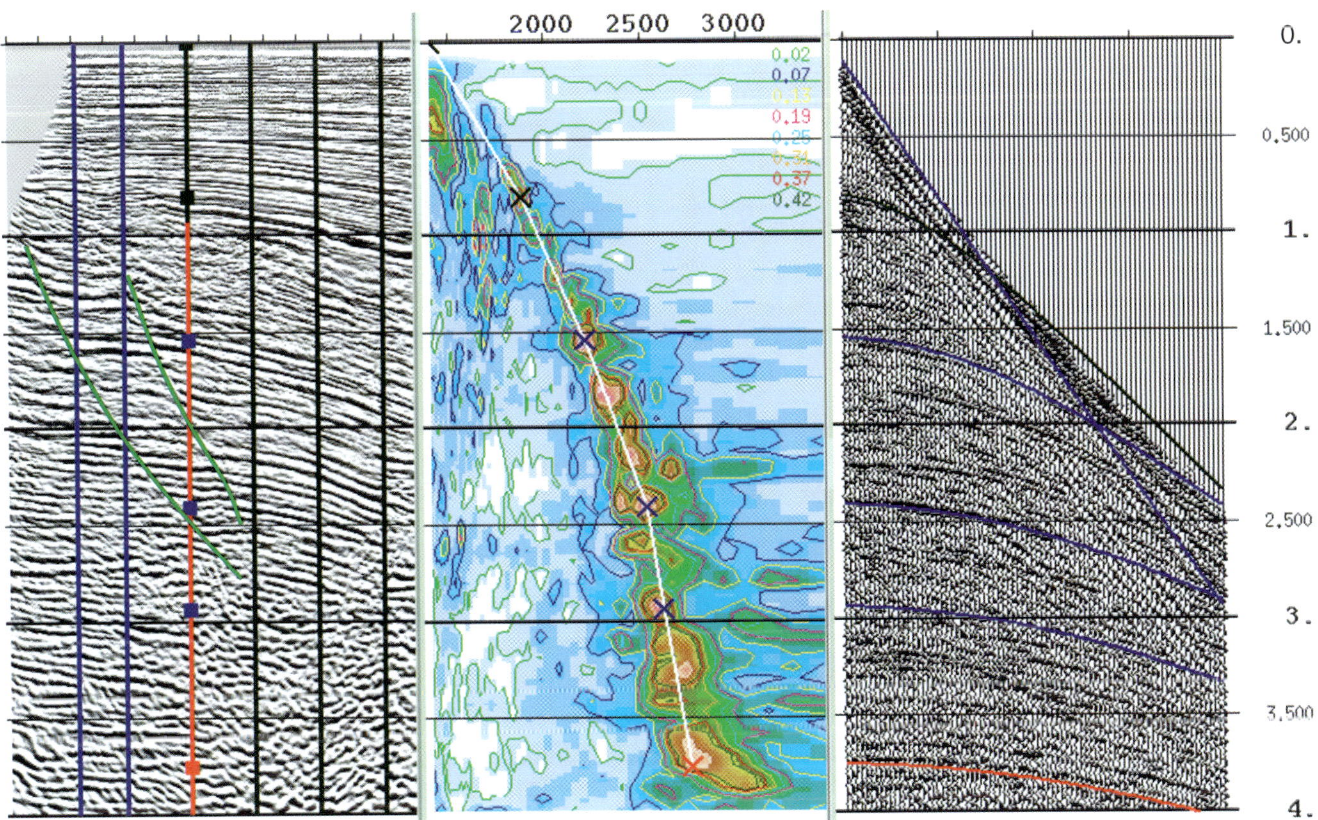

Figure 22. Seismic cross section, semblance panel, and CMP at overpressured well B location. Sparse picks indicate trend. Note change in slope of stacking-velocity curve at event just above 2.5 s corresponding to location of fault. Common vertical axes are two-way traveltime in seconds.

Figure 24 compares conventional Dix velocity profiles (Figure 24B) with gradient profiles (Figure 24C) for the six velocity functions that sample the structure around well B in Figure 21. Figure 24A identifies the horizons that were picked in the initial analysis along with zones where the velocity gradient might be expected to change, for example, fault blocks and major unconformities.

Figure 25 plots velocity data from the six locations in Figure 24 on common axes. Figure 25A and B show both the final stacking-velocity picks and the calculated Dix interval velocities color coded by location. Note that the software package used for the velocity analysis allowed interactive adjustment of both the stacking velocity and the interval-velocity values relative to preliminary picks at the other locations. In other words, the consistency of the interval velocities in the shallow section already reflects an interpretive bias based on tracking chosen horizons and multiple passes through the data.

Each stacking-velocity function in Figure 25A can be closely approximated by a linear function in time from the surface down to the individual slope breaks below 2 s. That would have the effect of replacing the stair steps in Figure 25B with a straight line at each location. Such action amounts to vertical smoothing of the stacking-velocity picks.

Figure 25C and D display the same data color coded by geologic unit. The interval-velocity steps from Figure 25B have been plotted as single midpoint values in Figure 25D. Both plots suggest that a single, laterally invariant function is appropriate for the first three layers, whereas a separate function is appropriate for layers 4 and 5. The derived interval velocities show considerable scatter in the low-velocity overpressure regions (layers 6 and 7). Part of that scatter is attributed to the poor seismic data quality in those areas. A linear fit to the stacking-velocity data in those zones produces a better approximation of the true velocity field. Such smoothing is appropriate provided the adjusted picks still flatten the observed moveouts. The gradient curves in Figure 24C were obtained by fitting straight lines to the stacking-velocity picks within each zone prior to Dix conversion to interval velocity.

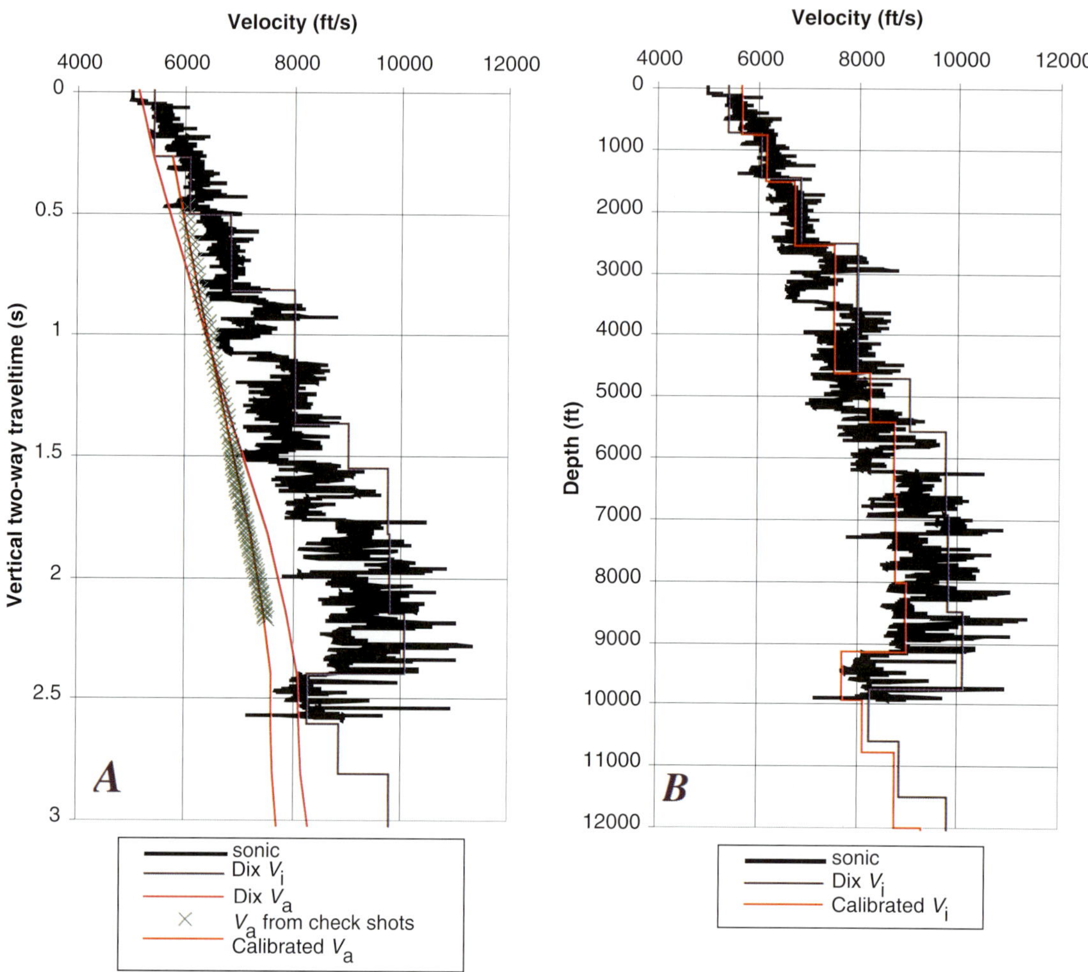

Figure 23. Comparison of seismic velocities with log data from well B. Time-depth information from well A used for calibration.

Figure 26 compares the calibrated Dix velocity profile from Figure 23 with the calibrated gradient profile from Figure 24 at the well B location. The gradient profile is obviously an acceptable alternative to the Dix display.

Figure 27 compares various ways to display seismic velocity information in cross section. Figure 27A shows the stacking-velocity field interpolated laterally. Such displays offer a few clues to the interval-velocity behavior. Constant thickness stacking velocity contours imply a uniformly increasing velocity with depth. In deep-water areas, those bands most likely follow the shape of the water bottom. An increase in the spacing of the color contours indicates a change in gradient of the stacking velocity.

If the stacking velocities are smoothed, either vertically and/or horizontally, as in the previous discussion, a display of the gradient of the stacking velocity commonly gives a rough indication of slow velocity zones that correlate with overpressure (Figure 27B).

Figure 27C and D are color analogs of the individual function displays shown in Figure 24C and B. The velocity field in Figure 27C is represented by vertical gradients that are allowed to vary according to geologic structure. Figure 27D imposes a laterally varying, constant vertical velocity within each layer. Both representations are correct in the sense that they produce the same vertical time-depth curve at the indicated boundaries. The gradient display, however, is probably a better representation of the true picture. Recall that the lateral consistency of average velocities from the check shots (Figure 11) implies that velocity contours commonly cut horizon boundaries and faults. Reilly (1993) discussed additional reasons for preferring the gradient view to the Dix conceptualization.

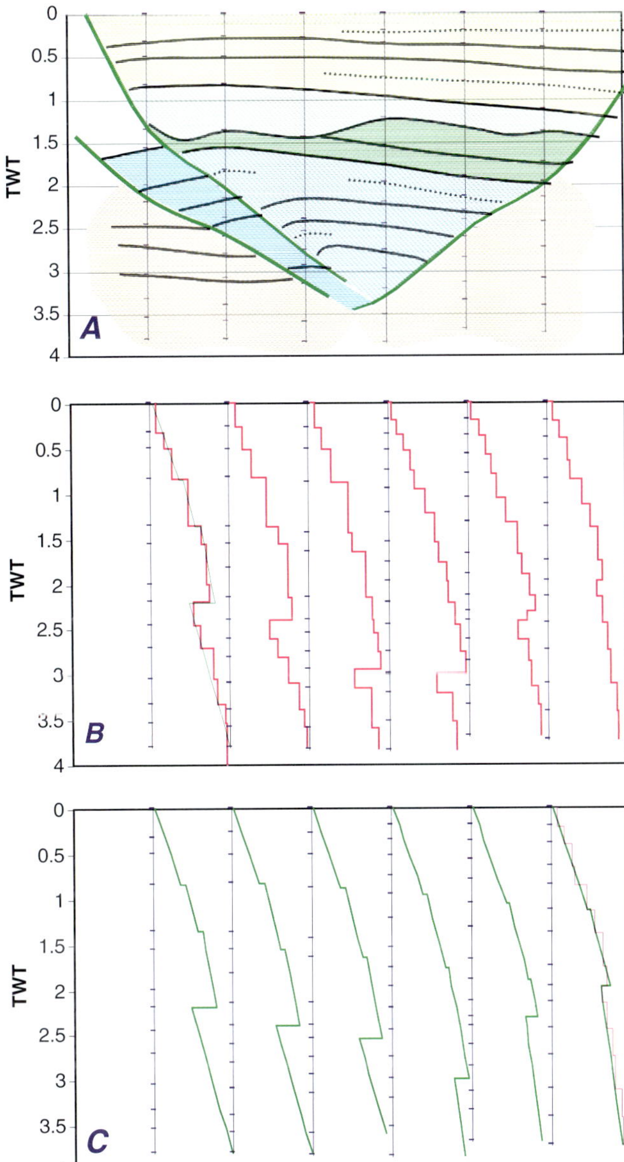

Figure 24. Comparison of velocity functions referenced to horizons. (A) Geologic units. (B) Dix interval velocities tied to horizon picks. (C) Gradient functions tied to unit boundaries. TWT = vertical two-way traveltime in seconds.

HORIZON-BASED SMOOTHING OF SEISMIC STACKING VELOCITIES

The concept of horizon-based smoothing of seismic stacking velocities is important enough to justify additional discussion. If used judiciously, it can improve confidence in seismically derived formation velocities. If misused, it can obscure or destroy information contained in moveout variations.

Figure 28 shows various ways to interpret five stacking-velocity profiles close to a proposed well location. Each point is within 0.5 km of the well, and given a 6 km cable, should sample a similar velocity field. The stacking velocities were generated by a reliable processing contractor. Analysis of the stacking velocities for time-depth conversion and regional pressure prediction was begun at the same time as the geologic interpretation. Four major horizon boundaries subsequently supplied by the interpreter are indicated on several of the plots.

A pronounced change in slope of the stacking velocities shown in Figure 28A suggests a possible pressure transition beneath horizon 2. Picks below the horizon show considerable scatter due to a poor signal-to-noise ratio and difficult imaging.

The processing geophysicist was asked to pick all reliable primary semblance events. Considerable scatter in the interval velocity estimates is expected when the Dix equation is used to convert closely spaced stacking velocities. Although each stacking-velocity pick represents a valid moveout velocity, the interval-velocity calculation is very sensitive to small time errors where the time interval itself is very small.

Figure 28B shows the raw Dix interval velocities. Several of the velocity values are physically unrealistic, and the large amount of scatter suggests that none of the deeper values should be used for pressure prediction without further analysis.

One way to reduce the sensitivity of the Dix equation to small intervals is to consistently make picks on horizons that are well separated in time. If that approach is taken, both well and seismic data should be reviewed beforehand to insure that the chosen interfaces are adequate to describe important variations in the velocity field. Figure 28C shows a limited choice of data points based on horizons that correspond to slope breaks in the original picks. The corresponding Dix interval velocities (Figure 28D) are now more consistent. Averaging the midpoint values of interval velocity from the decimated stacking velocities could result in an acceptable pressure prediction.

Note that the apparent improvement in the interval-velocity functions in Figure 28D did not arise from changing the individual stacking-velocity picks. The stacking-velocity picks that were rejected are probably just as valid as the ones retained. The final velocity interpretation should be better constrained if all the picks are considered.

One way to examine the additional data would be to choose other sets of similarly separated stacking velocities and then to interleave the resulting Dix interval velocities. Also, stacking-velocity picks for additional closely spaced horizons can be interactively tweaked to produce reasonable interval-velocity profiles. That was the procedure used to obtain the interval-velocity

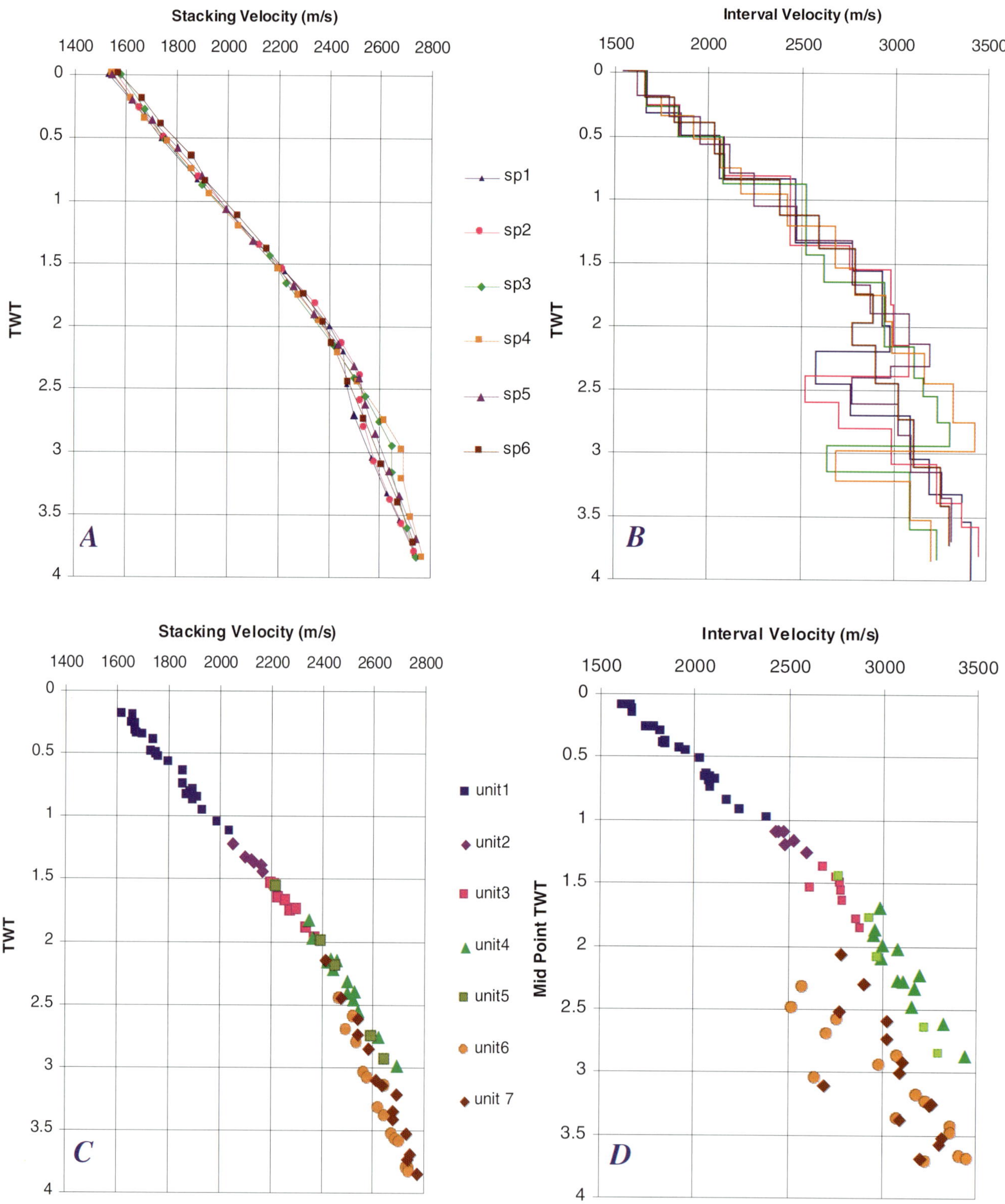

Figure 25. Comparison of (A) stacking velocity and (B) Dix interval velocity color coded by surface location with (C) stacking velocity and (D) midpoint interval velocity color coded by geologic unit. Curve fitting within units with a laterally constant velocity–depth relationship reduces uncertainty in individual Dix derived interval velocities. TWT = vertical two-way traveltime in seconds.

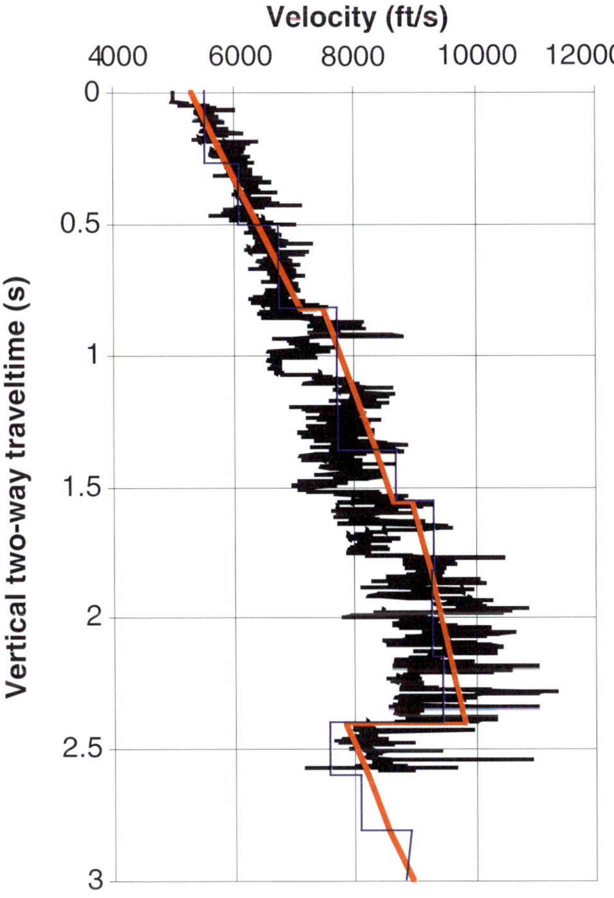

Figure 26. Comparison of Dix interval-velocity profile with gradient model.

curves shown in the previous example (Figure 25B). This technique, however, can be time consuming and subject to interpreter bias.

Also, it is possible to employ simple statistical techniques to extract robust interval-velocity estimates from closely spaced stacking-velocity picks. Simple curves may be fit over geologically consistent units, as was done to derive the gradient curves in Figure 24C.

Figure 28E demonstrates yet another technique for smoothing the stacking velocities. The individual stacking-velocity functions in Figure 28A were linearly interpolated in time to a uniform sampling interval. A median filter was then applied to the collection of five data points at each time sample. That operation produced a single stacking-velocity function smoothed over location. A running median filter was then applied along the time axis to smooth the data vertically. The final result is a smoothed version of the original data (Figure 28E). Median filters are better suited to such operations than averages. Median filters reject large spurious events that could skew an average. Median filters also preserve slope changes and distinct layers that are longer than half of the filter window.

Figure 28F shows the result of applying the Dix equation to the smoothed stacking velocities in Figure 28E. The result adds appropriate detail to the stair step function in Figure 28D and is easily reproduced by anyone starting with the same stacking-velocity picks. Anytime stacking-velocity picks are modified, however, it is appropriate, if possible, to use the smoothed stacking-velocity function to apply normal moveout to the original gather. If the seismic events are not reasonably flat, then further analysis is needed.

Before lateral smoothing is applied to stacking-velocity functions in regions of variable water depth, it is important to remove the influence of the water layer (recall Figure 10). Figure 29A shows three widely separated stacking-velocity functions from a three-dimensional survey. The water depth ranges from 480 to 919 m. Lateral smoothing of these functions is inappropriate. Figure 29B shows the same data after the effect of the water column is removed. Now, it is now apparent that the three functions follow the same initial trend but deviate from the trend at different times. Such information suggests that a single velocity-time curve can be used to aid velocity interpretation throughout the area.

Measured stacking velocities can be datumed to the mud line if the water velocity, V_w, and two-way traveltime to the water bottom, T_{wb}, are known:

$$V^2_{\text{stack new}} = (V^2_{\text{stack old}} T_{\text{old}} - V^2_w T_{wb})/(T_{\text{old}} - T_{wb}) \qquad (12)$$

$$T_{\text{new}} = T_{\text{old}} - T_{wb}$$

The previous equations cause stacking velocities measured from the surface to coalesce if the only difference is a constant-velocity medium above the compaction datum. They do not have the same effect if the velocity intercept changes due to uplift and/or erosion or if multiple geologic units have the same depth reference but different velocity gradients. Both time and depth equations can be derived to extract compaction curves where such effects are present. Changes in intercept between locations with the same velocity gradient can be used to infer the amount of erosion and/or uplift. (Magara, 1976; Poix, 1998) Such interpretations are beyond the scope of this chapter other than to suggest that it is best to work in the depth domain for such applications. Although mathematically equivalent expressions for velocity vs. time can be written, compaction trends are physically connected to depth rather than traveltime. Thinking in time for simple situations is advantageous because the initial seismic velocities are in that space. In complicated cases,

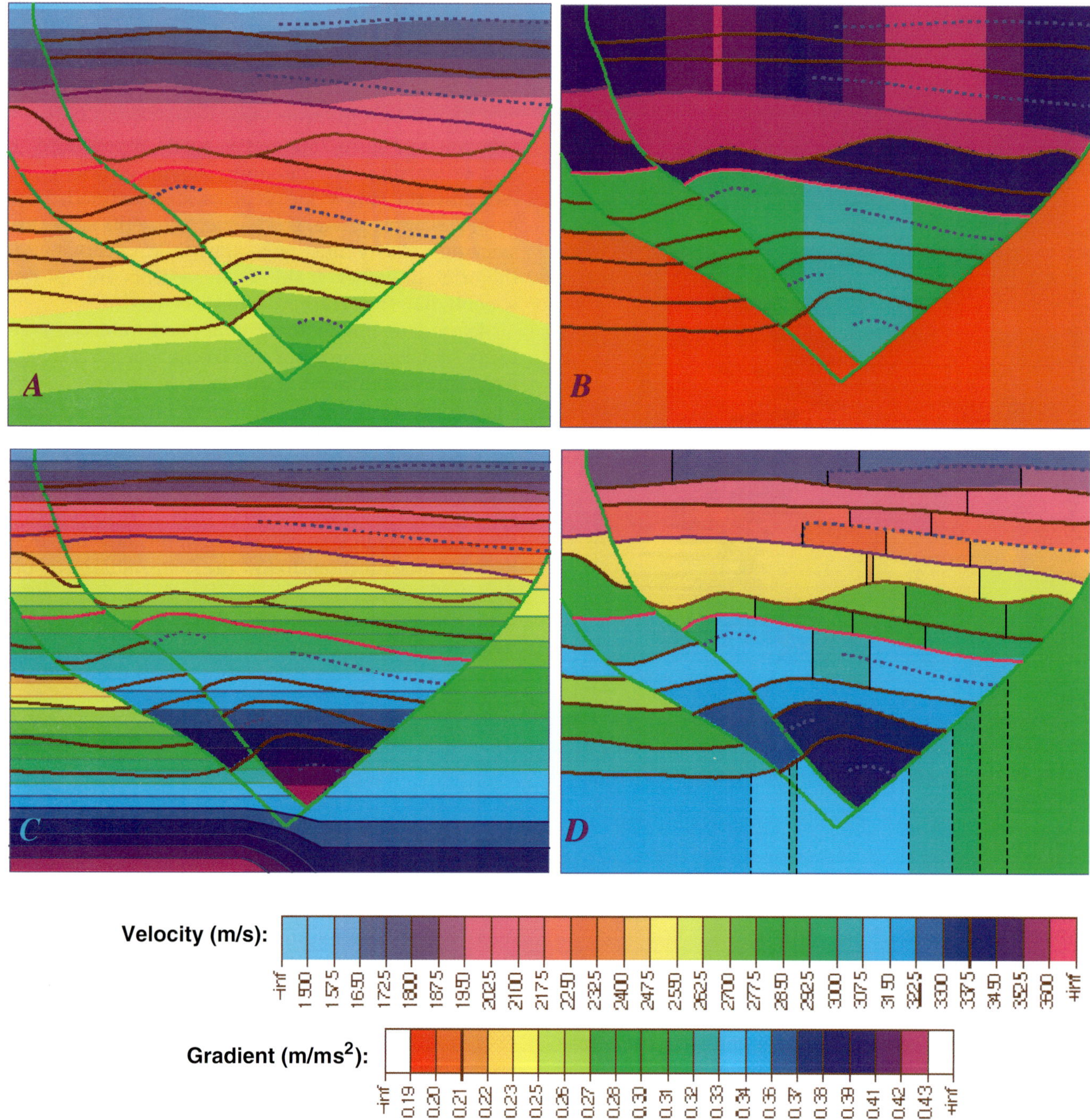

Figure 27. Comparison of velocity displays. (A) Stacking velocity. (B) Gradient of stacking velocity. (C) Layer velocities represented by gradients between horizons. (D) Layer velocities constant between horizons. Vertical traveltimes at boundaries are the same in C and D. Vertical axes are vertical two-way traveltime.

however, the time behavior is not intuitive and can be misleading.

The data analysis presented in this chapter was done using spreadsheets and generic statistical software. Commercial software packages are also available that are designed to apply such concepts. They combine statistical analysis of stacking velocities, sonic logs, check shots, and horizon picks to produce a velocity field for time-to-depth conversion. They have the ability to characterize the expected error in each data set and to determine if changes in analytic coefficients fit to the data represent reliable variations in velocity

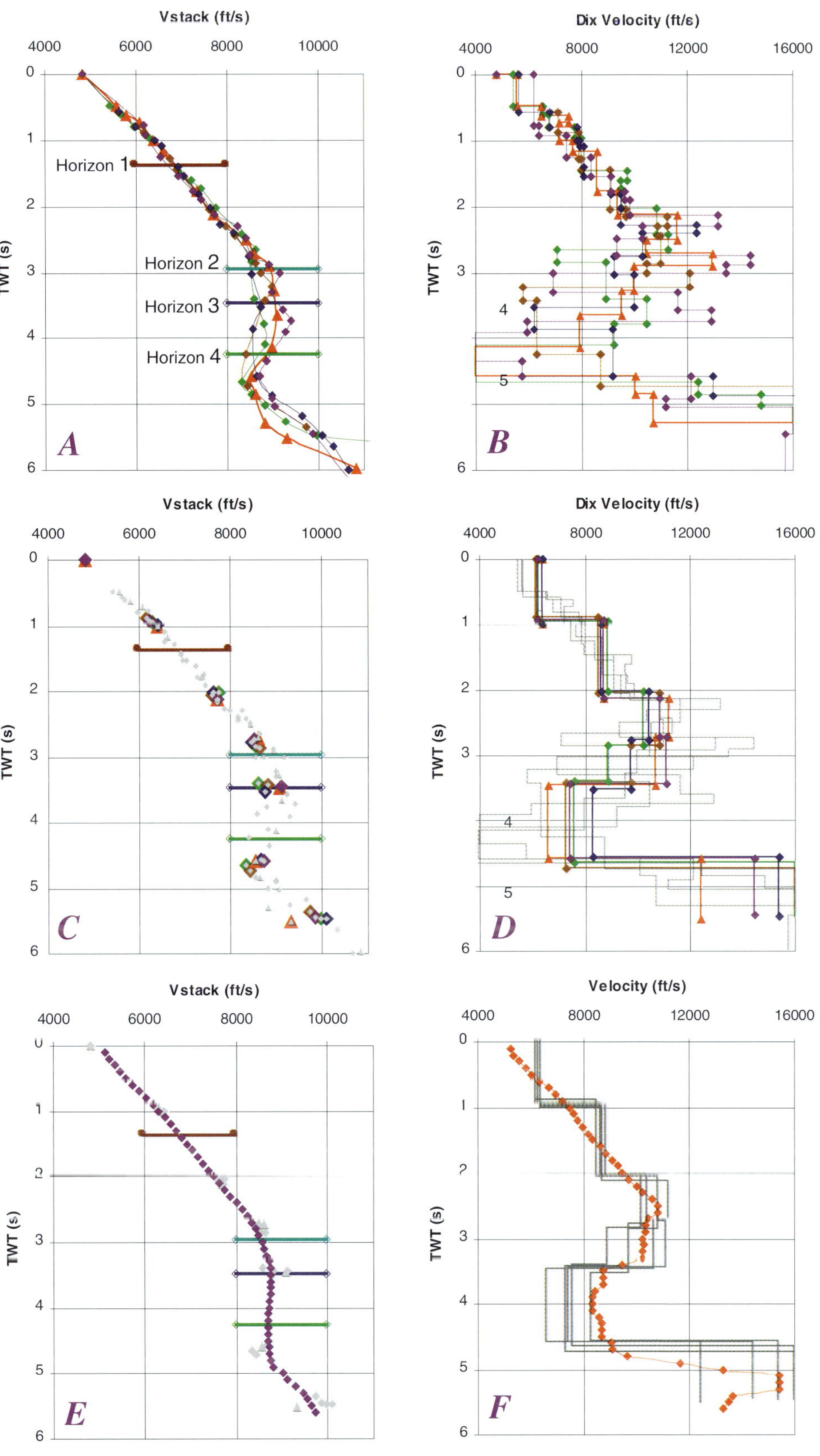

Figure 28. Effect of smoothing. (A) Raw stacking-velocity picks in red at a proposed well location along with the four nearest neighbor functions from a 0.5 km grid. Note that interpreted horizons do not correspond to slope breaks. (B) Resulting scatter in Dix interval velocities. (C) Picks decimated to major slope breaks. (D) Dix interval velocities from decimated picks. These are more consistent than those in part B (curves without symbols). (E) Result of linear interpolation followed by median filters in both time and position. The filtering effectively fits a smooth curve to the raw stacking-velocity picks. (F) The resulting interval-velocity profile compared with the interval velocities computed in step D (stair-step curves). TWT = vertical two-way traveltime.

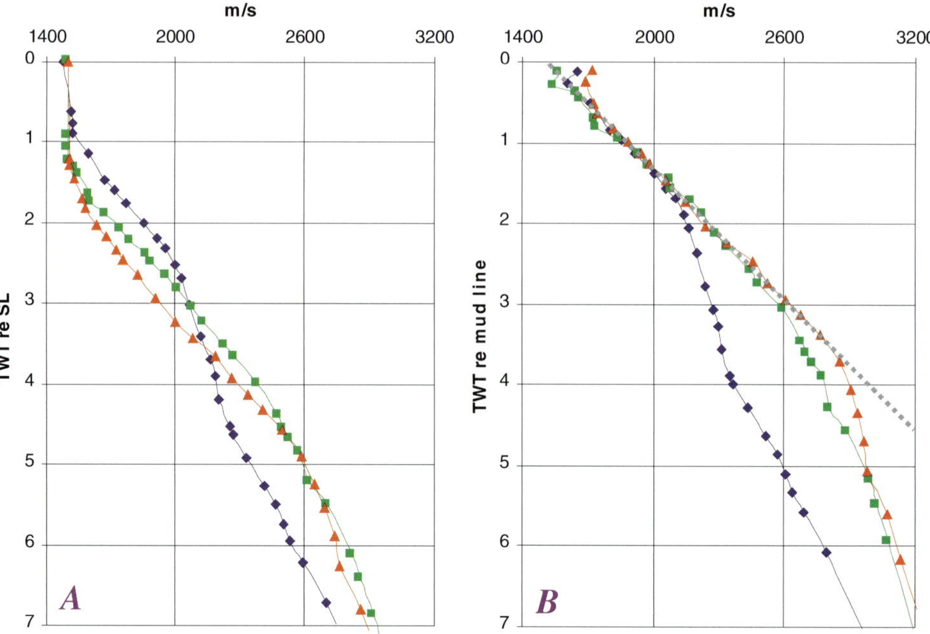

Figure 29. Removing effect of water depth on stacking velocity reveals common trend and aids analysis, particularly curve fitting in regions of similar gradient. (A) Three stacking-velocity functions with water depths of 480 m (blue), 714 m (green), and 919 m (red). Surface locations are separated by 10 km. (B) Same data after correction for the water layer. TWT = vertical two-way traveltime. SL = sea level.

or simply scatter in the measurements (See Al-Chalabi, 1997). Such modules should be used to refine velocity estimations for pressure work beyond the traditional reliance on a simple Dix inversion.

POSTSTACK INVERSION

The previous discussion concentrated on deriving formation velocities from differences in event moveout on offset seismic gathers. Velocity information can also be extracted from reflection amplitudes. Figure 30 reviews the basics of poststack inversion of seismic traces. A stacked seismic trace can be modeled as the convolution of a wavelet with a reflection coefficient series (Sheriff, 1991). The zero-offset reflection coefficient at each point in the series is given by the difference in seismic impedance divided by the sum of the impedances. Impedance is the product of velocity and density.

Figure 30A–F shows the forward modeling of a seismic trace. The goal of inversion is to reproduce the original impedance log. Various techniques are available to estimate the impedance, but they all have to contend with the band-limited nature of the seismic trace. The fine detail in the impedance log is not resolved at typical seismic frequencies and cannot be recovered unambiguously from the seismic data. Low-frequency data is similarly absent from the seismic trace but can be recovered from the stacking velocities or added from well control.

Figure 30G shows the result of inverting the modeled seismic trace in Figure 30F. Where the missing low-frequency information (Figure 30H) is added and adjustments are made between velocity and impedance, it is possible to reproduce much of the fine structure in the sonic log (Figure 30I). The inverted trace is obviously a more detailed representation of the sonic log than the previous results shown in Figure 26. Poststack, or newer prestack, inversions are recommended for detailed lithology interpretation based on seismic velocity and are also suited for detailed pressure predictions. The critical information for pressure work, however, is contained in the missing low-frequency signal. All of the preceding concerns are still pertinent and must be considered in the derivation and interpretation of an inverted trace.

SUGGESTED PROCEDURE FOR VELOCITY ANALYSIS FOR PRESSURE PREDICTION

By way of conclusion, the following 11 steps condense the previous material into a guideline for velocity analysis using seismic data. Application to pore-pressure prediction by tying existing well control to a proposed well location is assumed, but the recommendations are also appropriate with slight modifications to regional pressure studies without nearby well control.

1. Determine What Software Is Available for Velocity Analysis and Collect the Appropriate Data

The software available for velocity picking, display, and analysis influences the way data is manipulated

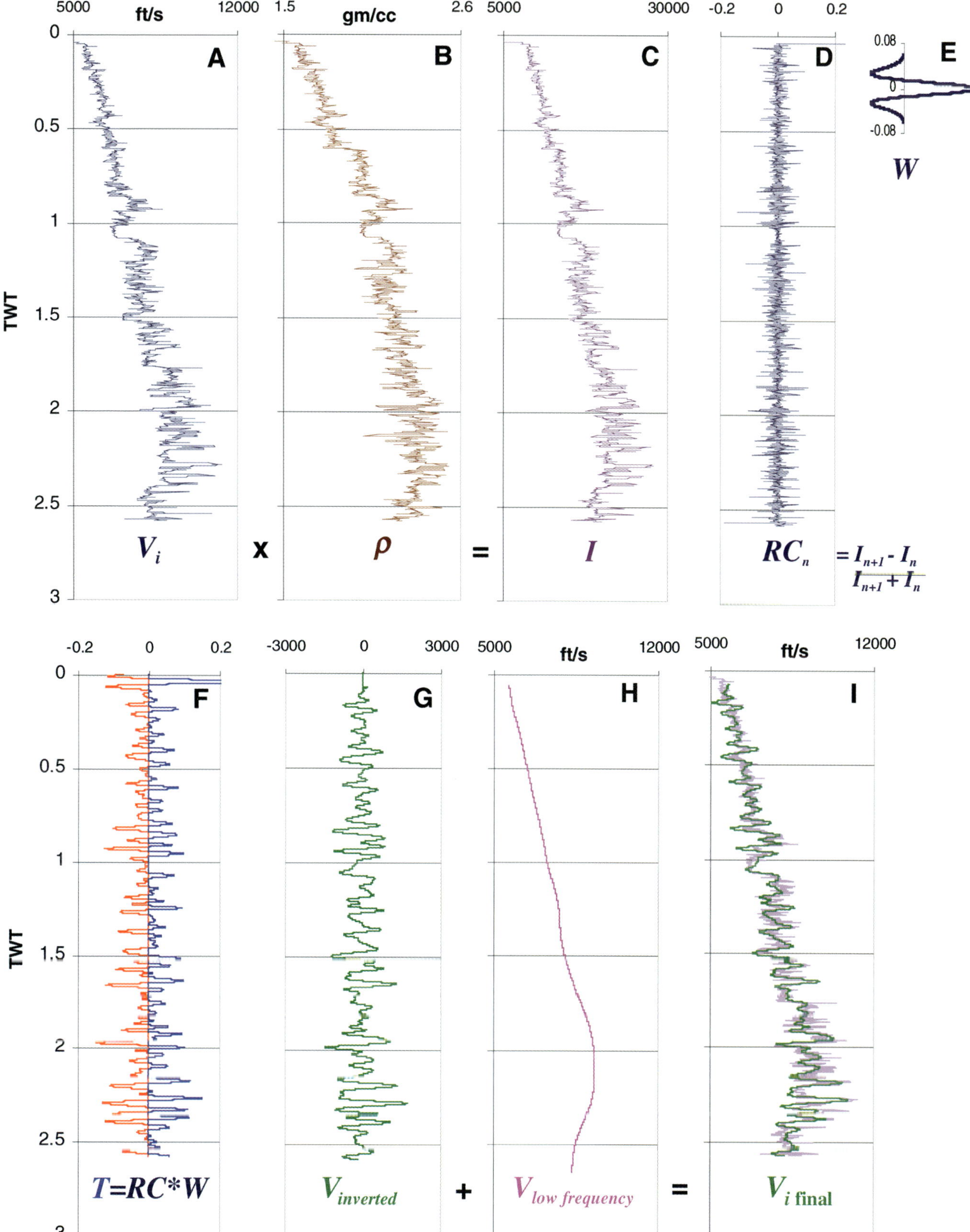

Figure 30. Review of poststack inversion. (A) Velocity multiplied by (B) density gives (C) impedance. (D) The reflection coefficient series is given by the difference in impedance at an interface divided by the sum of the impedances. Poststack inversion assumes that a (F) stacked seismic trace is the convolution of the reflection coefficient series with (E) a known wavelet. (G) Inversion of the seismic trace gives a band-limited estimate of the velocity after removing the density term. (I) The final velocity estimate requires addition of (H) a low-frequency trend obtained independently from well logs or stacking velocities. TWT = vertical two-way traveltime.

and the quality of the final result. Seismic-processing systems have basic routines for picking stacking velocity based on normal moveout of CMP gathers. Semblance panels and constant-velocity stacks are common tools. Such tools are appropriate in good data areas with simple velocity fields. The CMP traces are input at select locations. Simple calculations such as Dix conversion to interval velocity and interpolation are commonly included, but detailed analysis of the stacking velocities and calibration with well control normally require transfer of the basic data to a separate software package.

Spreadsheets or generic statistical packages are commonly adequate for further analysis of stacking velocities. Software designed for seismic time-to-depth conversion, however, may offer easier data input and simpler access to robust tools for statistical analysis and comparisons with sonic logs, check shots, and horizon times. Regardless of the software used, effort is required to collect and format the various data types for input. (Ideally, various modules will draw from the same database and share data via common formats, but that is not always reality, particularly where modules from different vendors are chosen for particular features.)

More sophisticated processing packages offer tomographic inversion, ray-trace modeling, and depth migration tools to account for structural complexity. Such programs work with an entire prestack data set and contain advanced tools for estimating interval velocity based on matching nonhyperbolic moveout. Interpreted horizons may be input into such systems or picked as part of the analysis. Calibration of the velocities with well control is still needed. Although such tools are based on advanced theoretical techniques, data quality in difficult areas may still be insufficient to provide velocities without a relatively large degree of uncertainty.

2. Select the Appropriate Seismic Lines and Velocity Analysis Locations

Any nearby wells available for velocity and pressure calibrations should be tied to the proposed drilling location via a series of velocity analysis locations along lines having the same acquisition geometry and processing history. Spacings of 0.5–1.0 km are appropriate. Additional velocity locations should be selected to optimize analysis near the wells, particularly if the traps are small fault blocks less than a cable length in extent. If possible, line ties should avoid poor record areas, such as salt diapirs, that prevent direct horizon ties between the wells. For two-dimensional (2-D) data, dip lines are preferred to strike lines (2-D DMO and migration cannot correct for out-of-plane dip effects). Velocity analysis at 2-D line crossings near the well locations helps quantify the potential error in the final velocity result. Ideally, gathers from separate lines that have the same midpoint location should sample the same velocity field. Differences between velocities at such locations can be diagnostic of problems at other locations that might otherwise escape notice.

The detailed velocity analysis needed for pore-pressure prediction is also ideally suited for time-to-depth conversion. Additional analysis points may be useful for that purpose.

3. Determine Whether Prestack Seismic Data Are Available

If prestack seismic data are available, stacking velocities should be repicked for the express purpose of pressure prediction and time-to-depth conversion. If prestack data are not available, then previous contractor stacking velocities should be critically examined to see if they contain the desired level of information.

The contractor stacking velocities should have been picked after DMO. Avoid picks that have been interpolated to a regular time interval because it is then difficult to determine which values correspond to actual moveout information. Processors sometimes make phantom picks deep in the section, particularly if the signal-to-noise ratio is poor. Such picks will adversely effect the analysis if they are used to interpolate intervening values. Also avoid picks that have been smoothed laterally, as is commonly done for poststack migration.

4. If the Stacking Velocities Are to Be Repicked, Sort the Data into Supergathers after Appropriate Preprocessing

Data at the chosen velocity analysis sites should be binned to include all available offsets. Velocity analysis should follow DMO and mild passes at removing multiple reflections. Strong multiples that obscure primary events should be suppressed. If the multiple removal is too robust, however, slope breaks indicative of velocity reversals may be eliminated.

Use of data after prestack time migration is also appropriate, or superior, provided moveout has been restored to reconstituted gathers. If prestack time, or depth migration, or tomographic techniques are to be used for the velocity analysis, then DMO should not be applied.

5. Examine Sonic and Check Shot Information

Sonic and check shot information should be examined to identify major velocity units and the expected imprint of pressure on velocity. Sonic logs should be check shot corrected and displayed in two-way traveltime for comparison with events picked on seismic time sections. Look for assumed compaction trends, offsets in the compaction trends, thick units with relatively constant velocity, velocity reversals associated with lithology, and velocity reversals associated with overpressure. Look for consistent velocity units at different well locations and units that change velocity with location. Display well data as average velocity vs. time to mimic the behavior expected of stacking-velocity picks. In particular, notice how changes in interval velocity correspond to changes in slope of the average-velocity curve.

6. Review Geologic Interpretations of Existing Seismic Data

Extrapolate the observations made at the wells to the proposed location based on the latest interpretation. The seismic interpreter, the seismic processor, the velocity analyst, and the pressure analyst (or at least as many of those who are separate people) should discuss expectations prior to the detailed velocity work. Typically, the interpreter concentrates only on a limited number of horizons of geologic interest. The velocity analyst needs to be aware of geologic changes in the units that most impact the stacking-velocity field.

7. Make a Preliminary Pass at Picking the Stacking Velocities

Getting an overall feel for variations in both stacking velocities and data quality, particularly if well control is limited or absent, is useful. Commonly, the area of exploration interest corresponds to a region of poor data quality. After the initial pass, it may be useful to rethink the preprocessing and selection of analysis locations prior to detailed work. Additional CMPs can be added to the bin gather to improve the signal-to-noise ratio if needed. If event moveout is not hyperbolic, depth migration should be considered.

8. Decide on Criteria for Picking Events

In good data areas, it is possible to minimize subsequent analysis if stacking-velocity picks are limited to a set of predetermined horizons that include the major velocity interfaces identified from the well logs. Tracking such events is easier if an interpreted seismic section is displayed concurrently with the semblance panel and CMP gather. Note that horizon times may not correspond to a maximum in the semblance. That is, it may be necessary to visually interpolate a pick to the selected horizon. Interference between reflection events in the semblance window may shift the maximum. More importantly, the event in the CMP may not necessarily be from the same subsurface point as the horizon pick on the migrated data. A first pass through the data should concentrate on flattening events in the gather. Subsequent passes should compare interval-velocity plots to either side of a given analysis location so that minor adjustments to the stacking velocity can be made to insure consistency in the final interval-velocity field. The last step is an interpretation rather than a deterministic exercise and may require several iterations around a loop tied to well control, particularly if the time separation between events is small. Phantom picks may be needed to maintain horizon consistency near faults and in poor data areas.

An alternative approach limits stacking-velocity picks to those events with unambiguous moveout but makes as many such picks as is reasonable, without concern for horizon consistency or the time difference between the points. The intent is to accumulate as much valid information as possible and to avoid making phantom picks that could later skew statistical analyses. Scarcity of picks in a given zone implies increased uncertainty in the final velocity interpretation. This approach initially avoids the problem of events that come and go or that are hard to correlate across faults. It is quicker than horizon-keyed picking and more reproducible. The procedure is appropriate if the velocity picking is to be done in a processing shop by someone other than the person doing the final velocity interpretation or if a geologic interpretation of the data has not been completed. Horizon consistency is imposed after the picks are made rather than as part of the picking exercise.

9. Analyze Velocities for Geologically Consistent Trends

Data redundancy and statistical analysis can be used in several ways to improve the accuracy of velocity predictions. Data should be analyzed for vertical trends and horizontal consistency within geologically similar units, for example, between major unconformities. The effort needed depends on the geologic setting and the quality of the seismic data. Working directly with the stacking velocities is generally better than beginning with the derived interval velocities.

Plot numerous stacking-velocity curves on the same velocity-time plot. Determine if a single compaction curve is appropriate for the shallow section. In deep water, first datum the stacking velocities to the sea floor. (Similar information can be obtained by plotting the Dix interval velocity vs. the interval midpoint time relative to the vertical two-way traveltime from the sea floor. The scatter in the Dix velocities, however, may not be representative of the quality of the moveout information if the events are close together in time.)

If there is a common trend in the shallow section, look for changes in the slope of the stacking-velocity trends that represent velocity changes with depth. If possible, correlate the slope change with a geologic horizon or fault. Fit stacking-velocity curves to points above the slope change. Check that curve coefficients at any one location are consistent with the statistics of fitting data from several locations at once. The stacking-velocity curve may then be converted to interval velocity either analytically or by applying the Dix formula at regularly sampled intervals along the curve.

If there is an areal variation in the stacking velocity in the shallow section, look for a plausible geologic explanation. A subtle slope change in the stacking velocity may have been overlooked. Variations can also arise from uplift and erosion (in which case the normal compaction trend has a different datum at each location) or from gradual changes in overpressure with depth (in which case there is no well-defined normal trend). Fitting a curve to several points within a unit at a given location, but not including points from adjacent locations, may be appropriate. Check that available well control is consistent with the assumed cause, but be alert to changes that are not represented by limited well sampling.

Apply similar analyses unit by unit for deeper events. Ideally, the standard deviations for a fitted curve should be within the uncertainty of the picks. Interpolated or smoothed stacking velocities should be used to apply normal moveout to the gathers at critical locations. If the interpreted velocity does not flatten distinct hyperbolic events (at least at near offsets), then the results are suspect. If distinct events are absent, that is, the data are poor, then a corresponding uncertainty should be placed on subsequent pressure predictions.

10. Calibrate Well with Well Control

At this stage, vertical two-way traveltimes of boundaries between major velocity units should match reasonably well with the check shot–corrected sonic logs displayed in time, but the velocities are probably too high to produce an accurate time-depth conversion. The seismic interval velocities should be converted to average velocity and compared with average velocities from check shots. A time-dependent calibration should be determined to correct the former to the latter. Ideally, a similar calibration should apply at all available wells. If not, geologic insight can be applied to determine a lateral variation in the calibration. The calibrated average velocities should then be converted back to interval velocity.

11. Apply Time-Depth Conversion

The final step is to convert the calibrated interval velocities from functions of time to functions of depth. Any remaining differences in interval velocity between the seismic functions and the well logs should be factored into the uncertainty in the pressure prediction at a proposed location.

APPENDIX

Definitions of some common velocity analysis terms (adapted from Sheriff, 1991; Bates and Jackson, 1980):

common midpoint (CMP)—Having the same midpoint location between the source and the geophone.

gather—A side-by-side display of seismic traces that have an acquisition coordinate in common; for example, a common midpoint.

geophone—An instrument that measures the passage of seismic energy. The output record is called a trace.

moveout—A change in the traveltimes recorded by different source/geophone locations for the same seismic event. Normal moveout is the two-way traveltime change that occurs as the distance between the source and the geophone is varied assuming a flat reflector model. Dip moveout is the two-way traveltime change that occurs assuming a reflection from a dipping plane.

multiple reflection—A seismic pulse that has been reflected more than once. Commonly referred to as a multiple.

offset—The distance between a source and a geophone.

primary reflection—A seismic pulse that has only been reflected once. Commonly referred to as a primary.

source—A device that releases a pulse of seismic energy.

stack—n. A composite record made by adding together several traces. v. Adding together several traces.

trace—A single seismic record, normally amplitude vs. two-way traveltime, that represents the subsurface seismic response for a real or hypothetical source/geophone geometry either before or after various processes are applied to improve the image.

traveltime—The time it takes a seismic pulse to travel from one reference location to another.

two-way traveltime—The time it takes a seismic pulse to travel from its source to a reflector and back to a geophone.

vertical two-way traveltime—The two-way traveltime calculated for a hypothetical seismic pulse that travels straight down and up.

wavelet—A seismic pulse.

ACKNOWLEDGMENTS

Alan Huffman provided encouragement to produce this chapter and useful suggestions during the editing. Bob Lankston edited the final draft. Glenn Bowers helped clarify concepts relating pressure and velocity. Rob Meek provided the example in Figure 20. The other real-data examples come from productive interactions with numerous Conoco colleagues who have provided me with challenging opportunities and guidance in learning a trade that continues to evolve. Thanks to all.

REFERENCES CITED

Al-Chalabi, M., 1997, Parameter nonuniqueness in velocity versus depth functions: Geophysics, v. 62, p. 970–979.

Alkhalifah, T., and I. Tsvankin, 1995, Velocity analysis for tranversely isotropic media: Geophysics, v. 60, p. 1550–1566.

Bates, R. L., and J. Jackson, eds., 1987, Glossary of geology, 3d edition: American Geological Institute, 788 p.

Bowers, G. L., 1995, Pore pressure estimation from velocity data: accounting for overpressure mechanisms besides undercompaction: Society of Petroleum Engineers Drilling and Completions, v. 4, no. 10, p. 89–95.

Deregowski, S. M., 1986, What is DMO?: First Break, v. 4, no. 7, p. 7–24.

Eaton, B. A., 1975, The equation for geopressure prediction from well logs: 50th Annual Fall Meeting of the Society of Petroleum Engineers, SPE paper 5544, unpaginated.

Heasler, H. P., and N. A. Kharitonova, 1996, Analysis of sonic well logs applied to erosion estimates in the Bighorn basin, Wyoming: AAPG Bulletin, v. 80, p. 630–646.

Hottmann, C. E., and R. K. Johnson, 1965, Estimation of formation pressures from log-derived shale properties: Journal of Petroleum Technology, v. 17, p. 717–722.

Kaufman, H., 1953, Velocity functions in seismic prospecting: Geophysics, v. 18, p. 289–297.

Korvin, G., 1984, Shale compaction and statistical physics: Geophys. J. R. astr. Soc. v. 78, p. 35–50.

Kumar, N., 1979, Thickness of removed sedimentary rocks, paleopore pressure, and paleotemperature, southwestern part of Western Canada basin: discussion: AAPG Bulletin, v. 63, p. 812–814.

Liner, C. L., 1999, Concepts of normal and dip moveout: Geophysics, v. 64, p. 1637–1647.

Lines, L. R., F. Rahimian, and K. R. Kelly, 1993, A model-based comparison of modern velocity analysis methods: The Leading Edge, v. 12, no. 7, p. 750–754.

Magara, K., 1976, Thickness of removed sedimentary rocks, paleopore pressure, and paleotemperature, southwestern part of Western Canada basin: AAPG Bulletin, v. 60, p. 554–565.

Marsden, D., M. D. Bush, and D. S. Johng, 1995, Analytic velocity functions: The Leading Edge, v. 14, no. 7, p. 775–782.

Neidell, N. S., and N. Berry, 1989, Documenting the sand/shale crossover: Geophysics, v. 54, p. 1430–1434.

Poix, O., 1998, Sonic anomalies, a means to quantifying overpressures, *in* Mitchell and Grauls, eds., Overpressures in petroleum exploration: Bulletin Centre Recherche Elf Exploration and Production, Memoir 22, p. 207–211.

Reilly, J. M., 1993, Integration of well and seismic data for 3D velocity model building: First Break, v. 11, no. 6, p. 247–260.

Sheriff, R. E., 1991, Encyclopedic dictionary of exploration geophysics, 3d edition: Society of Exploration Geophysics, 384 p.

Whitmore, N. D., and J. D. Garing, 1993, Interval velocity estimation using iterative prestack depth migration in the constant angle domain: The Leading Edge, v. 12, no. 7, p. 757–762.

19

The Future of Pressure Prediction Using Geophysical Methods

Alan R. Huffman
Conoco Inc., Houston, Texas

ABSTRACT

The technology of pore-pressure prediction has advanced significantly in recent years. In the future, new methods for pore-pressure prediction will routinely use shear-wave data gathered using multicomponent seismic technology. Overburden and fracture gradient will be predicted in three dimensions using gravity and magnetic inversion technology. Seismic inversion, both prestack and poststack, will provide refined estimates of the velocity field in the subsurface, and new seismic-processing methods will allow velocity anisotropy to be predicted accurately so that it can be used to predict both pore pressure and real triaxial stress fields in the earth. These new methods will be used to make advances in the prediction of pressures in nonclastic rocks and to extract information that can be used to accurately predict structural hyperpressuring in reservoirs to assist in drilling difficult wells. Pressure prediction will become a standard tool in basin-scale and prospect-scale evaluation of the hydrocarbon system and will be used to guide the exploration process. In the production environment, pore-pressure prediction will be used routinely to provide a three-dimensional model for the pressure regime in the subsurface that will be critical to effective reservoir simulation and reservoir management.

Despite all these advances, however, pore-pressure prediction will still be limited by the quality of seismic data acquisition and processing technology that is used to prepare the data and by the structural complexity of the subsurface that is to be imaged. Predictions will continue to be limited by the lack of predrill information about the state of compaction in the subsurface that is critical to a robust pressure prediction. Lastly, prediction accuracy will continue to be limited by the presence of secondary pressure in situations where velocity reversals are difficult to detect on seismic data.

INTRODUCTION

Predrill pressure prediction using geophysical data and methods has historically been done using very simple models and overly simplistic estimates of the Earth's velocity field. The methods commonly incorporate a locally calibrated set of curves for pressure that contained imbedded assumptions about the cause of pressure in the geologic section sampled by the control wells. The advent of the effective-stress concept and the pressure-prediction and the pressure-prediction methods that developed from that concept led to a much needed inclusion of fundamental physics into the art of pressure prediction. The use of effective-stress methods has become the standard for pressure prediction with many variants including the Eaton method (Eaton, 1975), the Bowers method (Bowers, 1994), and the Sperry Sun method (Holbrook and Hauck, 1987), to name a few. The range of software available for pressure prediction has grown significantly in recent years, along with the sophistication of the parameters used. Still, weaknesses remain because of (1) the limitations of the seismic velocities themselves, (2) the lack of understanding of the basic causes of pressure, and (3) the effects of pressure

Huffman, Alan R., 2002, The Future of Pressure Prediction Using Geophysical Methods, in A. R. Huffman and G. L. Bowers, eds., Pressure regimes in sedimentary basins and their prediction: AAPG Memoir 76, p. 217–233.

on physical properties, including velocity, density, and porosity, of the rocks. Despite the level of sophistication that is used in pressure prediction today, most practitioners are concerned primarily with predrill prediction and overlook the vast significance of pressure variations at the basin and prospect scale. This chapter discusses some of these issues and tries to put them into context.

EFFECTIVE-STRESS AND LOADING-PATH DEPENDENCY OF PRESSURE

The functional relationship between pore pressure and velocity has been generally recognized for many years. The level of understanding about the relationships between various physical properties and pore pressure, however, varies widely. Many people still are unaware that there is more than one cause of pressure and that present-day pressure regimes are the result of the complete loading path that a rock has undergone since its deposition.

The causes of abnormal pressure can be divided into two types. Undercompaction, also known as compaction disequilibrium, is related to the compaction process itself and occurs where the rates of deposition and burial are sufficiently great relative to the vertical permeability of the sediments. Where large loading rates are applied to rocks such as shales with relatively low-vertical permeability, the confined fluids in the rock mass cannot escape abruptly enough to maintain a hydrostatic fluid pressure gradient. This type of abnormal pressure is observed in many young Tertiary basins worldwide and is commonly recognized in seismic-velocity data by the slow decrease in the velocity gradient with depth.

The other class of abnormal pressure mechanisms is not associated with the compaction process, and occurs where the pressure of the fluid in the rock mass is allowed to increase relative to hydrostatic pressure through one of several mechanisms (Plumley, 1980). These mechanisms include (1) aquathermal fluid expansion (Magara, 1975), (2) hydrocarbon source maturation and fluid expulsion (Spencer, 1987), (3) clay diagenesis (Bruce, 1984), (4) fluid pumping from deeper pressured intervals during fluid migration, and (5) decreases in overburden caused by tectonic activity. Although each one of these mechanisms are distinctly different in their behavior, they all produce a similar effect in pressured rocks in that they work to cause a decrease in the effective stress on the formation for a given porosity. In particular, clay diagenesis, aquathermal expansion, and source maturation occur at elevated temperatures so that cold sediments should not be affected by these mechanisms. In contrast, fluid pumping and overburden decreases can occur in any sediments. Where secondary pressure occurs, it is commonly manifested through a reversal in the velocity trend with depth without an increase in porosity. Although not all velocity reversals are caused by secondary pressure, it is important to remember that reversals caused by secondary pressure contain some of the most severe pressure increases. Thus, it is prudent to treat velocity reversals as if they are caused by secondary pressure unless there is clear evidence from well data that the velocity reversals are due to undercompaction, lithology changes, or other possible causes. Also, velocity reversals, if not recognized as being due to secondary pressure, overestimate porosity if it is assumed that the rocks are simply undercompacted.

To understand the relationships between effective stress, porosity, and velocity, consider the concept of critical porosity that was first defined by Marion et al. (1992). Figure 1 shows the relationship between porosity and velocity for clastic materials from laboratory experiments. The boundary between Wood's equation behavior and the loading-bearing behavior occurs at porosities of 38 to 50% and is defined as the critical porosity. The trend of velocity with porosity shows two dominant trends that follow (1) Wood's equation (Wood, 1941) at porosities above the critical porosity and (2) a modified Voigt-Reuss behavior at porosities below the critical porosity (Nar et al., 1991). The Wood's equation behavior is characteristic of slurries, and the modified Voigt-Reuss behavior is characteristic of frame-bearing solids, including clastic rocks under significant effective-stress conditions. Figure 1 shows how velocity increases as porosity decreases. This trend correlates with the degree of compaction that the material has undergone and is part of the reason that some pressure methods use porosity as a proxy in determining the compaction state of a material.

We must also consider the relationship between velocity and effective stress that defines the normal compaction trend. Tosaya (1982) performed experiments on clastic rocks to demonstrate the critical factor of the effective stress. Figure 2 shows that the velocity of the material follows the effective stress nearly perfectly regardless of the total overburden stress that is applied. This experimental result is an excellent demonstration that Terzhagi's effective-stress relationship is valid and can be correlated with velocity changes in clastic materials.

One way to think about the two causes of abnormal pressure is to recognize that the velocity of any rock in the subsurface is a direct function of its depositional and burial history. Figure 3 shows a hypothetical loading path for a rock in a clastic basin in porosity–velocity–effective-stress space. This diagram is a three-

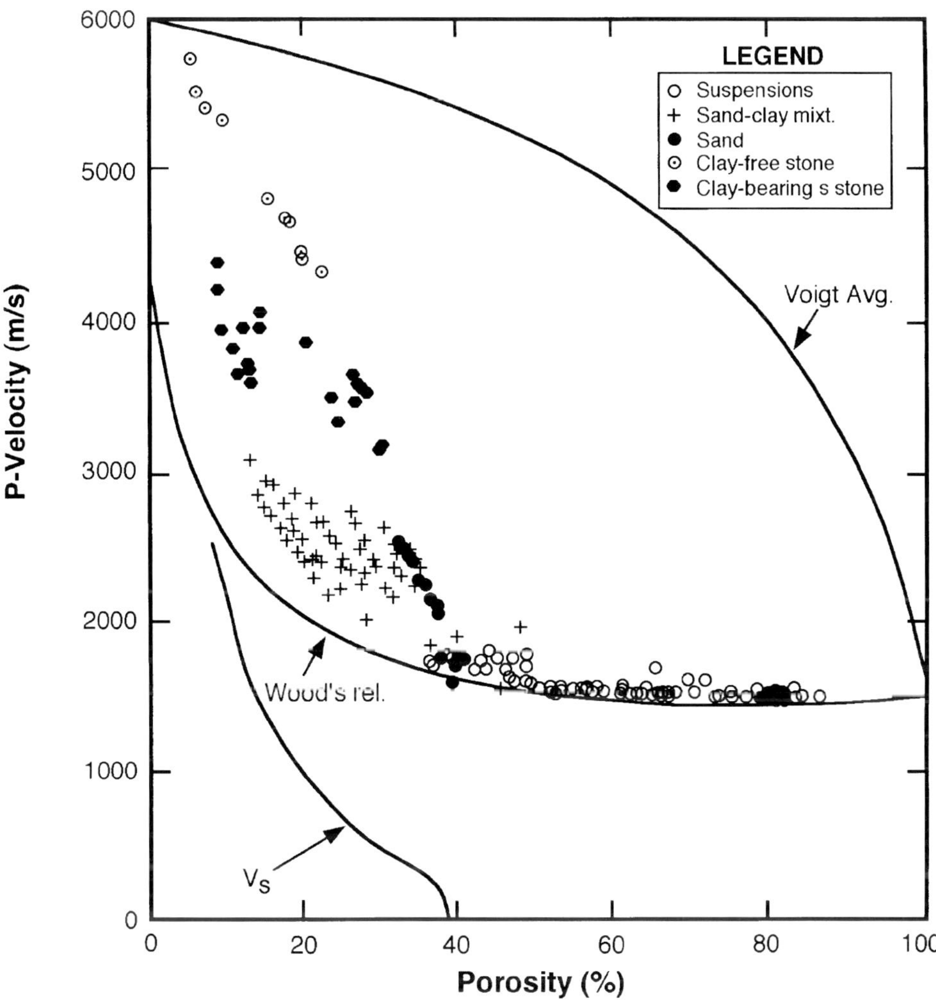

Figure 1. Velocity-porosity relationships in clastic sedimentary materials. Note the change in behavior caused by the inception of load-bearing capability at the critical porosity point (about 40% porosity). At porosities less than critical porosity, the material behaves like a Voigt-Reuss material. Also note the shear-wave behavior and how abruptly it changes from zero to nonzero as you pass the critical porosity. Figure modified from Marion et al. (1992).

dimensional composite of Figures 1 and 2 that demonstrates the interplay between velocity, porosity, and effective stress. The loading path starts at an effective stress of zero, and the velocity increases and porosity decreases until the material changes over from a Wood's equation material to a frame-bearing clastic rock that can support an effective stress on the grains. The Wood's equation part of the loading path (blue curve) occurs as the material is initially deposited and compacted near the surface. Once the critical porosity is reached, the material follows the primary compaction curve (black curve), achieving either a compacted or undercompacted state. If allowed to compact normally with fluid draining out of the pore spaces, a rock continues up the normal loading path, velocity increases, and porosity decreases. Both of these properties are dependent on the effective stress on the grains that are bearing the external load. If at some point the fluid is prevented from escaping, the rate of ascent up the normal pressure curve decreases so that the rock has a lower velocity and effective stress than would be expected at normal pressure conditions at a given depth of burial. This condition is known as undercompaction or compaction disequilibrium. The key to understanding undercompaction is to recognize that a rock under these conditions still remains on the normal compaction trend, only it is not as compacted as you would expect it to be at that depth of burial under normal hydrostatic pressure.

Unlike undercompaction, a rock subjected to secondary pressure cannot stay on the normal compaction curve. Where fluid is pumped into a rock or expands within the pore spaces in the rock, the compaction process is arrested, and the rock begins to display a form of hysteresis behavior in velocity–effective-stress space. Where this occurs, the porosity essentially does not change except for some minor elastic rebound (Moos and Zwart, 1998), and the velocity behavior is strictly controlled by the contact area and the grain-to-grain contact stresses in the rock. Because there is essentially no porosity change, the net effect is to flatten out the velocity–effective-stress trend and produce an unload-

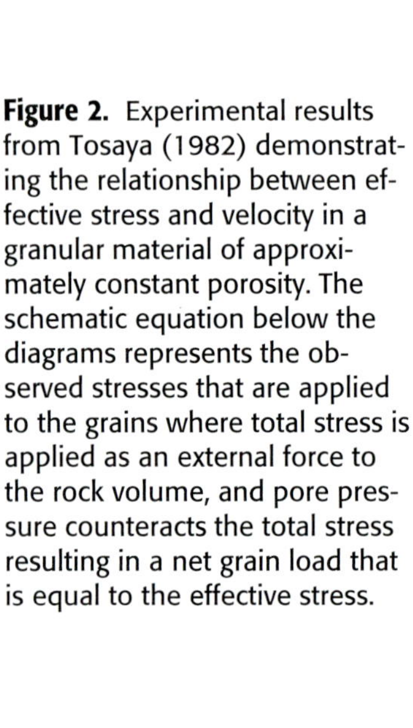

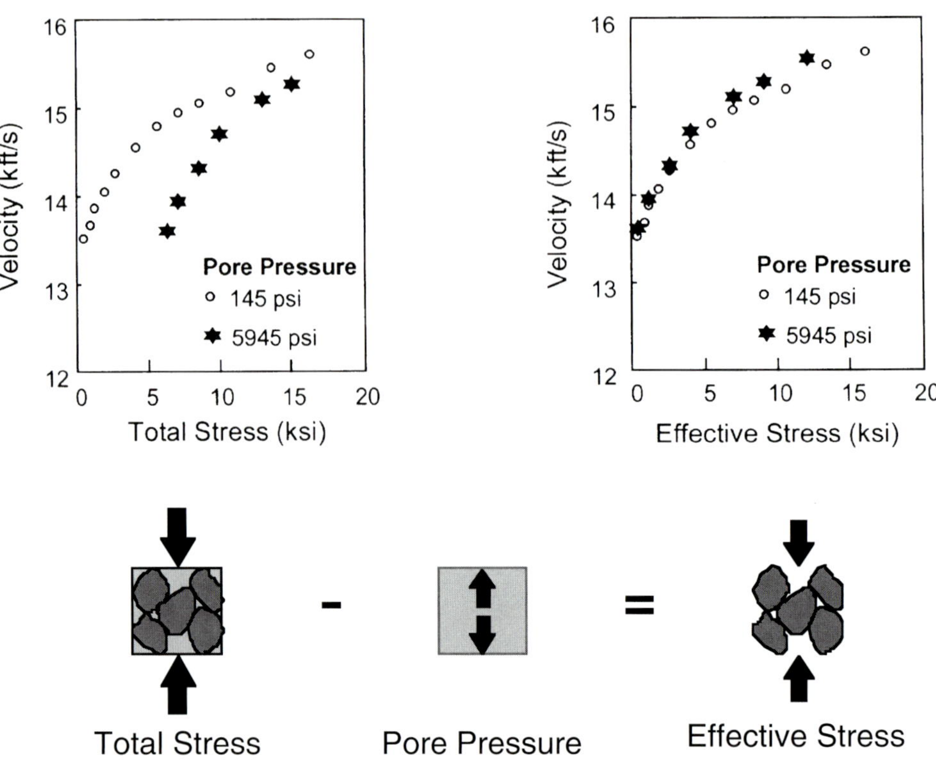

Figure 2. Experimental results from Tosaya (1982) demonstrating the relationship between effective stress and velocity in a granular material of approximately constant porosity. The schematic equation below the diagrams represents the observed stresses that are applied to the grains where total stress is applied as an external force to the rock volume, and pore pressure counteracts the total stress resulting in a net grain load that is equal to the effective stress.

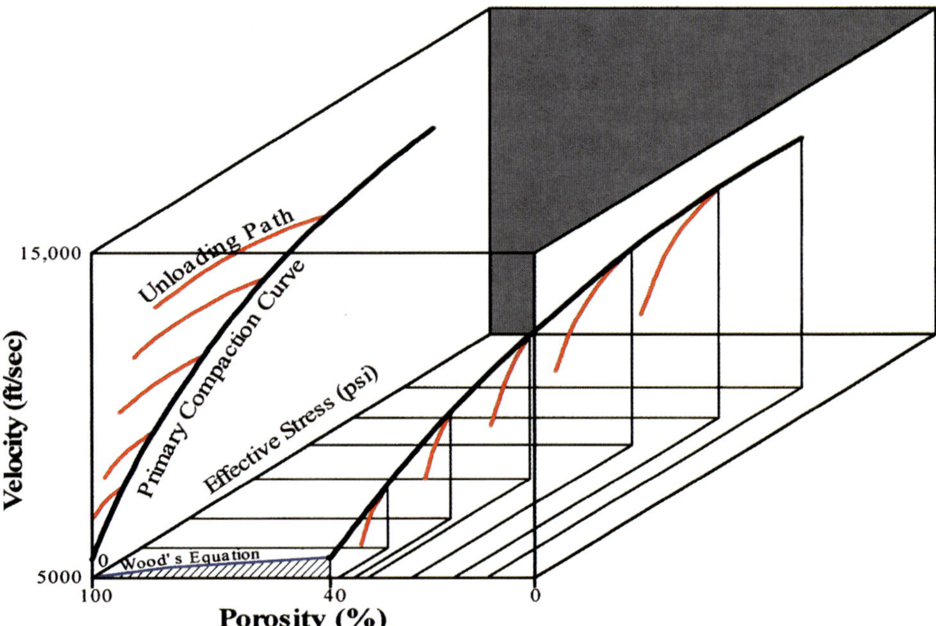

Figure 3. A three-dimensional diagram showing the loading history of a hypothetical shale material in terms of effective stress, velocity, and porosity. The actual three-dimensional normal compaction trend and unloading limbs are projected into the velocity–effective stress plane to the left, which is the same display shown in Figure 4B.

ing trend that is different from the primary compaction trend. The unloading curve must start from the velocity–porosity–effective stress point on the primary compaction curve where the unloading begins (Bowers, 1994). This is why unloading (red curves on Figure 3) always starts from a porosity–velocity–effective-stress point on the primary compaction curve. Note that the unloading paths occur essentially in the velocity–effective-stress plane as the porosity decrease associated with compaction is arrested during unloading and very little elastic rebound (less than 1 porosity unit) occurs during the unloading process. As the effective stress decreases because of higher fluid pressures at fixed overburden, the velocity decreases in direct relation to the stress change. Once a rock is on an unloading path, the rock does not change porosity unless other phenomena such as diagenesis or cementation are occurring concurrently with the pressure changes. For the rock to begin compacting again, the secondary pressures must first bleed off, or the overburden must increase sufficiently by additional sediment loading to counterbalance the secondary fluid pressures that were added within the rock mass. In either case, the rock responds to the change in effective stress and moves back up the unloading path until it contacts the normal compaction curve again. Once the effective stress has exceeded the value where unloading began, the rock can begin to compact again. If the stress never reaches this level, the rock remains on the unloading path indefinitely. Important to recognize in this context is that the normal compaction curve is also the maximum compaction, maximum velocity, and minimum porosity that a material can achieve at normal pressure for a given effective stress.

To properly predict pressure ahead of the bit, it is essential to know not only the normal compaction trend, but also the slope of the secondary pressure curve and the maximum stress-velocity state that was achieved before unloading began. Important to recognize is that the presence of two possible pressure mechanisms and a range of possible maximum velocities for unloading leads to a range of possible predicted pressures according to which mechanism is assumed to be at work and where unloading began. For any velocity (Figure 4), there are a range of possible pressures that are a function of the normal trend, the maximum velocity attained by the rock, and the unloading curve slope. In practical terms, for any observed velocity value, the minimum pressure case is represented by the effective stress at the equivalent depth of burial on the normal trend curve, and the maximum pressure case is represented by the greatest reasonable maximum velocity on the normal trend and the slope of the unloading curve from that point back to the observed velocity value. Thus, it is imperative that the pressure-prediction expert be aware of both causes of pressure and also recognize when and how to apply unloading corrections to the velocity data. Figure 4A is the velocity-depth plot that is used to predict pressure, and Figure 4B is the velocity–effective-stress plot that is used to show the normal compaction trend and unloading behavior. Point D in Figure 4A represents the velocity in the zone being determined in the analysis and correlates with the velocity D in Figure 4B. Point A represents the maximum loading case where the material has been completely unloaded from the normal compaction state. Point B represents the case where the velocity zone just above the unloaded interval represents the maximum compaction state. Point C represents the equivalent-depth point above the unloaded zone (point D) that would have to be used if the material displayed no unloading characteristics in the zone.

LIMITATIONS ON VELOCITY ANALYSIS WITH DIFFERENT METHODS

Velocity analysis has been used for many years to predict pressures prior to drilling. The level of understanding of the various velocity tools and their applicability to pressure prediction, however, is not always adequate to achieve the best results. A standard approach for pressure prediction is to use conventional stacking-velocity analysis and convert the stacking velocities to Dix equation–corrected interval velocities (Bell, 2002). Beyond this simple approach to velocity analysis lies a range of more sophisticated techniques including horizon-keyed velocity analysis, refraction and reflection tomography, and prestack inversion (Bell, 2002). These techniques can increase the accuracy of the velocity analysis but require additional analysis and processing, commonly at higher cost. The question is which technique is appropriate for a given situation. Also important to recognize is that geologic and interpreter input to the velocity analysis is essential to a good pressure prediction.

Preprocessing of Seismic Data for Pressure Prediction

As the complexity of the subsurface increases, the need for increased effort in velocity analysis becomes important. The simplified Dix model becomes progressively more problematic in the presence of steeply dipping and complexly structured geology. In addition, velocity analysis becomes difficult where multiple reflections of the same seismic pulse are severe enough to overwhelm first arrivals (primaries), and where anisotropy is present. Where large lateral and vertical velocity gradients and significant raypath distortion are encountered near salt bodies, volcanics, or other diapiric and nonstratigraphic features, it may become difficult to get any good velocity information.

Where structural complexity increases, it is also necessary to increase the amount of effort that goes into preparing the data for velocity analysis. Prestack time migration, demultiple, and other techniques are commonly required to clean up the seismic data so that it is suitable for analysis. These techniques improve the quality of the seismic data in most cases, but they also drive up the cost of the process significantly. The advanced methods for seismic imaging can improve the quality of the seismic data to facilitate pressure pre-

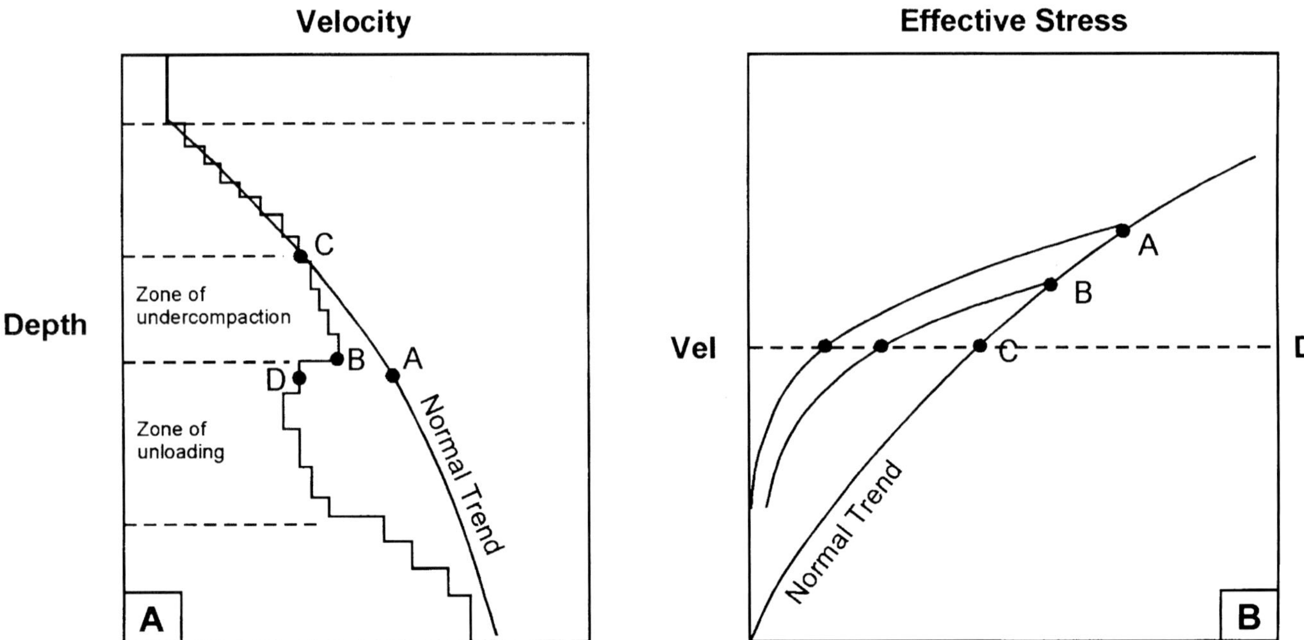

Figure 4. An example of the range of possible maximum compaction stresses that can result in the same velocity being tied to different effective stresses. (A) Velocity vs. depth; (B) effective stress vs. velocity.

diction. If applied carelessly or improperly, however, these methods can remove seismic events that are critical to a robust pressure prediction. Consider the case in Figure 4A and suppose that a velocity reversal occurs in the seismic data. If this seismic line has severe multiple reflection problems, a routine preprocessing stream might include a radon or frequency-wavenumber (f-k) demultiple step that removes or suppresses the multiple. If that multiple has the same or faster velocity as point D in Figure 4A, then the demultiple step removes both the multiple and the primary reflection event that allows the velocity in interval D to be picked properly. If this happens, the velocity analyst is not able to pick this event because it no longer exists in the data. In this case, the analyst would most likely make no velocity pick or would pick a faster velocity that will underpredict the actual pressure in the formation. All seismic data used for pressure prediction should be processed and quality controlled by someone with expertise in seismic processing specifically for pressure prediction. Standard processing streams may be poorly suited for the purpose, and advanced methods must also be used with caution.

The vertical resolution and accuracy of velocity analysis varies with the technique applied. For example, a sonic log has a resolution of around 1 ft (0.3 m), but seismic-velocity analysis at great depths may have a resolution of greater than 500 ft (152.4 m). Each coherent event on a seismic section provides an independent estimate of the average velocity where examined prestack and corrected for various geometric effects. Analysis of all reliable events helps to constrain the final velocity interpretation. Care must be exercised, however, where applying the Dix equation to derive an interval velocity between closely spaced events. Small timing errors inherent in picking each event can cause significant uncertainty in the interval velocity where the total time separation is small. The conventional approach to this problem is to use the Dix equation only for events greater than about 200 ms apart. A more robust approach involves careful smoothing of the stacking-velocity picks before the interval-velocity calculation is performed.

Poststack Inversion

Poststack inversion is one alternative to conventional velocity analysis that provides higher resolution by inverting for impedance from the reflection strength (Bell, 2002). Poststack inversion allows the analyst to separate the seismic wavelet from the reflection series represented by the geologic formations and results in an estimate of residual impedance for each layer. Poststack inversion can be applied using only the stacked seismic data or can be calibrated with well logs, check shot surveys, VSP data, and seismic-velocity data. Where calibrated properly, the analysis can be used to generate an estimate of the absolute impedance or its components of velocity and density. This requires a good set of density-velocity relationships for the lithologies encountered in the wells so that these two components can be separated effectively. This exercise is not trivial, however, because the poststack-inversion technique ignores the fact that offset-dependent behavior (amplitude vs. offset) is buried in the stacked response and can cause significant perturbation of the results. One way to overcome this limitation and also boost resolution of the results at the same time is to use only the near-offset traces for the analysis. This is a good method to use because it provides higher resolution results due to the removal of far-offset data that are degraded by normal moveout (NMO) stretch. Near-stack inversion also gives the most robust calibration to well logs that are essentially measuring the same vertical-incident information. Figure 5 shows an example of the difference in resolution between conventional velocity analysis and poststack inversion.

One of the challenges of poststack inversion is that the method does not normally include a low-frequency trend for velocity, but instead predicts variations in residual impedance that must be separated into velocity and density trends using well log data. Incorporating a low-frequency velocity trend in the analysis is possible, but it is commonly observed that the low-frequency trend, where combined with the residual impedances on the seismic data, does not match the predicted impedances from the well logs. Thus, the calibration of this method still remains problematic in many cases.

PRESTACK INVERSION IN PRESSURE PREDICTION

The techniques employed to get reliable velocity data for pressure prediction will undergo a revolution in the next 10 years. The availability of faster computers at lower cost will allow the use of prestack inversion for pressure prediction. Prestack inversion of seismic data will take us to the next level in accuracy and may allow us to predict pressure at a scale that was not achievable in the past. Prestack inversion can estimate the P wave velocity, shear wave velocity, and density simultaneously by using the near-offset reflectivity and amplitude vs. offset behavior of each reflection in the subsurface. This allows the user to estimate the overburden and effective stress from the same data set. The resolution limits for prestack inversion are approximately at the tuning thickness of the individual formations, so pressure data can be generated for layers on the order of 100–200 ft (30.5–61 m) at moderate depths in clastic basins.

The biggest drawback to prestack inversion at this time other than cost is its extreme sensitivity to data quality. A robust inversion requires data that are relatively clean, uncontaminated by multiples and noise, and in areas of little structural complexity. In addition, the resolution and accuracy of prestack inversion is only as good as the quality of the reflection events in the data. If there are pressure cells that do not have reflections associated with them, no velocity or reflection-amplitude technique including inversion will identify those zones.

The use of calibrated prestack inversion, especially where multicomponent (shear and compressional velocity) data are available, allows the estimation of density and velocity simultaneously along a line. This allows predictions to be made that consider the lateral variations in density that accompany changes in pressure. At present, most methods of pressure prediction assume that the overburden from the control well represents overburden globally, which is false in general. Prestack inversion helps to remedy this problem by predicting the density from the seismic data and allowing the pressure interpreter to make judgements about the density using the seismic and well data and his or her own intuition. Figure 6 shows a result from a prestack inversion of the same seismic line shown in Figure 5. Note that the analysis results in a P-wave section and a shear-wave section that can each be used to predict pressure in the presence of the proper calibration.

Prestack inversion allows isolation of velocities for individual sand packages so that the user can determine where disequilibrium may exist between the sand-bearing formations and massive shales and isolate the velocity and density effect of hydrocarbon-bearing reservoirs on the velocity field around them. At present, most methods lump these effects into thicker stratigraphic intervals that have a single velocity attached to them and hide the effect. These errors can cause predictions to overestimate or underestimate pressures significantly, which leads to less effective well planning. This problem is shown in Figure 7.

In the future, prestack inversion may allow the user to compensate for the effects of anisotropy using the initial velocity analysis and well data as constraints. At present, anisotropy is ignored in nearly all pressure analysis. In the future, prestack inversion will be able to separate the anisotropy effect from the isotropic Zoeppritz reflectivity effects (Sheriff, 1991) at the interface and allow a more robust estimate of the vertical velocity and density field.

In essence, prestack inversion with low-frequency information is the ultimate tool to allow the user to get the most out of the prestack seismic data for pressure prediction and other purposes. The greatest limitations to this powerful technique at present are the cost of application, the data quality limitations mentioned previously, and the continuing problem with

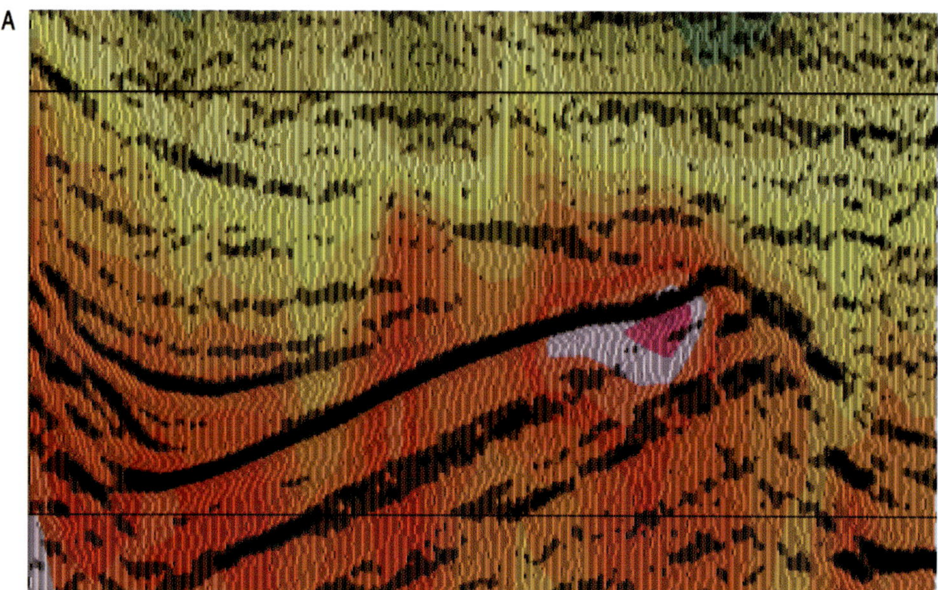

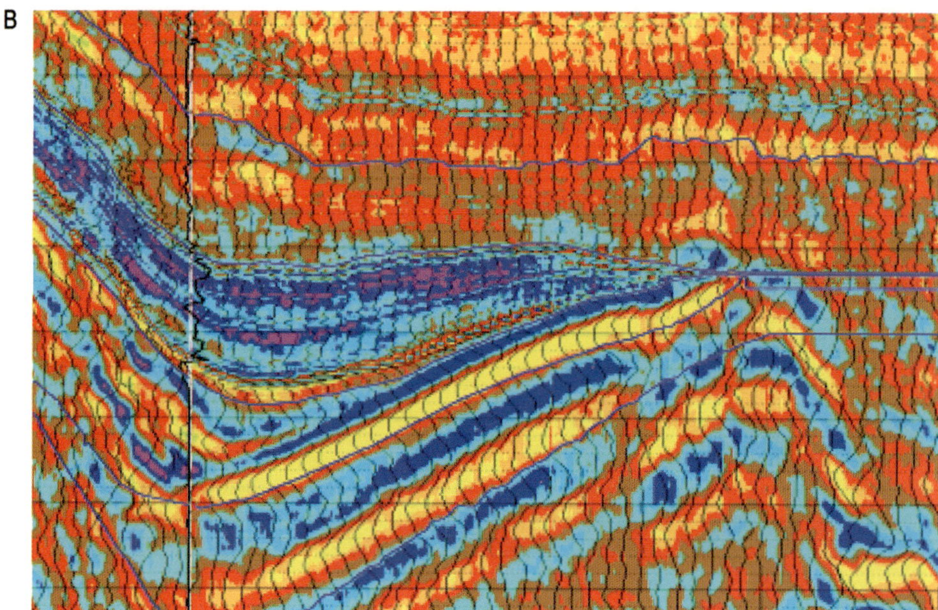

Figure 5. Comparison of conventional horizon-keyed velocity analysis (A) with seismic traces shown and poststack inversion (B). Note the higher resolution of the inversion results.

the low-frequency component. As computers improve in price performance, and we learn to deal more effectively with complex imaging problems, the use of prestack inversion for pressure prediction should increase dramatically.

PRESSURE PREDICTION USING MULTICOMPONENT DATA

Recent developments in multicomponent seismic acquisition and processing suggest that this new technology will be increasing dramatically in use in the future. This growth in multicomponent technology will create an opportunity to use mode-converted and direct shear-wave velocities for pressure prediction. Shear velocities provide another type of velocity data for pressure prediction that may turn out to be particularly valuable for shallow, grossly undercompacted sediments (Huffman and Castagna, 2000) and for zones of severe unloading where effective stress drops back to near zero. The variation of the shear-wave velocity with changes at low effective stresses has been demonstrated to be much greater than P-wave velocity variation, which may provide additional data and greater sensitivity for pressure prediction. Shear-wave data will also

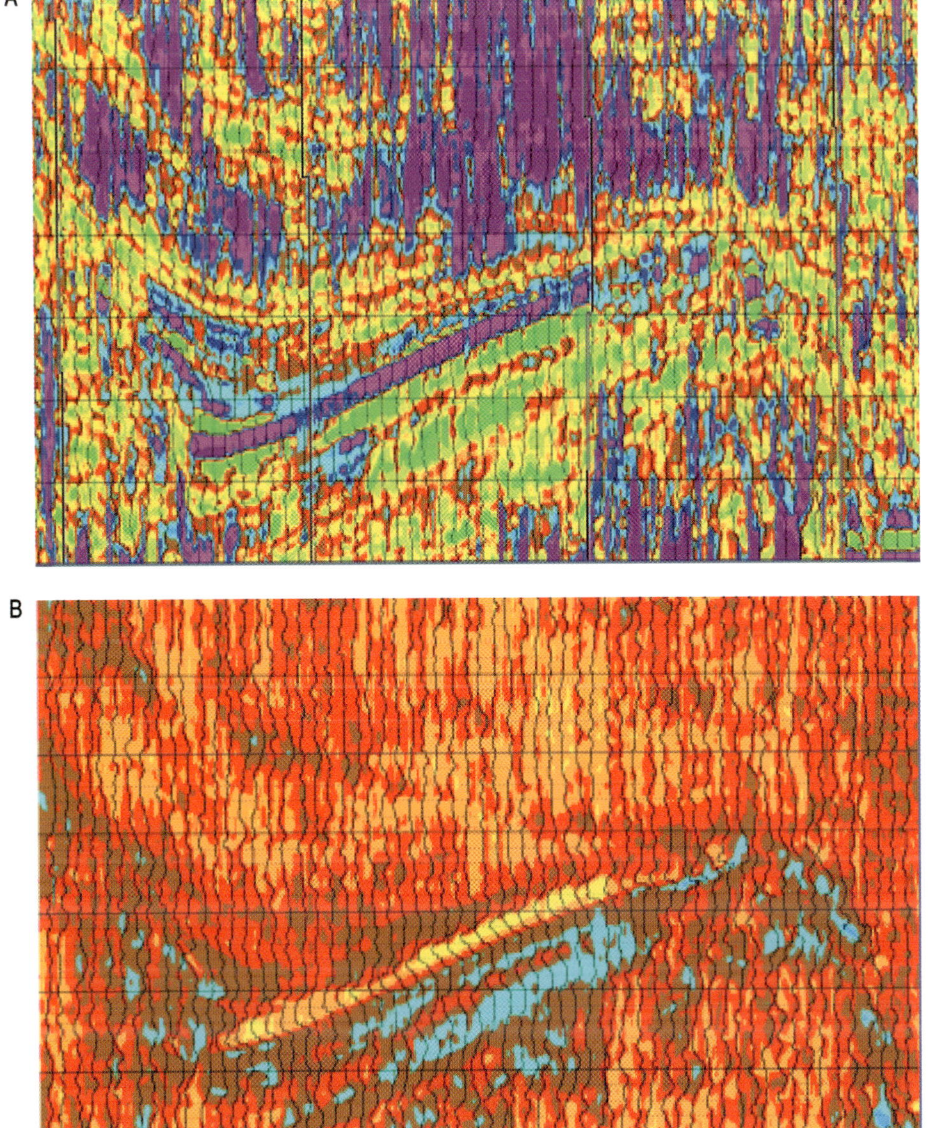

Figure 6. Prestack inversion results for the same seismic line shown in Figure 5. Note the differences in the (A) P-wave and (B) Poisson's ratio results from the inversion.

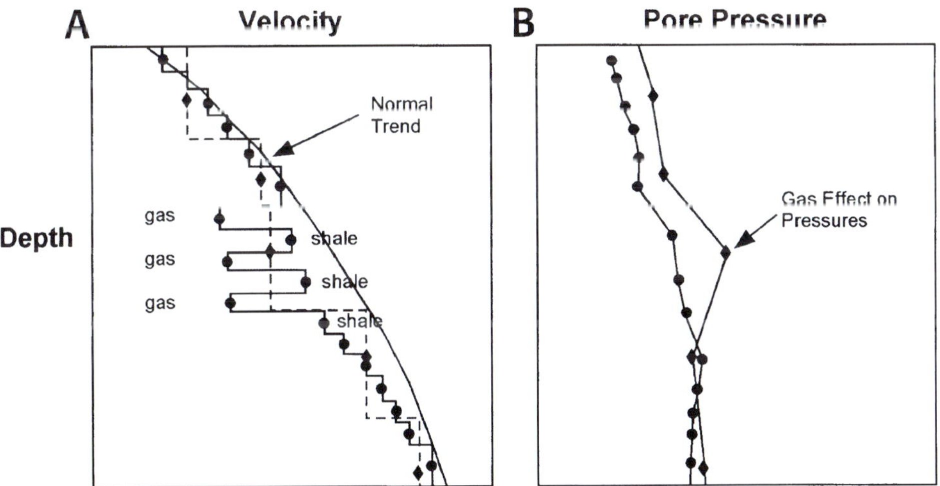

Figure 7. Comparison of velocity data from stacking velocity analysis (dashed line on part A) and prestack inversion with the low-frequency component from stacking velocities (solid line) in a hydrocarbon-bearing zone with interbedded shales. Note the temporal averaging in the stacking-velocity analysis that prevents the gas-bearing zones from being isolated in the pressure analysis on part B.

be valuable in areas where gas chimneys distort the compressional-wave velocity data and prevent robust velocity analysis. Shear-wave data also provides a second measurement of pore pressure in cases where hydrocarbons in a reservoir distort the compressional-wave velocity field. The fact that the shear waves are much less affected by fluid variations (Gregory, 1977) permits pressure prediction to be performed without worrying about how to correct for Gassmann effects in the reservoir as must be done for compressional-wave data.

OVERBURDEN PREDICTION USING GRAVITY DATA

One of the greatest limitations on robust pore-pressure prediction is the fact that the density data used to constrain the overburden and fracture gradient are commonly from one to two wells in the best circumstances. In some cases, little to no density data exist, and regional curves have to be used to predict overburden and fracture gradient. In all these cases, the pressure analysis relies on one-dimensional data to solve a three-dimensional problem.

One method for determining densities in the subsurface that has recently come into its own is gravity inversion. Recent developments in this area pioneered by Conoco (Jorgensen and Kisabeth, 2000) have revealed that gravity inversion using conventional gravity data and full tensor gradiometry data can be used to constrain the density field in three dimensions quite accurately for pressure prediction. Although the method was initially developed for determining the base of anomalous bodies such as salt and volcanics, it can also solve for lateral variations in sediment density due to variations in the compaction state. This method also allows the analyst to overcome one of the weaknesses of the velocity-based prediction methods in that it can see some of the density variations due to changes in pressure regime that can not be distinguished with velocity data alone because of the nonuniqueness of velocity–effective-stress relationships.

Figure 8 shows a model example of how critical gravity inversion results can be to a robust pressure analysis in areas where subsalt wells are to be drilled. The final density model from the inversion including the predicted anomalous salt bodies is integrated to get an overburden estimate. Note how the overburden stress decreases under the salt due to the effect of the lighter densities in the anomalous body. In many cases, a well drilled in the basin adjacent to the salt body might be used to predict the overburden gradient in planning for a subsalt well. If that approach is taken and the analyst does not correct for the density effect of the salt body, the well will be grossly overbalanced where it drills out of the salt. The gravity inversion result is especially suited to this problem because it allows a two-dimensional or three-dimensional prediction of the density and overburden that includes the variations due to the anomalous body and any variations in sediment density where no anomalous nonclastic bodies exist. In some cases, the sediments below the salt are observed to be anomalous relative to the regional density trend because of a local or regional salt seal. The method solves this problem by allowing the salt to be constrained from the inversion and seismic data along with the sediments everywhere but below the salt, so that variations related to pressure can be predicted from a second iteration of the gravity inversion. The gravity inversion results are also totally independent of the seismic-velocity information and thus provide additional data that can not be obtained by other means.

PRESSURE PREDICTION IN BASIN ANALYSIS

Today, only a handful of companies are using seismic-based pressure prediction as an end constraint in basin analysis and modeling. Pressure prediction at the basin scale can be very powerful in (1) determining where source rocks are actively maturing, (2) determining where large-scale fluid migration is occurring in a basin, (3) predicting the behavior of large regional faults and structures, (4) identifying the presence of secondary-pressured areas, (5) constraining the porosity model for the basin, and (6) evaluating the integrity of vertical seals in the basin. Such a case is shown in Figure 9.

Note that large-scale exploration already makes routine use of seismic-velocity data for time-depth conversion, but very few people are using these data for determining pressure and using this as an end constraint on basin models. This is truly ironic as it is one of the basinwide parameters that we can measure directly from seismic data.

Also, it should be possible to work the inverse problem and go backward from current-day conditions to the start of basin formation using the pressure prediction as the starting point. This approach would be similar to the current technology for palinspastic reconstruction used routinely by structural geologists but would require the added complexity of reverse engineering the basin loading history and compaction processes.

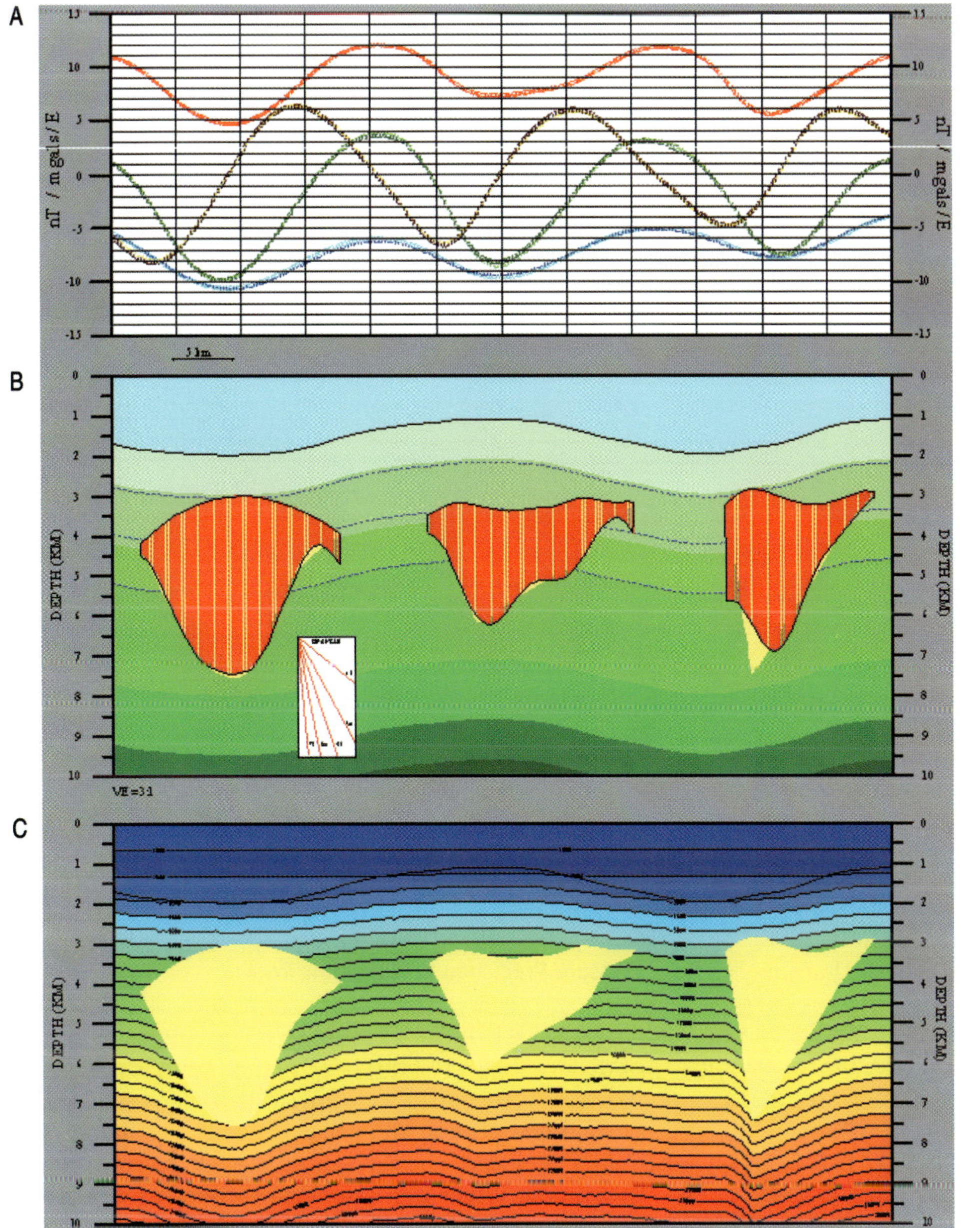

Figure 8. A two-dimensional gravity model for salt bodies imbedded in sediment showing (A) the calculated gravity and magnetic fields over anomalous salt bodies, (B) the final density model with the salt bodies, and (C) the overburden stress estimated by integrating the density function at each vertical location in the model.

PRESSURE PREDICTION AT THE PROSPECT SCALE

Like basin analysis, only a handful of companies today are using pressure prediction for prospect analysis. Considering the importance of pressure in establishing the fluid migration pathways and reservoir properties, it is quite surprising that this technique has not been more heavily used for this purpose. At the prospect scale, pressure prediction as currently applied can be used to (1) constrain the porosity and pressure regimes surrounding accumulations of hydrocarbons, (2) determine the sealing characteristics of faults, (3) evaluate vertical and lateral pressure seal properties, (4) evaluate the risk of structural effects on pressure in reservoirs, and (5) determine the production drive mechanism for a given reservoir based on its location relative to pressure. If prestack inversion is used at the prospect scale in pressure prediction, it may be possible to (1) isolate the pressure behavior of specific reservoirs and determine the centroid effect (Heppard and Traugott, 1998) in them, (2) identify cases where the sands are not in pressure equilibrium with the encasing shales, and (3) isolate and understand localized effects such as cementation, nonclastic rock units, and other factors on pressure prediction.

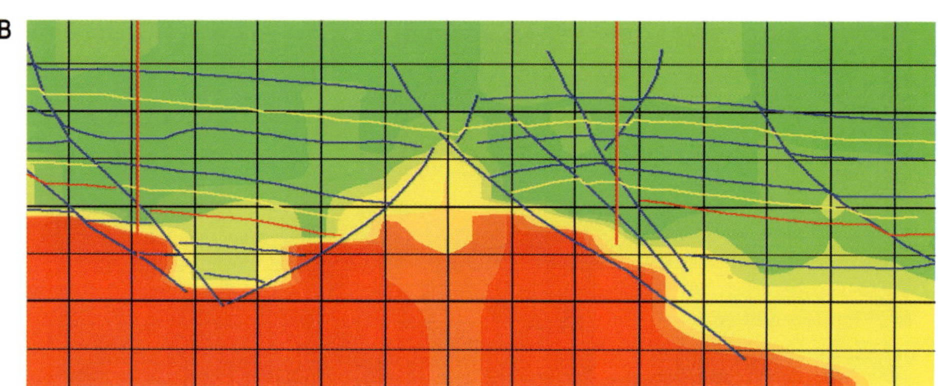

Figure 9. Example of (A) an interpreted seismic line and (B) the resulting horizon-keyed pressure prediction showing the sealing behavior of faults and pressure compartmentalization. Note the two faults on the far left that show fault seal failure in the deep section (red color) and compare this to the major fault in the center that acts as a pressure barrier between the red and yellow pressure isobars. The fault on the far right also appears to be a conduit for pressure pumping from deep in the section.

At the prospect scale, we are not currently using the results of pressure prediction in coordination with hydrogeologic modeling to predict fluid flow, migration pathways, and rates of hydrocarbon charging. A merger of hydrogeologic methods with pressure prediction could dramatically improve our ability to understand the hydrocarbon system at both the basin and prospect scale. Such an understanding would undoubtedly improve our exploration results by eliminating areas where migration and hydrocarbon charging are not viable. The combination of robust effective-stress methods, prestack inversion, and advanced hydrogeologic modeling may allow pressure prediction to become a routine part of the geophysical prospect evaluation process. Proper integration of pressure analysis with other geophysical tools and stratigraphic and structural analysis at the prospect scale should improve the industry's ability to successfully identify and produce economic accumulations of hydrocarbons in the future.

UNANSWERED ISSUES

Many issues still need to be addressed in pressure prediction. These issues have been viewed in the past as secondary concerns that were not critical to the analysis but are now becoming primary issues as we refine our methods for pressure prediction.

Pressure Prediction in Nonclastic Rocks

How do we predict pressure in nonclastic rocks ahead of the drill bit? Presently, the methods employed in the industry work fairly well for clastic rocks because the normal compaction trend has a sufficiently large gradient in velocity–effective-stress space to allow a robust calibration. Other rocks such as carbonates, however, are not as forgiving as clastic rocks. Carbonates in particular have a very flat velocity–effective-stress gradient so that there is very little velocity sensitivity to changes in pressure in these rocks (Figure 10). Furthermore, the hysteresis effect related to unloading in carbonates produces virtually no velocity change, which allows carbonates to sustain severe secondary pressure conditions with very little velocity change. At present, most workers use the encasing shales to infer the pressures in the carbonates, but this has proven to be very dangerous in many cases.

Pressure Prediction in the Presence of Velocity Anisotropy

Another issue is the effect of anisotropy on the measured velocity field. It was noted previously in this

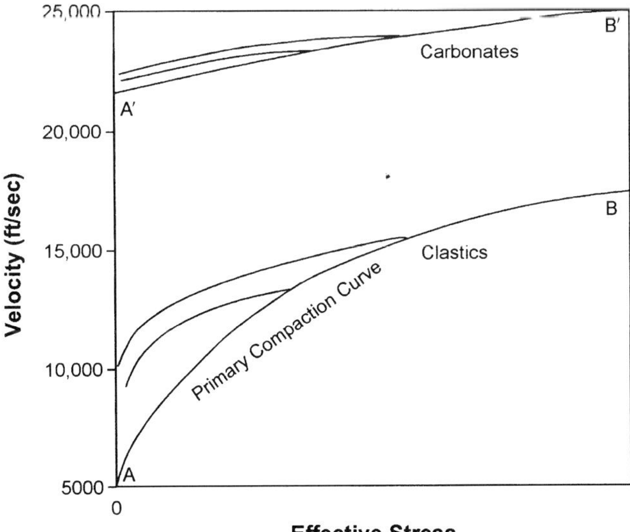

Figure 10. Velocity vs. effective stress plot showing normal compaction and unloading curves for clastic and carbonate materials showing the relative differences in their sensitivity. Note the large difference between points A and B for clastic rocks compared to the difference between A′ and B′ for carbonates. Also note the larger contrast between the unloading paths in clastic rocks vs. carbonates.

chapter that there are anisotropic migration and velocity analysis programs available to handle this issue. In the real world of clastic basins, anisotropy most commonly occurs as transverse isotropy (TI) but can sometimes occur as full anisotropy where the velocity varies in all three dimensions. Full anisotropy can be fracture induced, stress induced, and lithologically induced, singly or in combination. Distinguishing which type is operating can be important for pressure and fracture-gradient prediction. Once anisotropy is identified, the issue becomes one of determining the cause and then how to correct for it or use it properly in pressure work. In most of the young Tertiary basins that pressure prediction is applied to, the cause of anisotropy is commonly a mixture of thin-bed anisotropy, intrinsic shale anisotropy, and other forms.

Unfortunately, nearly all of the algorithms available today are designed to either ignore anisotropy or handle one form of anisotropy, TI, which is the case where the velocity field varies between the vertical and horizontal but does not vary with azimuth. Transverse isotropy requires two additional parameters to accurately produce vertical velocity as a function of depth from seismic moveout. Estimates of one parameter can be made from the seismic data itself and used to improve the seismic image. The moveout, however, is insensitive to the vertical velocity. Well data are needed to produce an accurate depth prediction. Transverse isotropy is one of several complications that can be handled if well data are available for calibration. Without such data, the ability of the seismic data to discern the anisotropy in the subsurface is limited (Bell, 2002). To make matters worse, the velocity effects related to anisotropy are commonly superimposed with other effects such as dip or cable feathering that makes the analysis even more difficult.

Velocity anisotropy that is related to fracturing presents an additional problem because it is not easily transposed into velocities that can be used for pressure prediction. In this sense, fractures are analogous to the problem caused by nonclastic rocks and hydrocarbons because it is an effect that must be removed from the velocity field before a pressure prediction can be performed. In theory, if a single set of oriented fractures exist in a formation, it should be possible to do a velocity analysis parallel with the fractures and get a velocity result that does not include the influence of the fractures. This is uncommonly the case, however, in real rocks. In practice, it is also very difficult to assure a priori that the seismic data are acquired in a way that meet this criterion for velocity analysis.

Stress State of Basins and Its Effect on Pressure Prediction

Large deviatoric or differential stresses in tectonically active areas can cause significant variation in horizontal stresses, and hence in fracture gradient, that can have a critical impact on the accuracy of a pressure prediction. The effects of the stress field are observed on both the velocity field through stress-induced anisotropy, and on the fracture gradient and induced fracture orientation for the basin. To consider these effects, we must look at simple models for the three dominant types of basin settings and their resultant stress fields (Figure 11).

In an extensional basin (e.g., Gulf of Mexico and Basin and Range province), the stress field is such that the minimum principal stress is horizontal and the maximum principal stress is vertical. The magnitude of the differential stress in extensional basins is commonly small, and the intermediate and minimum principal stresses are commonly close to the same value. In this case, the stress-induced maximum velocity is vertical, and the stress-induced minimum velocity is horizontal. This presents a significant problem in that the stress-induced anisotropy offsets some of the other forms of anisotropy acting in these basins as defined previously. This also implies that the velocity field measured from seismic data generally correlates reasonably with the velocities measured from vertical seismic profiles or check shots and from dispersion-corrected sonic logs. The degree of stress-induced velocity anisotropy in this type of basin is commonly sufficiently small that it is

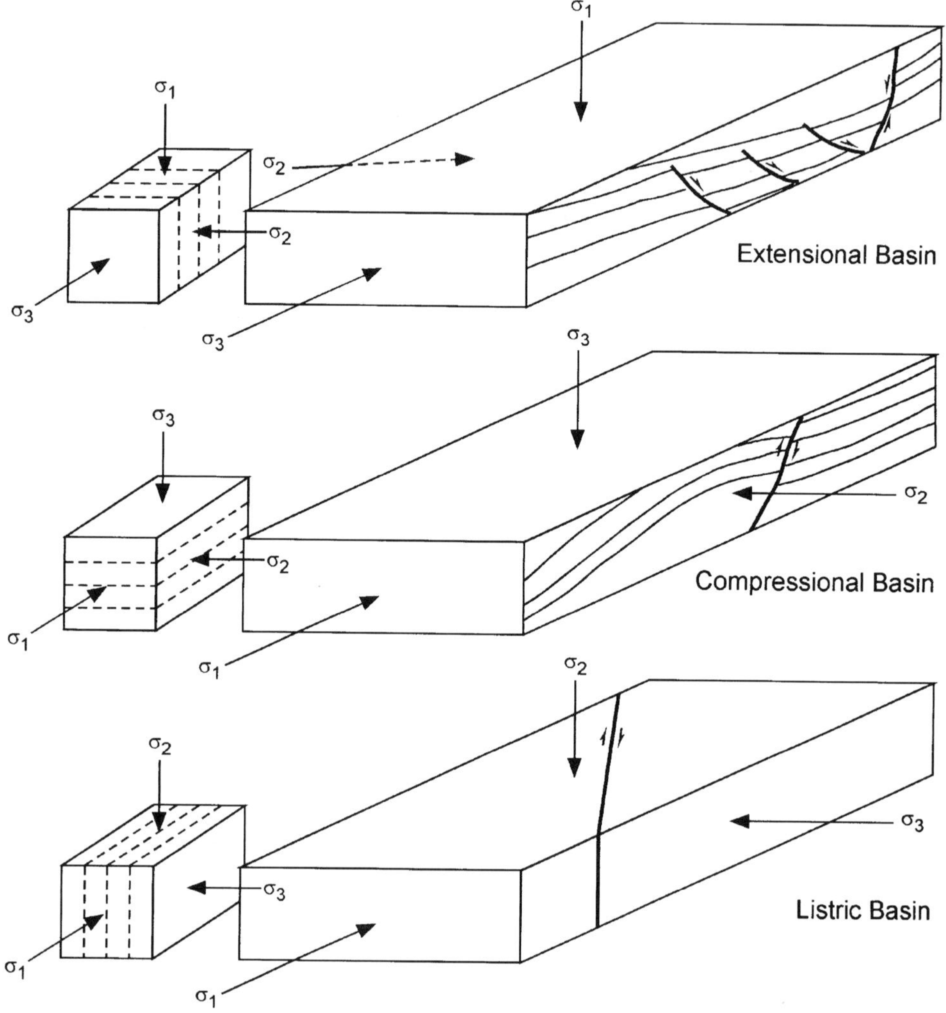

Figure 11. Models for stress regimes in basins showing the three principal stress orientations and their effect on the fracture orientation. Note that all of the stresses are positive so that true tensile stresses are not observed in nature.

within the measurement error of the velocity analysis methods and thus is not a significant problem in pressure prediction. Stress-induced velocity anisotropy is also offset by the presence of anisotropy (TI) that tends to occur in shale-prone basins. Fracture gradients in extensional basins are commonly slightly less than the overburden calculated from integration of the density logs. This is intuitively reasonable because the maximum principal stress is vertical and the mean stress (the average of the three principal stresses) is commonly not much less than the maximum principal stress. The fracture gradient is commonly controlled by the minimum principal stress, whereas the overburden is tied more closely to the mean stress. The orientation of induced fractures is normally parallel with the greatest principal stress and perpendicular to the minimum principal stress. For extensional basins, this implies that induced fractures will be vertical in orientation. As a result, severe induced fractures can propagate vertically through the formation and potentially undermine the cement job above the casing shoe.

In a compressional basin (e.g., Rocky Mountains, Zagros Mountains), the stress field is such that the maximum principal stress is horizontal and the minimum principal stress is vertical. The magnitude of the differential stress in compressional basins is significantly larger that in extensional basins due to the fact that rocks tend to be stronger in compression and thus support larger compressional loads. In this case, the stress-induced maximum velocity will be horizontal and the stress-induced minimum velocity is vertical. The degree of stress-induced velocity anisotropy in this type of basin can be significant and much larger than the measurement error of the velocity analysis methods. Fracture gradients in compressional basins can actually be greater than the overburden calculated from integration of the density logs. This can result in too conservative an estimate of the fracture gradient and lead to suboptimal well design. Obtaining good leak-off test (LOT) data in such areas to be sure that the fracture-gradient relationship to overburden is well understood before drilling additional wells is important. For compressional basins, induced fractures are commonly horizontal in orientation away from the wellbore. This orientation implies that fractures induced by drilling fluids pose less of a threat to casing and cement integrity because they will not propagate past the casing shoe. The local stress field around the wellbore, however, commonly induces a fracture that starts out vertical as in the other basin stress fields. This initial fracture can still pose a threat if it runs sufficiently far in the vertical direction along the wellbore to get past the cement job. If the fracture runs sufficiently outward from the well to get past the local stress field at the well, it can rotate to align itself with the regional stress field, which can add complexity to the interpretation of test data.

In a strike slip or listric basin (e.g., San Andreas fault region), the stress field is such that the maximum and minimum principal stresses are both horizontal and the intermediate principal stress is vertical. The magnitude of the differential stress in listric basins can also be significantly larger than in extensional basins. In this case, both the stress-induced maximum velocity and the stress-induced minimum velocity are horizontal. The degree of stress-induced velocity anisotropy in this type of basin can also be significant. This again presents a significant problem for anisotropy-processing algorithms because the velocity varies predominantly with azimuth in listric basins. This requires not only that the interpreter be able to process properly for anisotropy, but also that the subsurface be sampled with a range of azimuths, which has not been done historically except on some land seismic surveys. Fracture gradients in listric basins can vary over a wide range but are commonly found to be close to the mean stress, which is commonly close to the intermediate principal stress that is commonly vertical in this type of basin. For listric basins, induced fractures are commonly vertical in orientation. As with the extensional basin case, severe induced fractures can propagate vertically through the formation and potentially undermine the cement job above the casing shoe.

Structural Hyperpressuring in Reservoirs

Several authors (e.g., Heppard and Traugott, 1998; Stump et al., 1998) have addressed the centroid effect on pressures in structurally positioned reservoirs. This effect can be significant enough to cause seal failures and serious drilling problems if it is not recognized predrill. To date, however, the industry has not tackled the issue of how to detect this effect predrill using seismic data. Again, the use of poststack and prestack inversion may give us a significant advantage by being able to isolate the seismic behavior of a reservoir and evaluating it in detail as a function of structural position. This, however, requires sufficient knowledge of the physics of the process and its effect on both the reservoir and sealing rocks so that we can model the effect and interpret it in the seismic data. It also requires a good knowledge or definition of the reservoir geometry.

For water-bearing reservoirs, there are two factors that must be considered in dealing with structural hyperpressuring. The first factor is the possibility that the degree of compaction may change from the top to the bottom of the reservoir with large amounts of structural relief. In this situation, the centroid pressure profile observed in the velocity data may be magnified by the vertical change in compaction state of the reservoir. The second factor is the possibility that the reservoir has been breached by a fault or other conduit that allows fluids to escape from or recharge into the reservoir. In this case, it is possible that the centroid position in the reservoir may be shifted significantly. Also important to recognize is that the seal rocks adjacent to the reservoir can be affected by breaching a reservoir. For example, if a reservoir is drained slowly and then recharged, seal rocks adjacent to the reservoir may also undergo some compaction and reduction of fluid pressures. This results in a compaction halo whose thickness depends on the elapsed time for the drained condition. Once the reservoir is recharged, this halo stabilizes and preserves evidence of the drainage event that can be seen as a local increase in velocity and density in the seal rocks. This is important for inversion analysis because the local halo affects the reflection coefficients between the shale and reservoir and could result in misleading data. The halo also produces shale velocities and densities that are not the same as those further from the reservoir. This effect can be detected in inversion but would be missed by a conventional velocity analysis that measures the velocities more coarsely.

For hydrocarbon-bearing reservoirs, the effect of the hydrocarbons on the velocity field must also be considered in estimating the reservoir pressures. In this case, either the hydrocarbon effect must be calculated and removed by performing a Gassmann fluid replacement calculation, or the reservoir pressure calibration itself must be adjusted to include the hydrocarbon effect. In either case, enough must be known about the reservoir and its fluids to permit these adjustments to be made.

Recent work in deep-water drilling hazards has also suggested that shallow water flows (SWF) are the result of structural hyperpressuring (Huffman and Castagna, 2000). These shallow sands commonly occur within a few thousand feet of the mud line in abruptly depositing basins and exhibit pore pressures close to fracture gradient and near-zero effective stresses. They are probably the most significant hazard currently facing deepwater drilling and should be studied in the context of pressure prediction.

Identifying the Top of Secondary Pressure Zones

One of the most difficult issues in pressure prediction is knowing how to identify the correct maximum velocity–effective-stress point for a velocity reversal that is attributed to secondary pressure (see Figure 4). The technique for handling a velocity reversal requires that the maximum compaction state be known to estimate the correct unloading path for the interval. The selection of this maximum velocity determines the density and porosity in the unloaded interval, as well as the pressure that is attained for a given velocity (Bowers, 1994). The most common practice for determining the maximum velocity is to use the velocity of the interval directly above the unloaded interval. This is problematic, however, in that it assumes that there is continuity in loading history across this pressure boundary, which may not be correct. The maximum velocity that should be achievable for a given depth is the velocity that is on the normal pressure curve for the greatest depth of burial that the interval achieved because this is the maximum compaction state that any rock can attain. Selecting this velocity as the maximum velocity for any velocity inversion yields the maximum possible pore pressure that can be attained for that unloading condition. In most cases where the maximum velocity value is not well constrained, it is prudent to provide a worst case scenario using this approach. Overpredicting pressure is possible by using a secondary pressuring estimate where one is not needed. The minimum pressure case can be provided by simply assuming no unloading and treating the reversal as an undercompacted interval. The problem is knowing where unloading is actually occurring and what velocity point to tie it back to, which is a reflection of the loading history of the particular formation being analyzed.

This issue can be addressed in part by density data. If unloading is present, the compaction process is arrested by the secondary pressure. In that case, density should not continue to increase and porosity should not continue to decrease with depth. Therefore, collecting density and/or porosity data in real time can be invaluable in determining whether a velocity reversal represents secondary pressure conditions or undercompacted conditions and what maximum compaction point to tie the unloaded zone to. In the absence of unusual diagenetic alteration or cementation, the density of the formation gives a good indication of the maximum compaction achieved by the rocks. In general, if the velocity reverses and the density also decreases, this most likely indicates undercompaction as the cause. If the velocity decreases and the density and porosity remain the same, secondary pressure is likely to be the cause.

CONCLUSIONS

The future of pressure prediction will see dramatic changes in the purpose for which we use the technique, the types of data that are used, and the type of analysis that is employed to get more detailed high-resolution velocity data. Pressure prediction will become part of the holistic discipline of basin analysis and will be used to determine the present pressure state of a basin as a control on both forward and inverse basin modeling. Pressure prediction will also be used at the subregional to prospect scale to determine critical aspects about the hydrocarbon migration and trapping process, fault leak/seal, and source maturation.

Despite all of these new technologies and techniques for improving estimates of velocity and density, we will still be forced to cope with our lack of understanding of the nature of pressure and fluid movement in shale-dominated basins and the various physical processes that cause pressure. Future research will have to focus on these basic physical processes so that we can properly interpret the data that we are becoming so masterful at acquiring in large quantities.

REFERENCES CITED

Bell, D. W., 2002, Velocity estimation for pore-pressure prediction, *in* A. R. Huffman and G. L. Bowers, eds., Pressure regimes in sedimentary basins and their prediction: AAPG Memoir 76, p. 177–215.

Bowers, G. L., 1994, Pore pressure estimation from velocity data: accounting for overpressure mechanisms besides undercompaction: International Association of Drilling Contractors/Society of Petroleum Engineers Drilling Conference, p. 515–530.

Bruce, C. H., 1984, Smectite dehydration—its relation to structural development and hydrocarbon accumulation in northern Gulf of Mexico Basin: AAPG Bulletin, v. 68, no. 6, p. 673–683.

Eaton, B. A., 1975, The equation for geopressure prediction from well logs: Society of Professional Engineers Paper 5544.

Gregory, A. R., 1977, Aspects of rock mechanics from laboratory and log data that are important to seismic interpretation, *in* C. E. Payton, ed., Seismic stratigraphy—applications to hydrocarbon exploration: AAPG Memoir 26, p. 15–46.

Heppard, P. D., and M. Traugott, 1998, Use of seal, structural and centroid information in pore pressure prediction (abs.): American Association of Drilling Engineers Forum on Pressure Regimes in Sedimentary Basins and Their Prediction, unpaginated.

Holbrook, P. W., and M. L. Hauck, 1987, A petrophysical-mechanical math model for real-time wellsite pore pressure/fracture gradient prediction: Society of Professional Engineers Paper 1666.

Huffman, A. R., and J. P. Castagna, 2000, Shallow water flow prediction from seismic analysis of multicomponent seismic data: Proceedings of the 2000 Offshore Technology Conference, unpaginated.

Jorgensen, G. J., and J. L. Kisabeth, 2000, Joint 3-D inversion of gravity, magnetic and tensor gravity fields for imaging salt formations in the deepwater Gulf of Mexico (abs.): Society of Exploration Geophysicists Annual Meeting, unpaginated.

Magara, K., 1975, Importance of aquathermal pressuring effect in Gulf Coast: AAPG Bulletin, v. 59, no. 10, p. 2037–2045.

Marion, D., A. Nur, H. Yin, and D. Han, 1992, Compressional velocity and porosity in sand-clay mixtures: Geophysics, v. 57, p. 554–562.

Moos, D., and G. Swart, 1998, Predicting pore pressure from porosity and velocity (abs.): American Association of Drilling Engineers Forum on Pressure Regimes in Sedimentary Basins and Their Prediction, unpaginated.

Nur, A., D. Marion, and H. Yin, 1991, Wave velocities in sediments, *in* J. M. Hovem, M. D. R. Richardson, and R. D. Stoll, Shearwaves in marine sediments: Dordrecht, Kluwer, p. 131–140.

Plumley, W. J., 1980, Abnormally high fluid pressure: survey of some basic principles: AAPG Bulletin, v. 64, no. 3, p. 414–430.

Sheriff, R. E., 1991, Encyclopedic dictionary of exploration geophysics, 3d ed.: Society of Exploration Geophysicists, 384 p.

Spencer, C. W., 1987, Hydrocarbon generation as a mechanism for overpressuring in Rocky Mountain region: AAPG Bulletin, v. 71, no. 4, p. 368–388.

Stump, B. B., P. B. Flemings, T. Finkbeiner, and M. D. Zoback, 1998, Pressure differences between overpressured sands and bounding shales of the Eugene Island 330 field (offshore Louisiana, USA) with implications for fluid flow induced by sediment loading, *in* A. Mitchell and D. Grauls, eds., Overpressures in petroleum exploration: Bulletin Centre Recherche Elf Exploration and Production, Memoir 22, p. 83–92.

Tosaya, C. A., 1982, Acoustic properties of clay bearing rocks: Ph.D. dissertation, Stanford University, Stanford, California.

Wood, A. B., 1941, A textbook of sound, 2d edition: New York, Macmillan, 578 p.

Index